The Analysis of Actual Versus Perceived Risks

ADVANCES IN RISK ANALYSIS

This series is edited by the Society for Risk Analysis.

Volume 1   THE ANALYSIS OF ACTUAL VERSUS PERCEIVED RISKS
Edited by Vincent T. Covello, W. Gary Flamm,
Joseph V. Rodricks, and Robert G. Tardiff

A Continuation Order Plan is available for this series. A continuation order will bring delivery of each new volume immediately upon publication. Volumes are billed only upon actual shipment. For further information please contact the publisher.

The Analysis of Actual Versus Perceived Risks

Edited by

Vincent T. Covello
U.S. National Science Foundation
Washington, D.C.

W. Gary Flamm
Food and Drug Administration
Washington, D.C.

Joseph V. Rodricks
Environ Corporation
Washington, D.C.

and

Robert G. Tardiff
National Academy of Sciences
Washington, D.C.

Plenum Press • New York and London

T
174.5
.S615
1981

Library of Congress Cataloging in Publication Data

International Workshop on the Analysis of Actual Versus Perceived Risks (1981:
Washington, D.C.)
 The analysis of actual versus perceived risks.

(Advances in risk analysis; v. 1)
 "Proceedings of an International Workshop on the Analysis of Actual Versus
Perceived Risks, held June 1-3, 1981...Washington, D.C."—T.p. verso.
 Includes bibliographical references and index.
 1. Technology assessment—Congresses. 2. Risk—Congresses. I. Covello, Vincent T. II. Title. III. Series.
T174.5.I58 1981   363.1'0028'7   83-11071
ISBN 0-306-41397-3

Proceedings of the Society for Risk Analysis International Workshop
on the Analysis of Actual Versus Perceived Risks,
held June 1-3, 1981,
at the National Academy of Sciences Auditorium,
in Washington, DC

©1983 Plenum Press, New York
A Division of Plenum Publishing Corporation
233 Spring Street, New York, N.Y. 10013

All rights reserved

No part of this book may be reproduced, stored in a retrieval system, or transmitted
in any form or by any means, electronic, mechanical, photocopying, microfilming,
recording, or otherwise, without written permission from the Publisher

Printed in the United States of America

PREFACE

In 1980, a group of scientists from national laboratories, universities, and other research organizations gathered informally in a series of meetings to consider the state of research on risks to health, safety, and the environment. Each scientist had conducted research on the subject. All felt that the traditional disciplines and professional societies to which they belonged were neither adequate nor appropriate for addressing the extraordinarily complex problems of assessing the risks inherent in modern society. The consensus of the group was that a new society was needed to address these problems in a scientific and objective way. From these initial meetings, the Society for Risk Analysis was formed

The major aims of the Society for Risk Analysis, as stated in its constitution, are

- to promote knowledge and understanding of risk analysis techniques and their applications;

- to promote communication and interaction among those engaged in risk analysis; and

- to disseminate risk analysis information and promote the advancement of all aspects of risk analysis.

Members of the Society are drawn from a variety of disciplines, including the health sciences, engineering, the physical sciences, the humanities, and the behavioral and social sciences. An important function of the Society is the annual meeting, at which various aspects of risk analysis are discussed. The first annual meeting, represented by this volume, was the International Workshop on the Analysis of Actual vs. Perceived Risks, held from June 1-3, 1981, at the National Academy of Sciences in Washington, D.C.

The decision to hold a workshop on this theme was stimulated by the provocative research finding that technical experts and nonexperts differ substantially in their risk estimates. Risk estimates by technical experts tend to be closely correlated with annual fatality rates, whereas the risk estimates by nonexperts are only moderately-to-poorly correlated with annual fatality rates. In hope of clarifying this issue, the workshop organizers selected cases that represent some of the most important aspects and dimensions of risk: voluntary vs. involuntary; low vs. high probability of occurrence, exposure, and effects; and high vs. low health, safety, or environmental consequences.

The workshop was supported with funds from the Alfred P. Sloan Foundation, the Environmental Protection Agency, and the Nuclear Regulatory Commission. The meeting was co-sponsored by the World Health Organization and the U.S. National Academy of Sciences (the Board on Toxicology and Environmental Health Hazards, and the Assembly of Behavioral and Social Sciences).

Vincent T. Covello, Chairman

Publication Committee
Society for Risk Analysis

CONTENTS

I. AUTOMOBILE ACCIDENTS:
    THE PROBLEM OF PASSENGER RESTRAINTS

Effectiveness of Automobile Passenger Restraints
    H. C. Joksch..................................... 1

Public Perception and Behavior in Relation to Vehicle
    Passenger Restraints
    Leon S. Robertson................................ 11

Motor Vehicle Occupant Restraint Policy
    Joan Claybrook................................... 21

Summary of Panel Discussion and Commentary
    Christoph Hohenemser............................. 49

II. NUCLEAR POWER PLANT — EUROPEAN PERSPECTIVE

The Assessment of Nuclear Risk. Some Experiences
    from the Swedish Energy Commission
    Bengt Hansson.................................... 69

Nuclear Power Plant: West German Management
    of Risk — A Problem Analysis
    H. Paschen, G. Bechmann,
    and G. Frederichs................................ 81

Coping with the Risks of Nuclear Power Plants
    in the United Kingdom
    Timothy O'Riordan................................ 101

Risk Assessment Following Crisis in the United States:
    The Kemeny Commission
    Roger E. Kasperson and Arnold L. Gray............ 129

III. CANCER CHEMOTHERAPY

Role of Risk in Treating Advanced Lung Cancer
    Kenneth Stanley.. 157

IV. SMOKING CIGARETTES

Cigarette Smoke: Cancer Risk at Low Doses
    Charles E. Lawrence and Albert S. Paulson........... 169

Perceiving the Risk of Low-Yield Ventilated-Filter
    Cigarettes: The Problem of Hole-Blocking
    Lynn T. Kozlowski..................................... 175

Reactions to Perceived Risk: Changes in the Behavior
    of Cigarette Smokers
    Kenneth E. Warner..................................... 183

Perceived vs. Actual Risks: The Problem
    of Multiple Confounding
    Theodor D. Sterling................................... 203

V. NUCLEAR POWER PLANT — U.S. PERSPECTIVE

The Public Perception of Risk
    D. Litai, D. D. Lanning, and N. C. Rasmussen........ 213

Impact of the Three Mile Island Accident
    as Perceived by Those Living in the Surrounding
    Community
    Anne D. Trunk and Edward V. Trunk................... 225

"The Public" vs. "The Experts": Perceived
    vs. Actual Disagreements about Risks
    Baruch Fischhoff, Paul Slovic,
    and Sarah Lichtenstein........................... 235

Coping with Nuclear Power Risks: A National Strategy
    Chauncey Starr.. 251

VI. HEALTH IMPACT OF TOXIC WASTES

Health Impact of Toxic Wastes: Estimation of Risk
    Renate D. Kimbrough................................... 259

Perception of Risk: A Journalist's Perspective
    Joanne Omang.. 267

CONTENTS

VII. DEPLETION OF STRATOSPHERIC OZONE

Depletion of Stratospheric Ozone as a Result
  of Human Activities
    R. D. Hudson... 273

Relationship of Stratospheric Ozone Depletion
  to Risk of Human Skin Cancer
    Frederick Urbach..................................... 281

Depletion of Stratospheric Ozone: Impact of UV-B
  Radiation Upon Nonhuman Organisms
    Robert C. Worrest.................................... 303

Chairman's Summary
    Claud S. Rupert...................................... 317

VIII. PANEL DISCUSSION 323

APPENDIX

Actual vs. Perceived Risk: A Policy-Related
  Bibliography
    Vincent T. Covello and Mark Abernathy 351

INDEX 373

EFFECTIVENESS OF AUTOMOBILE PASSENGER RESTRAINTS

H. C. Joksch

The Center for the Environment & Man, Inc.

275 Windsor Street
Hartford, Connecticut 06120

1.0. Introduction

Occupant restraints are effective means to reduce the probability of death and injury, and the severity of injuries in motor vehicle crashes. Their effect depends strongly on the crash conditions, primarily direction and magnitude of the forces. The overall effectiveness of restraints also depends on how reliably they perform and, in the case of belts, how frequently they are used.

To estimate the effectiveness of restraint systems, actual crashes must be studied. One had to determine whether death or injury occurred and to quantify injury severity. Characteristics of the crash and of the victim which influence the fatality injury risks, and injury severity must be investigated. Then models are applied to separate the effects of the other factors from those of the restraint system used (if any). Because the crash conditions and their quantification are imprecise predictors of injury outcomes, such models are necessarily statistical.

2.0. Injury Severity

At first glance, death appears to be a well-defined consequence of an accident. However, two legal definitions are used in the U.S.: death within 30 days, or within one year from the date of the accident.

"Injury" is not as precisely defined. "Whiplash" can be a nontrivial medical problem, though it might not be easily observable. A minor bruise, though objectively identifiable, might be too trivial

Table 1. Approximate Distribution of OAIS Levels Among Injured Car Occupants (based on NCSS* data)

OAIS	%
1	72
2	15
3	8
4-6	5

*NCSS = National Crash Severity Study.

to be considered an injury. The most commonly used indicator of injury severity is the "police-scale":

K = killed
A = incapacitating injury
B = non-incapacitating, evident injury
C = possible injury
O = no injury.

This scale is widely used by police agencies in accident re-
This scale is widely used by police agencies in accident reports. Several weaknesses of this scale are that the "A" category have changed over time, and that, even if some definitions remained the same, there are great regional differences in their interpretation. Nevertheless, this scale has been successfully used in many studies and is still being used in studies of mass accident data.

Since 1974, police agencies in New York have used another description of injuries using three variables to describe different aspects of injuries. Other states have adopted the same or similar injury descriptions. However, these systems do not directly give a one-dimensional measure of injury severity.

For special accident investigations, the Abbreviated Injury Scale (AIS) was developed. Each injury (not injured person) is characterized on the following scale:

AIS 0 = no injury
AIS 1 = minor injury
AIS 2 = moderate injury
AIS 3 = severe, not life threatening
AIS 4 = serious, life threatening
AIS 5 = critical, survival uncertain
AIS 6 = currently untreatable.

An extensive catalog gives the AIS value for precisely described injuries.

To quantify the overall effect of several injuries, several measures have been developed. An overall value "OAIS" is a subjectively assigned a number between 0 and 6, corresponding to one injury which would have the same effect as all injuries together. A less subjective, but less comprehensive indicator, is "MAIS," the maximum of the AIS for all injuries. The Injury Severity Score (ISS) is the sum of the squares of the highest AIS values for the three body regions with the highest AIS. The ISS is closely correlated with the probability of death from the injuries.

An important point to consider is that the AIS quantifies injury severity according to threat to life. Long-term injury consequences are not considered. Injuries leading to blindness, e.g., can have a low AIS value; whereas, a person with higher AIS values can recuperate without long-term consequences.

Table 1 shows the approximate distribution of OAIS values among injured car occupants.

3.0. Factors Influencing Injury and Death in Accidents

Many factors influence the occurrence of death and injury (and their severity) in motor vehicle accidents. Therefore, one cannot simply compare the occurrence of death or injury among restrained car occupants with that among unrestrained occupants. Rather, it is necessary to account for differences in these factors when making comparisons.

The most important factor is crash "severity." One indication of crash severity is the vehicle damage. A simple description of crash damage is given by the "TAD" scale. It describes three aspects: type of impact, direction of force, and severity of damage. It can be used by a police officer investigating an accident.

A more detailed description is given by the Collision Deformation Classification (CDC). It describes damage in terms of direction of force, area of deformation (3 aspects), type of damage, and severity. To classify the damage, direct measurements are required.

These scales allow for the classification of accidents by severity and, thereby, more extensive comparison of injury and fatality risks both with and without restrains.

A quantitative measure of accident severity is the velocity change (ΔV) a vehicle experiences in an accident. It is the most important single variable on which the occupant injury and fatality risk depends. The fatality risk varies with the fourth power of ΔV (unpublished findings of the author). Velocity change can be estimated from vehicle damage, vehicle characteristics, and vehicle trajectory, if available.

In addition to ΔV, the area of impact on the vehicle is important. Up to $\Delta V = 40$ mph, the fatality risk in side impacts is 2-3 times as high as in frontal impacts.

The effect of a given ΔV depends on vehicle factors. If the vehicle structure is so that the occupant experiences the ΔV over a longer time period, injuries are less than when it is experienced over a shorter time period. This is one reason for the beneficial effects of restraints: a restrained occupant experiences the ΔV over the entire time the vehicle takes to come to a stop; an unrestrained occupant experiences much of the ΔV during the time the person impacts the vehicle interior.

Protrusions in the occupant compartment are factors that increase the injury and fatality risk. If the protrusion is not too large, restraints prevent the occupant from being struck by the protruding parts.

Ejection combined with subsequent striking of the ground or of objects is another factor increasing the fatality and injury risk. Restraints decrease this risk by reducing the probability of ejection.

In addition to the factors describing accident severity, the probability of death or injury also depends on occupant characteristics. The most important one is age: the risk increases substantially with age. For instance, with an ISS = 15, the fatality risk for persons 70 or older is 7 to 9 times as high as for persons under 50 (for higher ISS values - i.e., more severe injuries - the difference is less).

4.0. Methods to Estimate Restraint Effectiveness

Definition of effectiveness: If, under the same crash conditions, a fraction x of unrestrained occupants (of a specified seat position) and a fraction y of restrained occupants are injured, the effectiveness of the restraint is described as

$E = (x-y)/x.$

If the effectiveness of the restraint depends on crash conditions and driver characteristics, a weighted average can be used.

Comparing injury or fatality rates between vehicles equipped with certain restraint systems and those not equipped does not give

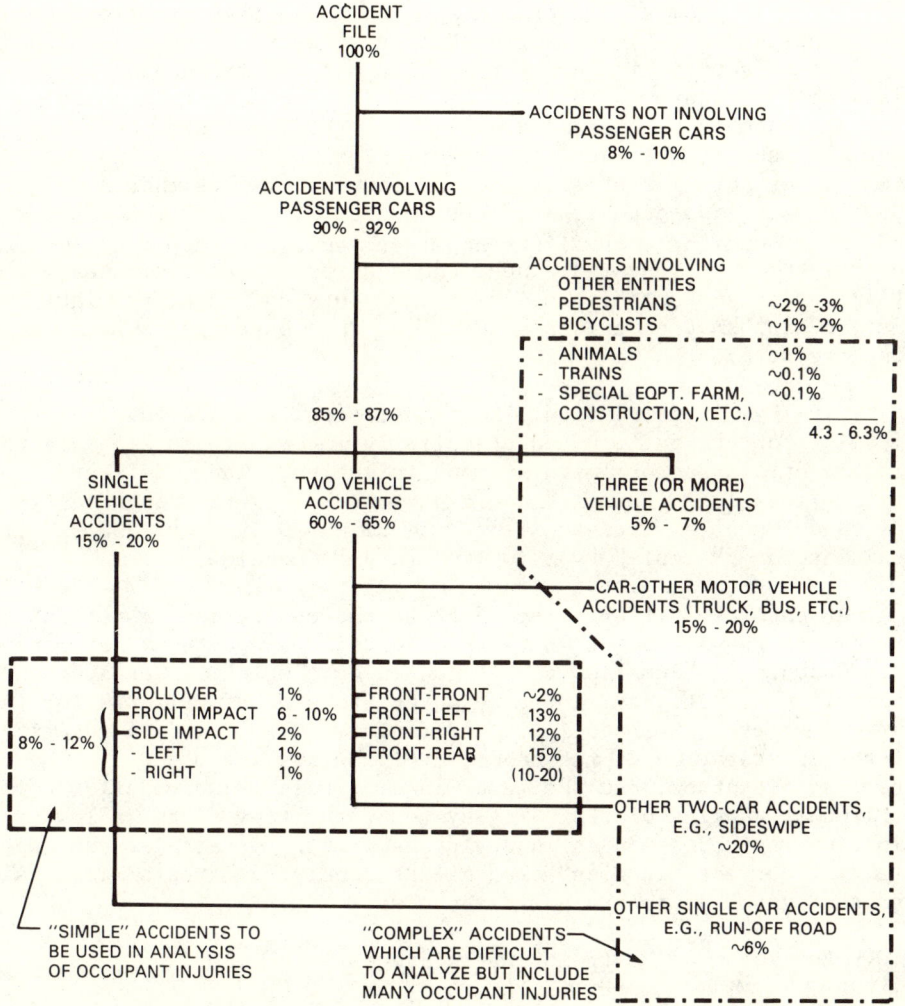

Fig. 1. Typical taxonomy of accidents in a state accident data file.

a valid estimate of restraint effectiveness because usage of restraints and accident conditions may differ between the vehicles. Comparing injury or fatality rates between occupants using and not using restraint systems avoids the problem of possibly differing restraint usage, but the problem of possibly different vehicle usage and accident conditions remains. Comparing the injuries (or occurrence of death) of restrained and unrestrained occupants in comparable crashes is valid but generally impractical because the numbers of accidents with sufficiently comparable conditions is very small.

The most promising approach appears to be the use of mathematical models describing the probability of death or injury related to various factors, including restraint use and the use of statistical techniques to estimate the coefficients of the models. This approach is conceptually valid, but there are some practical problems such as the fact that information on restraint use in an accident is not always reliable. Another problem is that ΔV differs between users and non-users of restraints (NCSS). Though the model should account for such differences, systematic errors may remain, because the differences are near the accuracy of ΔV estimates while still having an appreciable effect on injury and fatality risk. Also, ΔV information exists only in few data bases and only in a complete form.

Finally, it is difficult to mathematically model the influence of all important factors and practically impossible to estimate the effects of all factors with a known influence. The models are limited to relatively "simple" accident types. Figure 1 shows a typical taxonomy of a state accident data file. Only about 50% of all accidents are "simple" types which could be modeled.

An entirely different approach to the evaluation of restraint systems is the use of crash tests. For this approach, instrumented anthropomorphic "dummies" are placed in cars which are crashed under controlled conditions. Three measures are used as criteria for injury severity: the head injury criteria (HIC) which is a function of the acceleration of the head (involving its 2.5th power), the chest acceleration, and the femur load. These measures allow comparison or ranking of restraint systems, but they do not allow estimating the reduction in injury probability. Other disadvantages are that current "dummies" are realistic only for frontal impacts and that crash tests are very expensive.

5.0. Estimates of Restraint Effectiveness

Beginning with the 1964 model year, cars were equipped with lap belts; and from 1968, they were equipped with lap and shoulder belts for the outboard front seats. The buzzer/light warning that seat belts were not fastened was introduced in 1973; and in 1974, the ignition-interlock, which was abandoned in 1975, was introduced. Since 1974, lap and shoulder belts have been combined. Since 1964, about 150 million cars with seatbelts have been produced. Currently, over 97% of all registered cars have seatbelts.

Therefore, a large body of data on accidents involving cars with active restraints exists. The power of the analysis of active belts is not limited by the number of cars with belts, but by the number of occupants using them - recently only 11%. The number of police-investigated accidents in state data files, where seatbelt use is reported is of the order of 10 million. The reliability of

Table 2. Restraint System Effectiveness from the Restraint System Effectiveness Program. The Figures Indicate the Reduction of Injury Frequency in Percent

Restraint type	Injury severity		
	AIS ⩾ 2	AIS ⩾ 3	AIS = 6
Lap belt	31(+5)%	46(±10)%	71(±27)%
Lap and shoulder belt	56(±4)%	57(±9)%	55(±29)%

(Standard errors are given in parentheses.)

Table 3. NHTSA's Estimates of Passenger Car Restraint Effectiveness

AIS injury level	Lap belt	Lap and shoulder belt	Air cushion*	Air cushion* and lap belt	Passive belt and knee bolster	Knee bolster	Frequency of AIS level, %
1	0.15	0.30	0.0	0.15	0.20	0.06	72
2	0.22	0.57	0.22	0.33	0.40	0.10	15
3	0.30	0.59	0.30	0.45	0.45	0.15	8
4-6	0.40	0.60	0.40	0.66	0.50	0.15	5

*Overall effectiveness in all crashes, including those where the air cushion is not designed to deploy.

accidents which have been more thoroughly investigated and where seatbelt use is more reliably indicated is of the order of a few tens of thousands.

The situation is quite different with air cushion restraint systems. In 1973, 1000 Chevrolet Impalas were equipped with them; in 1974-76, 10,000 full size Buicks, Cadillacs and Oldsmobiles contained them; and in 1972, 800 Mercury Montereys were produced with them. About 200 accidents in which the air cushion deployed have been reported.

Many studies have been performed on the effectiveness of safety belts. The estimates of effectiveness are typically between 30 and 50%, but much smaller and much higher values (over 80%!) have been found. One hypothesis has been proposed to reconcile these differences: that the effectiveness increases with injury severity, reaching 70% for fatal injuries [6].

The most thorough study of seatbelt effectiveness, using specially collected accident data, is the Restraint System Effectiveness Program [8]. It reported the effectiveness estimates shown in Table 2.

The air cushion restraint system (ACRS) is designed to deploy in frontal impacts with a velocity change of about 11 mph or larger. However, in some cases, deployment has occurred at velocity changes of less than 11 mph. In some side impacts, the ACRS deployed, but there were side impacts with a velocity change of 25 mph where it did not deploy.

Since the ACRS has beneficial effect only in which it is deployed (for other accidents, occupants are expected to use the lap belts which are provided), direct effectiveness estimates are restricted to deployment accidents. One study [8] found that, for 180 front seat occupants, the ACRS reduced the frequency of injuries (AIS $\geq$ 2) by 19%. Another study (i.e., 60 front seat occupants in frontal impacts, where the vehicle deformation index was less than 6), found a 79% reduction of the fatality risk with ACRS and a 72% reduction with lap/shoulder belt combinations.

Table 3 shows the National Highway Traffic Safety Administration's estimates of effectiveness of several types of restraints.

6.0. The Effects on Deaths and Injuries

The beneficial effects of restraint systems depend not only on their effectiveness when in use in an accident but also on the usage rate for active restraint, the defeat rate for passive belts, and, for air cushions, on the fraction of accidents in which they are designed to deploy as compared with the frequency with which they fail to do so.

Reliable information on seatbelt usage can be obtained only by observation. Observations made in 1976/77 showed that only 19% of all drivers used seatbelts [9]. Usage varied with vehicle age, as shown in Table 4. The increase in 1974 due to the ignition interlock is obvious. Usage varied also with car size. Usage rate was 29% for subcompacts, 21% for compacts, 16% for intermediates, and 16% for full size cars, for the 1973 and later model years.

Observations in 1978 showed an overall usage rate of 13% and in 1979 of only 11%. Again, usage varied with the size of car: 18% for subcompacts, 11% for compacts, 10% for intermediates, and 9% for full size cars, of model years 1976-80.

With a usage rate of approximately 10%, and assuming that safety belts reduce the fatality risk by 50%, car occupant deaths would be 5% higher without seatbelts. With 30,000 car occupants killed each

Table 4. Observed Seatbelt Usage Rate (%) in 1976/77 by Model Year of Car

Model year	64-67	68-71	72	73	74	75	76	77
Usage rate	10	15	18	20	25	21	18	17

year, seatbelts save approximately 1500 lives per year. Increasing usage to 75% would save an additional 32.5%, or about 10,000 lives annually. If one assumes a reduction of the fatality risk of 70%, current savings would be 2100 deaths annually; and potential savings with 75% usage would be 14,000 additional lives.

The VW Rabbit is the only car equipped with passive belts in large numbers. Of the passive belts, about one quarter are defeated - whereas the overall seatbelt usage rate for the Rabbit is an unusually high one-third. One cannot assume that the defeat rate of one-quarter would apply if all cars were equipped with passive belts, because passive belts in the Rabbit are an option, though part of a package, presumably requiring selection by the owner. Therefore, one would expect a higher defeat rate if all cars were equipped with passive belts.

If air cushion restraints were installed in all cars, together with lap belts, and a 10% lap belt use is assumed, about 12,000 lives would be saved. This estimate includes the decrease in risk for those 10% currently using lap-shoulder belt combinations.

7.0. Reservations and Qualifications

The effectiveness of restraint systems to reduce the risk of injury or death for car occupants is established beyond any reasonable doubt. The quantitative estimates of the effects on injuries are reasonably accurately known. The quantitative estimates of the effects on fatalities, however, are not very precise: standard errors are about 40%. Estimates of the effects of air cushion restraint systems are even less precise because of the small number of accidents involving air-cushion equipped cars. Therefore, comparisons of the life-saving potential of different restraint systems have to be made with caution.

References

1. "Fact Book, Statistical Information on Highway Safety," U.S. Department of Transportation, National Highway Traffice Safety Administration, occasionally updated.
2. "Highway Statistics," published annually by the U.S. Department of Transportation, Federal Highway Administration.

3. "Accident Facts," published annually by the National Safety Council, Chicago, Illinois.
4. H. C. Joksch, "Design of Field Passive Restraint Evaluation," CEM Report 4250-694, March 1980.
5. D. W. Reinfurt et al., "A Statistical Analysis of Seat Belt Effectiveness in 1973-75 Model Cars Involved in Towaway Crashes," Highway Safety Research Center, University of North Carolina, 1976.
6. B. J. Campbell, D. W. Reinfurt, "The Degree of Benefit of Belts in Reducing Injury - An Attempt to Explain Study Discrepancies," SAE Paper 790684, 1979.
7. D. Mohan et al., "Air Bags and Lap/Shoulder Belts - A Comparison of their Effectiveness in Real World Frontal Crashes," Proceedings of the 20th Conference of the American Association for Automotive Medicine, 1976.
8. H. D. Pursel et al., "Matching Case Methodology for Measuring Restraint Effectiveness," SAE Paper 780415, 1978.
9. A. C. Grimm, "Use of Restraint Systems, a Review of the Literature." The HSRI Research Review, Vol. 11, Nos. 2 and 3, September-December 1980.

PUBLIC PERCEPTION AND BEHAVIOR IN RELATION

TO VEHICLE PASSENGER RESTRAINTS*

Leon S. Robertson

Yale University
14A Yale Station
89 Trumbull Street
New Haven, Connecticut 06520

With motor vehicle injuries currently occurring to 1 in 50 to 60 people per year in the U.S. [1], the objective risk of injury in a vehicle crash during a lifetime is high. But the probability of such a crash on any given trip is usually very low. The risk is known to vary substantially by time of day, day of week, characteristics of vehicles, characteristics of roadways, and weather conditions as well as driver's age and use of alcohol and drugs. Less well-known is the extent to which transient psychological and psysiological states of drivers increase or decrease their own, their passengers' and other road users' risk. Such factors as stress, anger, preoccupation with matters other than driving and numerous others, probably increase the risk. They are sometimes inferred based on post-crash interviews but are not measured precisely because of their transience and the questionable validity of recall.

In view of the low probability of injury per trip, and the variety of conditions that can change the risk and the perception of the risk, no one should be surprised that use of restraints that must be activated on each trip - seat belts and child restraints - were used by less than 15 percent of the population in recent surveys [3]. Despite a variety of efforts to persuade people to use seat belts (and, more recently, child restraints), use in the United States today is little different than it was more than a decade ago [4].

---

*Preparation of this paper was supported by a grant from the Henry J. Kaiser Family Foundation.

Persuasion and Belt Use

Prior to the late 1960s and early 1970s, estimates of belt use and research on correlates of belt use were based on claimed use in interview surveys [5]. These studies were brought into question when a team of researchers in North Carolina compared claimed belt use in interviews with actually observed use. Of those who claimed to use belts always when driving close to home, only 77% were actually using them when observed in their cars. The discrepancy was even greater among those who claimed to use belts always on long trips; only 46% of these drivers were actually using belts [6].

In 1970, a study compared belt users and nonusers observed at the same time, place, and moving in the same direction with respect to various attributes, experiences, and attitudes toward belts. This study design thus controlled as much as possible for objective risk which was probably similar in both groups. Users and nonusers were different on a number of dimensions. Nonusers had less formal education, rated belts more often as uncomfortable and inconvenient, and were more often smokers. Users were more likely to have had a friend injured in a crash [4].

Partly based on these results, a series of television ads were developed in which surrogates for a friend injured were displayed in a variety of situations directed at particular audiences. One of the ads won a prize and another was highly placed in judged competition among advertisers. A dual cable television system, in which the two separate cables are laid in a grid in a community, was used to test the effect of the ads on belt use. The ads were shown 943 times to audiences targeted by type of program on one cable but were not shown on the other during nine months. In contrast to much public service advertising, prime time programs were inclued. The campaign would have cost $7 million if done nationally.

Belt use was observed at various sites in the test community throughout the nine-month period as well as a month before and a month afterward. Vehicles from households on the separate cables were identified by linking the names and addresses of vehicle owners from the state license tag file to names and addresses on the cable television billing file. There was no significant difference in belt use by drivers in vehicles from the households on the separate cables or between these and other observed vehicles that were not associated with either cable [7].

Independent of this work, a study of belt use in three communities in California - one as a control and two exposed to varying intensity of multimedia advertising - found no effect on observed belt use [8]. A comparison of belt use in Grand Rapids, Michigan and Milwaukee, Wisconsin after an intensive ad campaign in Grand Rapids found no difference in use in the two cities [9]. Although the infinite variety of possible ad campaigns has not been tested, the available research suggests that substantial gains in belt use from advertising are unlikely to be achieved.

In response to the government's attempt to increase the criteria for automatic crash protection in the early 1970s, executives of motor vehicle manufacturing companies, in a White House meeting, apparently persuaded then President Nixon that alternatives should be tried [10]. Two were subsequently allowed - the buzzer-light system from January 1972 through the 1973 model year and the ignition interlock in the 1974 and part of the 1975 model year. With the exception of about 12,000 cars manufactured with automatic air bags, most manufacturers installed buzzers and interlocks. The buzzer and a light illuminating a "fasten seat belt" message on the dash were activated if one of the front seats was occupied by a certain weight, the transmission was in forward gear, and the belt was not extended from its stowed position. A large survey of observed belt use in 1972 cars with and without this system (so identified by linking license tag numbers to vehicle identification numbers in state license files) found little net difference in use [11]. Belt use increased in some groups but decreased in others. Some car occupants tied the belts in a knot to keep them extended from the retractors which rendered them useless without a substantial struggle to untie them.

The ignition-interlock, which would not allow the car to start unless belts were extended or latched when the seat was occupied by a certain weight, increased belt use to about 60% in those vehicles for a short time [12]. However, periodic surveys found a sharp decline in a few months as people learned that the interloack could be disconnected without damaging the car [13]. The interlock also provoked outraged letters to Congress by drivers who could not start their cars when any cargo above the specified weight was placed in the seat - groceries, dogs, or whatever [10]. Legislation was passed prohibiting the Department of Transportation from further requirement of interlocks.

A more personalized approach has been used in experiments attempting to persuade parents to place their children in child restraints. In-hospital counseling of mothers of newborns by a health educator had no effect and, in one group, giving them child restraints to remove any financial barrier to use resulted in only a slight increase in restraint use observed at follow-up visits [14]. Intense instruction and demonstration of use of child restraints by pediatricians in their offices increased restraint use observed in subsequent visits. However, after four months these gains were eroded to about the same use as that in a control group [15].

Belt Use Laws

While these attempts to increase belt and child restraint use were occurring in the U.S., a number of states, provinces and countries in other parts of the world took a different tack - the requirement that drivers and passengers, with some exceptions, be required to use belts by law. The first such law in an industrialized country, in the State of Victoria in Australia, was reported as a great success in the early 1970s [16]. Following this lead, a few jurisdictions each year around the world have adopted similar laws.

The Australians had reason to be proud of their law. Belt use increased initially to about 70% [16] and had reached 80% in the major cities some five years after the laws were adopted throughout the country [17]. A study of fatal injuries in Victoria before and after its law compared to other states that had not as yet adopted the law found a 20% death reduction in urban crashes and a 10% reduction in rural crashes that could be attributed to the law [18].

Similar high belt use was found in New Zealand [17] after passing its law and has been reported by researchers in other countries [19]. However, not all jurisdictions have found the laws that effective. When the Province of Ontario, Canada, adopted a belt use law in January 1976, initial use increased to 55% lap and shoulder belt use and an additional 11% lap belt use. As in the U.S., lap and shoulder belts in Canada could be latched separately in vehicles manufactured before the 1974 models. Within two months of the Ontario law's adoption, a public reaction to uncomfortable shoulder belts in pre-1974 cars resulted in an exemption from required shoulder belt use in 1973 and earlier models. This was apparently interpreted by the public as a failure of resolve to enforce the law because use of any belts declined to about 50% and remained there at last count [20]. Quebec adopted Ontario's weakened law and belt use was measured at 50% after the law [17].

No U.S. state had required seat belt use of adults; but use of child restraints, with exemptions, has been made law in Tennessee and Rhode Island at this writing. The Tennessee law exempted children riding on an adult's lap - a hazardous practice. Nevertheless, child restraint use by children subject to the law increased from 8% before to 29% two years afterward with little change in on-lap travel [21]. The more recent law in Rhode Island requires child restraint use in front seats only. Four months after the law restraint use in front seats increased from 22 to 35% and rear seat travel increased from 49 to 62% [22].

Although seat belt use is increased when required by law, the reduction in deaths has not been as large as expected from the known effectiveness of belts used voluntarily. The expected percentage change in death rate as a function of change in belt use is:

$$Pe = 100 \frac{E(U_2-U_1)}{1-U_1 E} \tag{1}$$

where Pe = expected percentage change in death rate; U_1 = proportion of vehicle occupants using belts at time t; E = effectiveness of belts in death reduction when worn.

In Australia, with one-piece lap and shoulder belts, E equals 0.6 [23] and the belt use rate increased from 20 to 70% from before to after the law. Applying Formula 1, the expected death reduction would be 35%; yet the actual reduction was 20% in urban areas and 10% in rural areas [18]. The apparent reason for the less than expected effect is that belts were less often used by those more likely to be involved in serious crashes. Research from various countries that have enacted belt use laws has found that younger drivers and drivers with high blood alcohol concentrations in crashes are less often using belts [17, 24].

Belt Use and Driving Behavior

Behavior that increases risk is associated with immaturity in judgement in the case of the young and with impaired judgement in the case of the alcohol impaired. Nevertheless, an economist has spun a theory saying that drivers have a set level of risk that they are willing to accept and are able to adjust their behavior accordingly. However, the problem of estimating objective risk, even by sober, mature drivers, much less the impaired, was not adequately addressed. According to the theory, a driver whose crash protection is increased will supposedly take more risks to reduce driving time in order to use the time to maximize other wants. It is doubtful, given the cost of auto repairs and insurance surcharges for crash involvement, that anything would be gained by such behavior. Nevertheless, the theory's proponent claimed that his time series regression model showed that increased motor vehicle crashworthiness caused greater hazard to pedestrians. Adjusted "pedestrian" death rates were higher than the model projected in the years following federal motor vehicle safety regulations [25]. However, a reanalysis of the data found that the variables in the model did not accurately project death rates prior to regulations and, therefore, were inappropriately used to estimate effects of regulation. Unregulated trucks were not separated from regulated cars; and motorcyclists were treated as "pedestrians" despite the facts that their rider death rates are about four times those of car occupants, that more than 30% of motorcyclist deaths do not involve other vehicles, and that motorcycle registrations were doubling every five years during the 1960s and early 1970s. Corrections of these and other errors in the analysis resulted in fewer than expected nonoccupant as well as occupant deaths attributable to regulated automobiles

[26]. Actually, the federal standards included crash avoidance as well as increased crashworthiness standards. A study of all the vehicles that actually hit other road users (pedestrians, motorcyclists, and pedalcyclists) during 1975-1978 found fewer deaths of occupants of regulated cars and of other road users struck by those cars [27].

An economist in Australia did a time series regression analysis of aggregated variables in relation to death rates and claimed that the belt use laws decreased deaths of occupants but increased deaths of pedestrians [28]. However, he failed to acknowledge the finding of a quasi-experimental design, comparing death rates in Victoria after its belt use law with other states without the law. That study found no increase in pedestrian deaths in Victoria other than that would have been expected from the experience in the other states without the belt use law [18].

Recent studies of observed risks in driving behavior by belted and unbelted drivers suggest that belt use is associated with other types of risk aversion rather than risky behavior. One research team measured the gap between leading and following vehicles and found that a greater gap was allowed by belted compared to unbelted drivers [29]. A second group counted shoulder belt use of drivers who ran through intersections after lights turned red compared to the drivers who were the first to stop after the light changed. Only 1% of those who ran the red light was belted compared to 8% belted of those drivers who were first to stop at a red light [30].

In sum, the studies with the more adequate research methodologies find no support for the theory that more risks are taken by drivers protected by restraints. The aggregated time series analyses that supposedly suggest the opposite actually illustrate the intellectual poverty of that methodology.

Public Preferences

The apparent resistance to belt use in the U.S. leads to speculation in some circles that the risk in currently manufactured vehicles is acceptable to the public. To the contrary, research for a decade has found that U.S. new car buyers (and the public generally) desire greater automatic protection in cars. Marketing research by General Motors Corporation (GM) in 1971 (not revealed publicly until 1979) found such a desire.

GM researchers showed new car buyers air bags that inflate automatically in severe frontal crashes and automatic seat belts that wrap around outboard front seat occupants as the doors are closed. Crash test film of air bags and automatic belts as well as favorable and adverse publicity on air bags were shown to the 630 respondents. The marketing report states, "consumers are overwhelmingly in favor of some kind of occupant restraint system... (are) not scared of the Air Cushion concept and that the Air Cushion Restraint concept is a viable one to the consumers... After seeing prices quoted, which were higher than those which respondents expected, and had a slight dampening effect - the Air Cushion still maintained half of all preference votes..." [31]. Other private GM studies and a number of public opinion polls found strong demand for air bags throughout the 1970s but thus far they have been offered to the public on only the most expensive models of 1974-1976 cars. They were not advertised, and buyers encountered dealer resistance and delays in filling orders for them [32].

The extent to which the willingness to buy and pay for increased crash protection is related to perceived personal vulnerability to injury was explored in a national random sample of 1,017 people who planned to buy new cars. To measure perceived personal vulnerability, the new car buyers were asked whether they thought their "chances of being killed or injured in a car crash" were greater than, the same as, or less than "people like yourself." Only 6% said "greater than" compared to 40% who chose "less than."

The perceived relative vulnerability to crash injury or death was not correlated with preference for crash protection or amounts the new car buyers were willing to add to their new car payments to save specified numbers of statistical lives. Only 15% chose exclusively "protection that you and your passengers must activate every time you travel." In contrast, 39% chose "protection so that you and your passengers do not have to do anything," and 38% chose "both types of protection." The remaining 8% expressed no opinion. These new car buyers expressed willingness to add to new car payments an average of $12.09 per month to save 6,000 lives per year, $16.69 per month to save 12,000 lives per year and $19.92 per month to save 18,000 lives per year [33].

Conclusions

The widespread lack of use of seat belts and child restraints is partly a result of the discomfort and inconvenience involved and, perhaps, the imprecise but relatively accurate perception of low probability of injury during any given trip. However, the disproportionate numbers of people who believe that others are more vulnerable than themselves suggests that convincing people of the objective, high lifetime probability of injury while traveling in motor vehicles applies to them may be impossible. It is conceivable, though difficult to prove, that the act of using a restraint is a confession of vulnerability that many people are unwilling to make. If that is so, neither argument based on rational and/or other grounds, nor law is likely to convince them otherwise.

Marketing studies clearly indicate that people desire, and are willing to pay for, automatic crash protection; but the market has not provided such protection except in vehicles that have the most important unsolved mysteries of our age is why the supply of increased automatic crash protection has not risen to satisfy the demand without government intervention.

References

1. Health Resources Administration, Current Estimates from the National Health Survey, United States - 1978, Washington, D.C., U.S. Government Printing Office, 1979.
2. W. Haddon, Jr., E. Suchman, and D. Klein (eds.), Accident Research: Methods and Approaches, New York, Harper and Row, 1964.
3. Insurance Institute for Highway Safety, Seat belt use continues to decline, Status Report, 15:6, 1980.
4. L. S. Robertson, B. O'Neill, and C. W. Wixom, Factors associated with observed safety belt use, Journal of Health and Social Behabior, 13:18-24, 1972.
5. E. g., D. I. Manheimer, G. D. Mellinger, and H. M. Crossley, A follow-up study of seat belts usage, Traffic Safety Research Review, 10:2-13, 1966.
6. P. F. Waller and P. Z. Barry, Seat belts: a comparison of observed and reported use. University of North Carolina Highway Safety Research Center, Chapel Hill, 1969.
7. L. S. Robertson, A. B. Kelley, B. O'Neill, C. W. Wixom, R. S. Eiswirth, and W. Haddon, Jr., A controlled study of the effect of television messages on safety belt use, American Journal of Public Health, 64:1071-1080, 1974.
8. G. A. Fleischer, An experiment in the use of broadcast media in highway safety, Department of Industrial and Systems Engineering, University of Southern California, Los Angeles, 1972.
9. L. S. Robertson, Auto industry belt use campaign fails. In: Background Manual on the Passive Restraint Issue, Washington, Insurance Institute for Highway Safety, 1977.
10. Motor vehicle and schoolbus safety admendments of 1974. Congressional Record, August 12, 1974, H 8119ff.
11. L. S. Robertson and W. Haddon, Jr., The buzzer-light reminder system and safety belt use, American Journal of Public Health, 64:814-815, 1974.
12. L. S. Robertson, Safety belt use in automobiles with starter-interlock and buzzer-light reminder systems. American Journal of Public Health, 65:1319-1325, 1975.
13. U.S. Department of Transportation belt use surveys done by various contractors, 1974-1978.
14. K. S. Riesinger and A. F. Williams, Evaluation of programs designed to increase the protection of infants in cars, Pediatrics, 62:280-287, 1978.

15. K. S. Riesinger et al., The effect of pediatricians counseling on infant restraint use, Pediatrics, 67:201-206, 1981.
16. D. C. Adreassand, The effects of compulsory seat belt wearing legislation in Victoria, Proceedings of the National Road Safety Symposium, Canberra, 1972.
17. L. S. Robertson, Automobile seat belt use in selected countries, states, and provinces with and without laws requiring belt use, Accident Analysis and Prevention, 10:5-10, 1978.
18. L. A. Foldvary and J. C. Lane, The effectiveness of compulsory wearing of seat-belts in casualty reduction, Accident Analysis and Prevention, 6:59-81, 1974.
19. F. G. Fisher, Effectiveness of safety belt use laws, Springfield, Virginia, National Technical Information Service (PB 80209-901), 1980.
20. L. S. Robertson, The seat belt use law in Ontario: effects on actual use, Canadian Journal of Public Health, 69:154-157, 1978.
21. A. F. Williams and J. K. Wells, The Tennessee child restraint law in the third year, American Journal of Public Health, 70:579-585, 1980.
22. Insurance Institute for Highway Safety, Rhode Island child restraint law evaluated, Status Report, 16:2:5-6, 1981.
23. L. S. Robertson, Estimates of motor vehicle seat belt effectiveness and use: Implications for occupant crash protection, American Journal of Public Health, 66:859-864, 1976.
24. J. B. Dalgaard, Experiences with the new seat belt law on fatal lesions of car occupants in Denmark, in: Proceedings of the Sixth International Conference of the International Association for Accident and Traffic Medicine, Royal Australasian College of Surgeons, Melbourne, 1977.
25. S. Peltzman, The effects of automobile safety regulation, Journal of Political Economy, 83:677-725, 1975.
26. L. S. Robertson, A critical analysis of Peltzman's "The effects of automobile safety regulation," Journal of Economic Issues, 3:587-600, 1977.
27. L. S. Roberston, Automobile safety regulations and death reductions in the United States, American Journal of Public Health, in press.
28. J. A. C. Conybeare, Evaluation of automobile safety regulations: the case of compulsory seat belt legislation in Australia, Policy Science, 12:27-39, 1980.
29. C. R. von Buseck et al., Seat belt usage and risk taking in driving behavior, Society of Automotive Engineers Paper No. 80038, 1980.
30. D. Deutsch, S. Sameth, and J. Akingemi, Seat belt usage and risk-taking behavior at two major traffic intersections, Proceedings of the Annual Meeting of the American Association for Automotive Medicine, October 1980.
31. Advertising and Merchandising Section, Consumer opinions rela-

tive to automotive restraint systems, General Motors Corporation Report No. 71-27P, 1971.
32. A. R. Karr, Sage of the air bag, or the slow deflation of a car-safety idea, Wall Street Journal, November 11, 1976.
33. L. S. Robertson, Car crashes: perceived vulnerability and willingness to pay for crash protection, Journal of Community Health, 3:136-141, 1977.

MOTOR VEHICLE OCCUPANT RESTRAINT POLICY

Joan Claybrook

2034 37th Street NW
Washington, D.C. 20007

INTRODUCTION

In 1980, more than 51,000 Americans were killed on our nation's highways, approximately the number of Americans killed in the entire Vietnam War. About 34,890, or about 70%, of these highway fatalities were occupants of cars, light trucks, or vans [1]. About half of these vehicle occupants could have been saved from death (and many more from serious injury) by properly using vehicle occupant restraints [2]. Other vehicle design changes could prevent additional deaths and mitigate the severity of injuries. This has been demonstrated by the National Highway Traffic Safety Administration's (NHTSA) Research Safety Vehicle Program and New Car Assessment Program [3].

The automotive manufacturers have known for 30 years that use of an occupant restraint system could save many lives in automobile crashes. During this period, more than 650,000 Americans have died whose lives probably would have been spared with the use of an effective crash protection restraint system [4]. During much of this time, members of Congress, state legislators, the U.S. Department of Transportation (DOT), the insurance industry, consumer organizations, and the United Auto Workers have debated with the automotive manufacturers about installation of occupant crash protection systems in new cars. Ford and Chrysler offered lap belts as optional equipment in 1955. But the U.S. auto companies resisted standard equipment installation, first of lap belts, then shoulder harnesses, then air bags, and, most recently, automatic (or passive) belts. The automotive companies eventually installed manual (active) lap belt systems on new cars as standard equipment after a few state laws were enacted and public pressure was exerted in the early 1960s.

The federal government required shoulder belt installation beginning in January 1968 [5].

During the last decade the arguments have focused on automatic restraint systems - air bags and automatic belts - designed to hold occupants in place during a violent crash without any action by the user. Automatic restraints, particularly the air bag, are attractive because they significantly increase restraint usage and do not depend on decisions by the more than 100 million motorists to use the system on each trip. The effectiveness of the belt restraint systems depends on usage. By comparison, the air bag is unobtrusive, conforms with the stylistic preference of most manufacturers for uncluttered vehicle interiors, and gently spreads the crash forces to protect the most critically vulnerable parts of the human body, the head, neck, and chest.

The effectiveness of designed-in automobile crash protection was documented by the General Accounting Office (GAO) in report CED-76-121 in 1974 [6]. Based on the GAO methodology used in this report, NHTSA estimated in 1980 that approximately 70,000 lives have been saved since the installation in 1968 of the minimal crash protection systems in accordance with the federal safety standards [7].

In spite of the documented payoff of automotive restraint systems, safety belt use by the public has been declining since 1974, when the interlock requirement was revoked. Average usage in cars of all sizes was observed at just under 11% in 1979 [8].

The risk to the public of being harmed in a motor vehicle crash is well known and has been documented by NHTSA. The death rate per 100,000 vehicles and per miles traveled has dropped dramatically in the last 50 years because of increases in travel and the vehicle population as well as vehicle and highway safety improvements.

When measured for public health analyses, however, the highway death rate per 100,000 persons has remained relatively constant, even though our knowledge of how to prevent deaths and injuries in crashes has grown enormously. The population figures indicate that while a citizen, on the average, can expect to drive farther without suffering a fatal crash, the probability of being killed in a crash is about the same today as it was some 50 years ago [9].

While the likelihood of being involved in a crash in any given trip is not high, especially for adults above age 25, the harm which occurs in many crashes is devastating. The total number of Americans injured each year is enormous, and the opportunities for mitigating the casualties are readily available at a modest cost.

In terms of frequency, each person in the nation, on the average, can expect to be involved in a crash once every 10 years. The question, then is how serious will one's crashes be?

In terms of severity, the odds are that one of every 40 perons born this year will die in an auto crash and will die young, and one in every 20 will suffer a severe injury. On the average, each day of the year there is a fatality every 11 minutes and an injury every 9 seconds [10]. Approximately 1.5 million Americans suffer disabling injuries each year, requiring medical treatment that restricts their activity. Auto crashes are the largest single cause of parapelgia and quadriplegia and a major cause of epilepsy; they are the largest killer of Americans under age 34. Today's motor vehicle death rate is of epidemic proportions, comparable to the typhoid and diptheria death rates of the early 1900s - and it is rising [11].

With the current shift from large to small cars, the number of deaths per year is estimated to increase by about 10,000 to 15,000 per year by 1990, if no additional safety provisions are built into the new smaller cars. Unfortunately, most Americans making this switch do not know they are probably doubling their risk of death and injury in a crash. The 1979 fatality figures document the problem. In crashes between subcompact cars and full-size cars, eight occupants of the smaller car were killed to one in the larger. The ratio for subcompact- and compact-size cars is 3.5 to 1. While 38% of the cars on the road in 1979 were subcompact, occupants of these small cars accounted for 55% of the deaths in two-vehicle crashes and 51% of the single-vehicle crashes. By comparison, 62% of the fleet in 1979 were large- and intermediate-size cars, but they accounted for 45% of the occupants killed in two-car crashes [12].

There are numerous well-known vehicle design remedies to reduce the risk of death or injury in a motor vehicle crash. Thirty years ago the public focus on vehicle design was reinforced when researchers pointed out that we could prevent death and injury, even if we could not prevent the crash. The auto manufacturers had known this for years. In 1940, a Dodge car ad suggested:

> Instrument panel is smooth and flush. Back of front seat is heavily tufted to protect rear seat passengers in case of an emergency stop. Door handles are smooth, rounded and curve inward for safety [13]!

The important occupant safety features, in addition to smooth and soft interiors, are 1) structural designs for improved crash energy management to absorb, control, and reduce the crash forces on occupants; 2) structural integrity to prevent occupants from being ejected, trapped, burned, or crushed by the steering column or collapse of the occupant compartment; and 3) most critical, effective occupant restraint systems designed to prevent or soften the second collision of the occupant with the vehicle's interior [3].

HISTORY

The history of the past 30 years is not encouraging for assessing what circumstances or incentives might have caused known restraint system technology to be built into automobiles to protect U.S. car occupants or for predicting what might achieve this in the future, but a short look at the past might help to suggest the course, however uncertain, for the future.

The value of belt systems has been known since World War I when they were used in the early years of military aviation. By the late 1920s, federal regulations required installation and use on all civilian passenger aircraft. The work of Hugh DeHaven at Cornell Aeronautical Laboratories (CAL) in the 1940s and Colonel John Paul Stapp for the Air Force in the early 1950s convinced car racing associations to require racing drivers to use belts by the late 1940s and early 1950s.

Research information released from CAL in 1954 and 1955 showed the high death rate from ejection and the dangerous trajectory of and damaging impact on children in crashes. In 1955, Chrysler and Ford announced lap belts would be available as options, but it was not until 1964 that lap belts became standard equipment. Until that time, no anchorage attachments were available in vehicles sold without belts installed as options, making belt installation extremely difficult [14].

In 1959, State Senator Edward Speno, chairman of the New York State Joint Legislative Committee on Motor Vehicles and Traffic Safety, decided seat belts should be factory installed, as are other critical vehicle systems. After negotiations in 1961 between Senator Speno and the manufacturers to get standard installation of at least lap belt anchorage systems for all front seats, New York passed such legislation. In 1961, Wisconsin and, in 1963, New York enacted laws requiring front seat lap belts installed in all new cars.

In 1963, one auto manufacturer broke out of the opposition phalanx. Sherwood Egbert, president of Studebaker, announced that front seat lap belts would be installed in all Studebakers beginning February 15, 1963. Mr. Egbert stated, "It is our feeling - a strong feeling - that safety measures in motor cars should not come by petition from motorists but that automobile manufacturers should lead in safety equipment" [14].

Following further pressure from New York's Senator Speno, the other auto manufacturers announced in August 1963 that front seat lap belts would be made standard equipment beginning January 1, 1964 [15].

The enactment of the new auto safety law in 1966 focused national attention on the potential for vehicle safety design. The government was required to issue initial safety standards by January 13, 1967 [16]. One of the obvious priorities was the shoulder harness (rear seat lap belts were also mandated). The manufacturers naturally objected, claiming lack of lead time to design properly for these new systems. The new auto safety agency nevertheless required their installation in 1968 models, but that was not the end of the battle [5].

In the summer of 1967, just a few months before the beginning of the new model year, General Motors (GM) brought a series of crash test films to the Department of Transportation to show that the shoulder harness might be dangerous. The films showed unbelted rear seat occupants coming forward and hitting the heads of front seat occupants held in place by shoulder belts, potentially causing injury [17].

While this type of event appeared unlikely to occur, Dr. William Haddon, Jr., M.D., the first director of the auto safety agency, broadcast an appeal to experts world-wide for data on the performance of shoulder belts. A thorough response came from Volvo, a company that prides itself on building safety into its vehicles and that had been installing shoulder harnesses as standard equipment since 1959. Volvo presented data from a study of 28,000 crashes investigated in Sweden. There were no deaths in crashes under 60 mph when the belts were used, but there were fatalities at speeds as low as 12 mph for unrestrained occupants [18].

It was not until later that Dr. Haddon's agency obtained from the GM Truck Advance Design Group a secret GM report entitled "Reduction of Automotive Accident Injury" and dated February 14, 1964. This report, contradicting the message in the GM films and presentation of mid-1967, suggested design methods for reducing the likelihood of injury in 1967 GM light truck designs. It stated:

> Considerable passenger car controlled-impact test data is available. Results obtained from instrumented impact tests indicate nearly all impacts are survivable, even those against fixed barriers. The deceleration of the occupant depends on a number of factors; but the most important is effective restraint.
>
> When effectively restrained, the occupant can make full use of the decelerating properties of the collapsing structure; and is prevented from impacting against the interior or being ejected.

The GM report, after assessing survey information from Cornell automotive crash injury research, concluded:

1. Protection of the head and chest from sudden impact against the car interior in front-end collisions has the greatest potential for reducing serious injury to occupants;

2. Retaining the occupants within the compartment during impact and rollover is the next most important factor in reducing injury [19].

Motor Vehicle Safety Standards 208 and 210, requiring lap/shoulder belts and anchorages, took effect on January 1, 1968. (They did not, incidentally, require shoulder belts in light trucks and vans.) Just three years later, in early 1971, at the annual meeting of the Society of Automotive Engineers, W. D. Nelson of GM's engineering staff research and development laboratory presented a paper entitled "Lap-Shoulder Restraint Effectiveness in the United States" that concluded:

> Lap-shoulder belts, although infrequently used by vehicle occupants, are demonstrating a remarkably high reduction of injury in collisions where they were used. A search was made of all collisions in the GM files where at least one occupant was wearing the lap-shoulder belt combination restraint. Of the 160 cases found for this study, 60% of the vehicles had heavy damage of the type often associated with occupant injury; however, 99% of the lap-shoulder belt users either had no injury or only minor injury. The only two fatalities found in the study involved accidents occurring under unusual circumstances [20].

Mr. Nelson did not mention, of course, that his employer was responsible in large measure for the unfavorable public reaction to safety belts. Not only did GM fail to urge its customers to use belt systems, but its design and marketing representatives characterized shoulder harnesses as spaghetti, which perhaps was a fair description of their crude and uncomfortable design. The easier-to-use and more conformtable three-point belt systems produced by many foreign manufacturers were not installed in most U.S.-made cars until an amended federal safety standard required it in 1974.

In contrast to most of the decisions made on the installation of belt systems, the history of the air bag has been more public, more political, and more disappointing. Although invented and developed in America and funded mostly by the Detroit automakers, General Motors and Ford have bobbed and weaved for the past 12 years to avoid manufacture and sale of this magnificant space-age technological vaccine. The first federal proposal for installation of automatic restraints was issued in August 1969, by which time significant design and development work had been completed by the Detroit companies and their suppliers.

In August 1970, General Motors, in the first of many similar propositions to forestall government action, pledged voluntarily to install air bag systems as standard equipment in all 1975 models and before that as options in some vehicles, if the effective date of the proposed safety standard were delayed from 1973 to 1975 models [21].

The standard was first delayed until the 1974 models. Then, in 1971, the Nixon Administration responded to complaints by Ford Motor Company and delayed the standard for several more years, substituting instead, at Ford's suggestion, the ignition interlocking seat belt [22]. In August 1973, General Motors announced it would make no more than 150,000 air bag-equipped vehicles in 1974 and 1975 [23]. In 1974, the Congress revoked the interlock provision, returning the standard to a requirement for manual belts only [24]. After not promoting the availability of the air cushion restraint system [25], and having sold only 10,000 cars in 1974-1976 with air cushion restraint systems, GM decided in 1976 to discontinue the option.

In a January 1977 agreement with Secretary of Transportation William Coleman, however, Ford, General Motors, and Mercedes-Benz contracted with the DOT to manufacture between 40,000 and 440,000 air bag-equipped intermediate and compact 1980 and 1981 model cars on the condition that a federal regulation would not be issued [26]. In June 1977, Transportation Secretary Brock Adams issued an automatic restraint standard with 4 to 6 years lead time, attempting to resolve the matter once and for all by accommodating the industry lead time request [27]. He urged the companies to voluntarily carry out the Coleman agreement in selected 1980 and 1981 models before complying with the standard in all large cars in 1982. But this did not happen. No company would agree to fulfill the Coleman contract, although GM and Ford pledged in congressional hearings in the fall of 1977 to offer air bags on some 1981 large cars instead [28].

In October 1979, GM decided not to offer air bags as options in 1981 model cars because animal tests of the GM design suggested possible injuries to out-of-position children, possibilities that GM hypothesized might occur in actual use [29]. General Motors' announcement raised doubts in the minds of the public and had a severe dampening effect on the air bag work of other companies. Although GM had spent millions on its test program and implied that the problem was generic, it soon became obvious that it was readily controlled by the air bag designer with proper selection of height and width of the dashboard design, speed of inflation, bag shape, etc. [30].

Just two months later, in December 1979, GM announced it had solved its problems with design changes and would offer inflatable restraints as an option at the beginning of the 1982 model year [31].

Three months later, in March 1980, GM said it would not offer inflatable restraints on medium and small cars in 1982 through 1986 models but might still offer them on 1982 full-size cars [32]. That same month Mercedes-Benz announced it would equip all 1982 models with air bags as standard equipment [33].

Then, in June 1980, GM announced it had canceled plans to offer air bags as options on 1982 large cars because it had decided to redesign its large cars for 1983 [34], and the tooling for air bags was too costly for one model year. In July 1980, GM urged that the standard be delayed until 1983 and agreed to sign a contract with the DOT for production of some air bag systems in 1983 models, if small cars rather than large cars were covered by the standard in that year [35].

In August 1980, the GM proposals was incorporated instead into an enforceable legislative package and made applicable to GM, Ford, Toyota, Nissan, and Volkswagen (VW). It passed the Senate but failed to pass the House of Representatives in December by three votes [36].

In April 1981, the Reagan Administration delayed Standard 208 by one year and simultaneously proposed to eliminate or rearrange the 1983 and 1984 effective dates [37]. Shortly thereafter, Mercedes-Benz announced it would delay installation of air bags until the standard took effect [38]. In May 1981, General Motors and Ford canceled all air bag research and development work completely.

CRITERIA

In assessing the optimum choice in restraint systems to protect the public, there are basic criteria against which each option should be comparatively measured. In order of societal importance, they are:

1. Public acceptance

2. Performance in the crash

3. Cost

4. Time lag for taking effect

There are, of course, other issues of importance to manufacturers and the public, such as life span of the system, repairability, manufacturer capacity to mass produce, and environmental effects in production, operation, and disposal. But for seat belts and air bags, these issues present no significant problems, although the question of environmental effects has been raised occasionally during the course of various political debates. (See Appendix A for a discussion of environmental effects, life span, and repairability of restraint systems and notes on child restraints.)

The two basic restraint systems offered for sale commercially to date are belts and air bags (although no automotive company has offered air bags for sale since 1976). Each of these systems will be discussed in reference to the societal criteria, with special comments directed to the differences, where appropriate, between manual and automatic belts [39].

1. Public Acceptance

Public opinion polls and marketing studies conducted since 1971 show a strong customer interest in built-in crash protection and inflatable restraint systems. A 1978 survey for Amtrak by Louis Harris found that the overwhelming majority of respondents (83%) believed improvement in automobile safety was the most important change needed in transportation systems [40].

Two surveys in 1978, one by Volvo and one by Peter D. Hart Associates under contract with the NHTSA, found that air bags were the overwhelming favorite compared to automatic belts. Consumers reported they do not wear manual belts because of discomfort and inconvenience. Thirty-seven percent said nothing could be done to make them wear their belts most of the time [41].

General Motors conducted inflatable restraint marketing studies in 1971, 1975, 1978, and 1979. All showed strong consumer demand for air bags in new cars. The 1978 study of more than 1000 GM car owners concluded, "The air cushion restraint system....received the highest ratings on all operation, comfort, and appearance items evaluated."

The 1979 study, based on in-depth interviews with recent Chicago purchasers of large GM cars, stated, "With passive belts and an Air Cushion Restraint System available, 70% of the total principal-driver sample selected the air bag even when they were told its cost would be more than four times the cost of a passive belt system." The results of the GM studies were not made publicly available until requested by Representatives John Burton of California in December 1979 [42].

Attempts to significantly increase the usage of manual belt systems through various mechanisms, including buzzers, interlocks, and mandatory laws, have had no long-term effectiveness in the United States, as has been discussed in more detail at this conference by Leon Robertson of Yale University. Usage rates of automatic belts have been shown to increase dramatically when voluntarily purchased [43], but there is skepticism about whether these high usage rates would continue with automatic belts installed in all new cars sold [37].

There are a number of reasons cited by the public for failure to use existing automotive belt systems. In several in-depth NHTSA studies, the primary reasons cited are lack of comfort of the system when worn and substantial inconvenience in fastenging the system [44]. To remedy these discincentives, the NHTSA issued a requirement for improved belt system comfort and convenience to take effect in 1982. Because of the great variety of vehicle types and sizes and the difficulty of defining performance specifications for belt comfort and convenience, there were numerous industry objections to the original proposal. The final rule issued in December 1980 addressed fewer problems [45], but would have pressed the manufacturers to improve their designs. However, the Reagen Administration has now announced that the scope of the standard will be drastically reduced, retaining only the shoulder belt pressure limit provision. This recent decision is rationalized on the basis of cost and lack of correlation between comfort and convenience surveys and observed usage [46].

In the fall of 1978, I wrote every governor a latter asking if they would work with the DOT for the enactment of mandatory adult seat belt use legislation in their states, recognizing that less drastic measures had failed. While the governors were sympathic to the problem, none of them would support such legislation [47].

Publicity and advertising have not been any more successful. As I explained in a letter to Representative John Dingell of Michigan in July 1979, approximately $10 million has been spent by the government and additional millions by private industry during the last decade to encourage and enhance seat belt usage - and it hasn't worked [48].

In summary, there is every indication that inflatable restraints are highly acceptable to the public, and the accompanying lap belt is the least obtrusive restraint system to provide protection for some side impacts and for rollover crash protection. In contrast, belt systems are not well liked by the public and have never been worn in large numbers despite serious scientific attempts to increase usage through buzzers, interlock systems, and sophisticated industry and government advertising. Since public acceptance and use are essential to crash performance effectiveness, inflatable restraints must be rated far superior to belt systems for this most critical criterion.

2. Performance in a Crash

The NHTSA has assessed the performance of air bag systems and belt systems - when used - and has concluded that they both supply a high level of protection in crashes. The air cushion restraint systems is designed to protect occupants in frontal crashes. Approximately 55% of the occupant deaths occur in this mode of crash.

The lap belt, which is installed with the air bag, is important for protection in rollover crashes and certain opposite-side impact crashes. (The seat structure and head restraint supply protection in rear crashes; the side structure and interior padding are the sources of protection for most side crashes.) The air bag/lap belt system, according to th NHTSA analyses, provides somewhat superior protection to the lap belt/shoulder harness. In crashes above 35 mph, the air bag is superior in that it distributes the crash energy evenly without harm to the occupant [49].

The automatic belt, installed in approximately 400,000 VW Rabbits since 1975 [50], has shown nearly a 50% reduction in fatalities compared to otherwise identical manual-belt Rabbits. The General Motors air bag cars, which have now traveled more than 800 million miles, have also worked magnificently, with more than a 50% reduction in fatalities and serious injuries compared to otherwise identical GM cars [51]. There have been two inadvertent deployments in moving vehicles (design corrections would prevent recurrence in new generation air bag designs), and in no case have the air bags failed to deploy in a crash. They have worked well in multiple impact crashes [52]. Various reports document the payoff of both of these systems [51, 53].

In May 1981, two women in Utah, a 53-year-old driver and an 81-year-old front seat passenger, in a 1975 air bag-equipped Toronado crashed head-on with a tank truck at a closing speed of 80-100 mph. The air cushion performed effectively. The women suffered only repairable injuries, leg fractures and a ruptured spleen, in this life-threatening crash [54].

As for manual belts, NHTSA crash tests at 35 mph during the past 2 years have revealed that most current manual belts do not protect occupants as well as they should. In some cases, the belts reeled out too much when they were loaded by the instrumented test manikin's forward movement in the crash, allowing the manikin's head to strike against the steering wheel or the instrument panel. In some cases, the belt loads were excessively high. According to biomechanics experts, loads in excess of 1500 pounds are likely to cause unnecessary rib fractures and other thoracic unjuries. Inexpensive webbing locks, if used, could prevent some of the problem. It is also possible to install more sophisticated systems for belt pre-tensioning and load limiting devices.

Careful attention to the design of restraint systems, whether belts or inflatable restraints, is necessary to achieve the best performance. Most current belt systems in virtually all cars on the road are not designed to encourage use and when worn do not provide the level of protection they should.

3. Cost

Manual belts are significantly less expensive than inflatable restraints and are effective for the roughly one out of 10 persons who use them. But this lower price must be measured against the costs of massive nonuse of belt systems. The major advantage of inflatable restraints in comparison with belts are automatic operation, and thus high overall effectiveness, and their daily comfort and convenience value.

Over the lifetime of a car, the driver enters the vehicle somewhere between 20,000 and 40,000 times. If the owner puts a price tag on comfort and convenience, how much would he pay per trip to secure superior crash protection without having to bother with the belts? If the answer were one penny per trip, that would be sufficient to pay for a several-hundred-dollar air bag.

The cost of restraint systems, whether belts or inflatable, is highly volume sensitive. Front seat manual belt systems now cost approximately $75. Automatic front seat belts add between $50 and $100 to the consumer price above existing manual systems [40, 55]. Some automatic belt systems, such as the luxury system produced by Toyota with a small motor to move the belt automatically, could add as much as $350 to the price of a new car [56].

Recent statements out of Detroit about $1100 inflatable restraints refer to limited production systems. In mass production, the price of inflatable restraints is competitive with numerous other and far less critical automotive systems sold as optional equipment, such as vinyl roofs for $120, AM/FM radios for $200, air conditioning for $600, and automatic transmission for $400. The average car buyer spends approximately $800 on optional equipment. The average working person spends $500 per year on health insurance.

Numerous estimates have been made of the price of inflatable restraints, ranging from DeLorean's $122 mass production figure in 1976 [57] to GM's $1100 limited production price in 1981 [58]. In 1977, GM estimated the price to be $193 in mass production [59]. Recent inflatable restraint supplier estimates indicate that in mass production of approximately two million units, the price to the consumer should be less than $200 in today's cars and less than $250 for one million units [60]. A recent study for NHTSA by the industry cost estimating company, Pioneer Engineering, found that a $375 price tag for the gold-plated Lincoln Continental air bag with 300,000 units produced would cover costs and profit [61].

Two other cost figures should be mentioned. First, the insurance industry has indicated it will reduce automobile premiums by $40 or more per year for owners who buy automatic restraint cars. At $40 a year in discounts, even a $350 air bag would be paid back during the lifetime of the car.

Second, because of the enormous savings consumers will accrue from the fuel economy standards now in effect, the average buyer of a 1985 car will save about $2800-$3000 net over the life of the car compared to the buyer of a 1977 car. This will be a net savings after subtracting the cost of the fuel economy technology improvements and the safety standards planned by the Carter Administration to take effect by 1985 (primarily improvements in automatic restraints, side impact protection, and pedestrian protection). Under this plan, the consumer gets a more fuel efficient car that is smaller but has compensating safety features added at a significant net savings.

When considered in these terms, the price for mass-produced automatic crash protection that will save thousands of lives with the convenience of inflatable restraints is one of the consumer's best buys.

4. Time Lag for Taking Effect

If installation of inflatable restraint systems began today on all cars produced, it would take 10 years before they would be installed in most cars on the highway. But in 1 year they would be in 10% of the cars, approximately the same percentage as the population now protected by manual belts. Thus, in just over 1 year the percentage of the population protected by existing belts would be exceeded if there were full scale introduction of inflatable restraints.

Although manual lap/shoulder belts are now installed in virtually every car on the highway, because they are not used, there is an indefinite time lag for achieving their potential effectiveness, and there is no known acceptable political method for achieving high manual belt usage.

In summary, while it would take 10 years for automatic restraints to be installed in most vehicles on the highway, there is every reason to believe this approach would achieve higher usage in the short and the long run then more futile attempts to increase usage of currently installed manual belts.

CONCLUSIONS AND SUMMARY

There is an enigma in this variegated history of government's pushing industry to install automotive restraints and industry's manipulating government to avoid it. For the last decade, we have seen one of the most powerful and most profitable industries in the world steadfastly refuse to sell a proven remedy for a massive public health epidemic in compliance with a federal safety standard. This experience has taught us several lessons which are useful for determining future strategies.

1) The first lesson is that the inflatable restraint technology is so meritorious and has such potential that it will not just die and fade away because General Motors tries to bury it. When General Motors announced in late April 1981 that it was closing down all research and development work on inflatable restraints, it was attempting to preempt any further decisions in this area by the Department of Transportation and the Congress.

But a historic parallel depicts a longer term perspective. About 100 years ago, before the automobile, train crashes were killing enough people each year to be a serious national problem. An inventor, George Westinghouse, developed a new safety device - the air brake - and obtained a patent on it in 1869. For 24 years the railroad companies vigorously opposed the use of this advanced technology. Commodore Vanderbilt, the railroad tycoon, typified the industry attitude when he told Westinghouse, "Do you pretend to tell me that you could stop trains with the wind? I'll give you to understand, young man, that I am too busy to have any time taken up in talking to a damned fool" [62]. Despite these rebuffs, Westinghouse persisted, and in 1886 Iowa passed legistlation requiring air brakes. Federal legislation followed in 1893.

2) Because the U.S. auto industry no longer completely controls the U.S. auto market, it can no longer decree that innovative technologies will be banned by refusing to manufacture them. Competition from the Japanese has even forced the traditional free trade advocates in the auto industry to seek government assistance in limiting auto imports. The competition stems from the refusal over the years of the U.S. industry to manufacture classy, smaller, more fuel efficient cars, and the shift in the last 2 years of U.S. public preference toward fuel efficiency. Where General Motors and Ford previously controlled over 75% of the market with their sales of large- and intermediate-size cars, importers now sell to almost 30% of the market, with the smaller Japanese cars controlling most of this share. With consumers, who for years have been used to buying larger, safer cars at the urging of the manufacturers, deciding now to buy more fuel efficient vehicles, there is an obvious market for the seller of an unusually safe, smaller car. If any one of the larger Japanese manufacturers, or even Mercedes-Benz, decided to sell inflatable restraints, General Motors could not ignore that challenge.

3) The societal benefits of most safety systems do not accrue to the manufacturers. Manufacturers measure payoff from either technological or stylistic investments in new cars in terms of their influence on sales, not in lifesaving potential.

Where there is no obvious sales benefit, safety systems are viewed on manufacturer balance sheets strictly as a cost. In contrast, the American public has a strong financial and humane interest in effective, fully used vehicle restraints. Professor William Nordhouse of Yale University recently estimated that the net benefits to society of Federal Motor Vehicle Safety Standard 208 in the 1982-1985 models is approximately $10 billion.

A direct reward and penalty system for manufacturers might be useful as incentives to enhance the public welfare. One future public policy option to integrate manufacturer design decisions and their lifesaving potential might be the adoption of far more sophisticated vehicle insurance rating schemes by make and model. These could be based on the make/model crash testing activities and fatal crash data analyses of the NHTSA.

An insurance company-sponsored property damage research program to assist in setting insurance rates has been successful in Thatchem, Great Britain, and U.S. insurance companies have begun to more aggressively surcharge and discount particular model cars because of property damage characteristics. Alternatively, a manufacturer could be encouraged to improve his vehicle designs by being made responsible for the first five years of the vehicle liability insurance as part of the purchase price of the new car, a system that has been used with some success in Sweden for vehicle property damage insurers.

Because of this requirement, for many years both Volvo and Saab have done investigations of accidents involving their vehicles in Sweden. It was this work that permitted Volvo to prepare the Bolin Report in 1967 for the U.S. Department of Transportation documenting the payoff of shoulder harnesses in 28,000 crashes. At Saab, the data base from such investigations caused the company to redesign its steering column, steering wheel, and rear shoulder harness anchorage points.

4) One remedy for manufacturer reticence to build technological improvements into new cars, particularly safety technologies, is consumer knowledge of vehicle safety performance by make and model.

U.S. vehicle manufacturers avoided sales programs boasting about technological accomplishments until fuel efficiency dominated the market in 1979 and they had no choice. Although for years they advertised a new model each year, they did not really introduce a new model with new technology every year. That would have been too expensive and not justified by the snail's pace of their technological progress. What they did each year was to introduce superficial cosmetic changes which led the public to think there was something new to buy. But, in fact, significant design changes occurred in any given vehicle only every 4 or 5 years. Even today most people do not realize that many cars with different names, such as the Citation and Omega, have the same chassis/platform with slightly different sheet metal and interior designs.

The U.S. manufacturers have avoided selling new technologies because they view this as too risky and costly. They have been able to avoid it because consumers have not had information about the comparative differences in technological achievements, or lack thereof, by make and model. Without this knowledge, car buyers have had no ability to exert pressure on the manufacturers to take technological initiatives.

With the purchase of smaller, more fuel efficient cars, the comparative crash safety of a car has become far more important. Beginning in 1978, the NHTSA crash-tested a large number of foreign and domestic cars to compare their safety quality and found that U.S.-made small cars were superior to the competitive Japanese models.

Safety information should be routinely developed for make and model by the manufacturers at the beginning of each new model year, or when new models are introduced, so that the public will have the ability to buy a car on the basis of this technical knowledge. A proposal to make safety information mandatory is pending in the NHTSA. Without such information, the consumer has no way of deciding which car has superior safety protection. With this information available to consumers, manufacturers are more likely to find it advantageous to develop crashworthy cars, including ones with publicly acceptable automatic restraints.

5) While much blame for the failure of U.S. manufacturers to install inflatable restraints has been laid on the adversarial system for issuance of regulations, this system has, in fact, successfully pushed the U.S. auto industry, with a seriousness of purpose, to develop innovative technologies that have subsequently been copied by industries of other countries. On an individual basis, in a speech here and a seminar there, top level engineers in the U.S. auto industry have commented about the excitement in their profession during this recent period in which government demands were made on their companies to produce effective technologies to make the automobile more socially responsible.

The U.S. rule-making process, with its opportunities for participation by the public as well as the industry, has been a little bit noisier and a little bit more difficult than secret decision-making, but, on the whole it has been far more creative and effective in assuring the production of safer cars. A recent report for the DOT by the Center for Policy Alternatives at the Massachusetts Institute of Technology has documented the view of other countries that, "The American insistence on 'defensible' positions in the regulatory debate (whether in Congress, the agencies, or the courts) has led to a level of analysis that far surpasses any other country. Indeed, other countries often await American research results before embarking on their own regulatory initiatives" [63].

6) Where government safety standards are implemented by the manufacturers, they have the capacity to make them work and guide public acceptance, or to undermine their purpose and to delay them. Where public acceptance is critical to the effectiveness of the standard, as is the case with vehicle occupant restraints, the government should design the safety standard to limit a manufacturer's opportunity to negate its intent. Under the National Traffic and Motor Vehicle Safety Act, the NHTSA has attempted to issue safety performance standards. This has worked well for most crash safety systems which do not interact on a daily basis with the car user, but in the case of occupant restraints, there is a strong argument, given the history of the past 12 years, for the government to exercise more definitive control over manufacturer options in complying with the standard to assure the systems produced are publicly acceptable.

For the past dozen years, the motor vehicle manufacturers have continued to sell the Model A seat belt instead of the preferred space-age inflatable occupant restraints, thus denying the buying public a choice of systems for protection in motor vehicle crashes which kill another American every 10 minutes and injure one every 9 seconds.

In view of manufacturers' symbiotic relationship with the nation's current political leaders who are now "unleashing" the business community, it appears likely this resistance will continue in the near future in spite of continuing exposure in the media and the persistent criticism of the scientific and medical communities. It appears our industry will fulfill the expectations of economist George Gilder who correctly noted in Public Opinion magazine, "Large corporations don't introduce many truly pioneering products in our economy."

The fragility of the U.S. manufacturers, however, in their continuing resistance to change, has become all too evident in the last several years, and it has hurt the entire nation. The pressures on this industry to adopt technological advancements, including inflatable restraints, will continue as the deaths and injuries continue to mount.

NOTES

1. The other major categories of fatalities were: pedestrians (8071); motorcyclists (5143); bicyclists (964); large truck occupants (1287). National Center for Statistics and Analysis, National Highway Traffic Safety Administration, U.S. Department of Transportation (hereafter referred to as NHTSA).

2. The preamble to the final automatic restraint rule as well as supporting documents discuss the life-saving value of restraint systems. 49 CFR 571.208, originally published as 42 FR 128, p. 34289, July 5, 1977; "Automobile Occupant Crash Protection, Progess Report No. 3," NHTSA, July 1980, DOT/HS 805 474, p. 3; "Occupant Protection Program, Progress Report No. 2," NHTSA, April 1979, DOT/HS 804 418, p. 12.
3. Letter from Administrator, NHTSA, November 28, 1980, to presidents of General Motors, Ford, Chrysler, Toyota, Nissan, Volkswagen, Fiat, Renault, American Motors; Small Car Safety in the 1980's, NHTSA, December 1980, DOT/HS 805 729, pp. 117-147; "Test Results of the Minicars Research Safety Vehicle," Goto, Japanese Automobile Research Institute (JARI), Proceedings of the Eighth International Technical Conference on Experimental Safety Vehicles, NHTSA; "The Safe, Fuel-Efficient Car, A Report on Its Producibility and Marketing," NHTSA, October 1980; M. Pavlick and W. T. Hollowell, "Improvements to Vehicle Side Structures for Side Impact Protection," Proceedings of the Eighth International Technical Conference on Experimental Safety Vehicles, NHTSA, October 1980.
4. National Center for Statistics and Analysis, NHTSA.
5. 49 CFR 571, originally published as 32 FR 123, pp. 2408-2421, February 3, 1967.
6. General Accounting Office Comptroller General's report to Senate Commerce Committee, "Effectiveness, Benefits and Costs of Federal Safety Standards for Protection of Passenger Car Occupants," July 7, 1976, CED-76-121.
7. The major federal crash safety standards include such items as laminated windshields, collapsible steering assemblies, removal of sharp dashboard protrusions, improved door locks, shoulder harness and lap belt systems, strengthened side doors, leak resistant fuel tanks, etc., "The Contributions of Automobile Regulation," NHTSA, Office of Plans and Programs, December 1979, DOT/HS 805 501.
8. Benjamin M. Phillips, Opinion Research Corp., "Safety Belt Usage Among Drivers," May 1980, for NHTSA, DOT/HS 805 398.
9. "Automobile Occupant Crash Protection, Progress Report No. 3," NHTSA, July 1980, p. 13; Small Car Safety in the 1980's, NHTSA, December 1980.
10. Congress of the United States, Office of Technology Assessment, "Changes in the Future Use and Characteristics of the Automobile Transportation System," Vol. II, Technical Report, February 1979, p. 185.
11. National Center for Statistics and Analysis, NHTSA; William Haddon, Jr., M.D., "Quadriplegia and Other Motor Vehicle Injuries: Some Implications and Choices for Motor Vehicle Manufacturers," Insurance Institute for Highway Safety, Automotive News World Congress, Detroit, July 25, 1978; "Automobile Occupant Crash Protection, Progress Report No. 3," NHTSA, July 1980, pp. 1-24; Small Car Safety in the 1980's, NHTSA, De-

cember 1980, p. 47; Hartunian, Smart, and Thompson, "The incidence and Economic Costs of Major Health Impairments," Insurance Institute for Highway Safety, 1981; Klauber, Barrett-Connor, Marshall, and Bowers, "The Epidemiology and Head Injury, A Prospective Study of An Entire Community - San Diego County, California, 1978," American Journal of Epidemiology, 113(5) (1981).
12. Small Car Safety in the 1980's, NHTSA, December 1980.
13. "Automobile Occupant Crash Protection, Progress Report No. 3," NHTSA, July 1980, p. 65.
14. Ralph Nader, Unsafe At Any Speed, Grossman, New York, 1965, pp. 96-110.
15. Ibid., p. 106.
16. 15 USC 1381, et seq. Originally passed as Public Law 89-563, Sec. 103(h), September 9, 1966.
17. Personal experience of the author, GM 1967 films are in FMVSS 208 docket, NHTSA.
18. N. I. Bohlin, Passenger Car Engineering Department, AB Volvo, "A Statistical Analysis of 28,000 Accident Cases with Emphasis on Occupant Restraint Value," 11th Stapp Car Crash Conference, October 1967, Anaheim, California, SAE Paper #670925.
19. Truck Advance Design Group, "Reduction of Automotive Accident Injury," General Motors Corporation, Detroit, February 14, 1964.
20. W. D. Nelson, Safety, Research, and Development Laboratory, General Motors Corporation, "Lap-Shoulder Restraint Effectiveness in the United States," Society of Automotive Engineers Congress, Detroit, January 1971, SAE Paper #710077.
21. GM pledged voluntarily to provide air bags, first as options and then as standard equipment, on all its cars by the 1975 model year. "Occupant Crash Protection," submission by General Motors Corporation to NHTSA docket 69-07, #4-085, August 3, 1970, p. 6.
22. 120 CR, H 8129 (Daily Edition), August 12, 1974.
23. Edward Cole, president of General Motors Corporation, letter to Secretary of Transportation Claude S. Brinegar, August 10, 1973; General Motors press release, November 29, 1973, confirmed this decision.
24. 15 USC 1409 note, originally passed as Public Law 93-492, October 27, 1974, the Motor Vehicle and Schoolbus Safety Amendments of 1974.
25. Statement of E. M. "Pete" Estes, president of General Motors Corporation, to Phil Donohue on The Today Show, NBC, September 27, 1979.
26. "The Secretary's Decision Concerning Motor Vehicle Occupant Crash Protection," U.S. Department of Transportation, December 6, 1976.
27. 49 CFR 571.208, originally published as 42 FR 128, July 5, 1977, p. 34289.

28. Secretary Brock Adams, U.S. Department of Transportation press release and statement, Washington Press Club, June 30, 1977. "Passive Restraint Rule," Hearings before the Subcommittee on Consumer Protection of the Committee on Commerce, Science and Transportation, United States Senate, 95th Congress, 1st Session, Department of Transportation's June 30, 1977, Passive Restraint Rule, September 8, 9, 14, and 21, 1977, No. 95-126; "Installation of Passive Restraints in Automobiles," Hearings before Subcommittee on Consumer Protection and Finance of the Committee on Interstate and Foreign Commerce, 95th Congress, 1st Session, H.R. 1019, A Bill to Amend the National Traffic and Motor Vehicle Safety Act of 1966 to Provide for Installation of Passive Restraints for Occupant Crash Protection in New Cars and for Other Purposes, and H. Con. Res. 273 (and all identical concurrent resolutions), Resolved that the Congress Disapproves the Federal Motor Vehicle Safety Standard transmitted to Congress on June 30, 1977, September 9 and 12, 1977, No. 95-89.
29. NHTSA, October 1, 1979, press release on the Department of Transportation reaction to the General Motors decision to postpone air bag installation in 1982 models; articles in The Wall Street Journal and The Washington Post on October 2, 1979, quoted GM officials as saying they would not offer air bags in 1981 models because of problems with out-of-position children; letter from Betsy Anker-Johnson, GM vice president, to administrator Joan Claybrook, September 27, 1979.
30. Biss, Fitzpatrick, Zinke, Strothers, et al., "A Systems Analysis Approach to Air Bag Design and Development, 8th International Experimental Vehicle Conference, Wolfsburg, Germany, October 21-24, 1980.
31. "GM Expects to Offer Air Bags in Some '82 Cars," The New York Times, December 9, 1979. The article quotes David Potter, GM vice president, as saying, "We believe we've got the air bag on tract," "GM to Offer Air Bags as Extra Cost Option on Some Cars in 1982," The Wall Street Journal, December 10, 1979, p. 48. This article quotes company officials as saying that design changes had been made which solved the problem of out-of-position children and that "unless some unforeseen developments occur during completion of a few more tests," air bags would be offered in drive and front seat passenger positions in 1982; in mid-November 1979, GM president E. M. "Pete" Estes told the author at a meeting in Detroit that this decision had been made by General Motors.
32. AP Wire Service, March 19, 1980, story saying GM indicated it would offer air bats on 200,000 large-size 1982 models; "Comments of the General Motors Corp. Regarding NHTSA's Evaluation Plan for FMVSS 208, Occupant Protection," March 1980, docket 74-14, No. 15, GM informed NHTSA that it "does not plan to offer inflatable restraints on medium- or small-size cars" in the 1982-1986 model years, but they may still offer them on full-size 1982 models cars.

33. The Wall Street Journal, September 26, 1979, article indicated Mercedes-Benz would "probably" offer air bags as standard equipment along with safety belts in its U.S. cars and as options in the higher priced models in Germany; the author was told in Stuttgart, Germany on March 4 and 5, 1980, during a visit to the company, that Mercedes would offer air bags as standard equipment in 1982 models; "All Mercedes Benzs to Offer Air Bags in '82," Automotive News, p. 2 (March 31, 1980).
34. "GM Drops Plans to Put Air Bag in Big 1982 Cars," The Wall Street Journal, June 4, 1980, p. 2; Automotive News, June 23, 1980, quoted Thomas Murphy, chairman of General Motors Corporation, as saying at a press briefing at the National Press Club, Washington, that the company would not offer air bags as options on any 1982 models and would not offer air bags until 1983; for a concise summary of the auto industry positions on the air bag: A. B. Kelley, "GM and the Air Bag: A Decade of Delay," Business and Society Review (35): 54-59 (Fall 1980).
35. E. M. "Pete" Estes, president of General Motors Corporation, letter to Neil Goldschmidt, Secretary of Transportation, July 24, 1980; see docket on FMVSS 208 at NHTSA.
36. Senate action: September 25, 1980, 126 CR, S. 13506. House action: October 1, 1980, 126 CR, H. 10206, by a vote of 209 yeas to 192 nays (insufficient for two-thirds needed on suspension calendar vote); December 4, 1980, 126 CR, H. 11919, by a vote of 186 yeas and 189 nays, 56 not voting.
37. 46 FR 68, pp. 21172-9, April 9, 1981.
38. "Air Bag Rules a Sore Spot with Car Makers," The New York Times, May 26, 1981, p. B-6.
39. It should be noted at the outset that U.S. manufacturers' dislike of automatic belts has resulted in designs which would narrow the differences between automatic and manual belts. Indeed, it could be argued that the push button release, the retractor system, and the absence of an ignition interlock with some automatic belts transform them into manual belts after the first time they are disconnected.
40. Louis Harris and Associates, Inc., "The Continuing Public Mandate to Improve Inter-City Rail Passenger Travel," Final Report, conducted for Amtrak, The National Railroad Passenger Corporation, March 1978, Study No. P2814T.
41. Peter Hart Associates, "Public Attitudes Toward Passive Restraint Systems," Summary Report, for NHTSA, August 1978; The Volvo study is discussed in Status Report, 13(13) (September 20, 1978).
42. John Burton, U.S. House of Representatives, press release, December 6, 1979; Larry Kramer, "General Motors Buyers Want Air Bags Studies Show," The Washington Post, December 7, 1979, p. E-1; Larry Kramer, "General Motors Planning to Offer Air Bags in 1982 Large Cars," The Washington Post, December 8, 1979, p. A-5.

43. Benjamin M. Phillips, "Auto Safety Belt Systems, Owner Usage Attitudes in General Motors Chevette and Volkswagen Rabbits," for NHTSA, May 1980, DOT/HS 805 399, and February 1981, DOT/HS 805 797.
44. William Ellis, "An Examination of Comfort and Convenience of 1979 Safety Belt Systems," for NHTSA, January 1979, DOT/HS 803 887; Jonathon Tom, "An Evaluation of the Comfort and Convenience of Safety Belt Systems in 1980 and 1981 Models," for NHTSA, March 1981, DOT/HS 805 860.
45. 46 FR 2064, January 8, 1981.
46. "Actions to Help the U.S. Auto Industry," The White House, April 6, 1981.
47. Joan Claybrook, Administrator, NHTSA, letter to governors of every state, August 22, 1978.
48. Administrator, NHTSA, letter to U.S. Representative John Dingell, July 18, 1979.
49. "Safety: Report of a Panel of the Interagency Task Force on Motor Vehicle Goals Beyond 1980," Office of the Secretary of Transportation, March 1976, TAD-443.1; "Occupant Protection Program, Progress Report No. 2," NHTSA, April 1979; "Automobile Occupant Crash Protection, Progress Report No. 3," NHTSA, July 1980.
50. Philip Hutchinson, Jr., director of government and industry relations, Volkswagen of America, Inc., statement before the Subcommittee on Telecommunications, Consumer Protection and Finance, Committee on Commerce, U.S. House of Representatives, April 30, 1981.
51. "Automobile Occupant Crash Protection, Progress Report No. 3," NHTSA, July 1980.
52. "Executive and Tabular Summary of Air Bag Field Experience," NHTSA, April 1977, Vol. 1, No. 1, DOT/HS 802 301; May 1978, Vol. 2, No. 1, DOT/HS 803 416; August 1979, Vol. 3, No. 1, DOT/HS 805-062; Vol. 4 (in publication).
53. "Injury and Fatality Rates for Equivalent Cars With and Without Air Bags," November 9, 1979, NHTSA docket 74-14. Passive Restraint Rule, U.S. Senate, Hearings before the Committee on Commerce, Science and Transportation, Subcommittee for Consumers, 95th Congress, 1st Session (1977), Serial No. 95-126. Installation of Passive Restraints in Automobiles, U.S. House of Representatives, Hearings before the Committee on Interstate and Foreign Commerce, Subcommittee on Consumer Protection and Finance, 95th Congress, 1st Session (1977), Serial No. 95-89. The Department of Transportation Automobile Passive Restraint Rule, U.S. House of Representatives, Report of the Committee on Interstate and Foreign Commerce, Subcommittee on Consumer Protection and Finance, 95th Congress, 1st Session (1977), Committee Print No. 95-23. Automobile Crash Protection, U.S. Senate, Committee on Commerce, Science and Transportation, 95th Congress, 1st Session (1977), Report No. 95-481. Decision of the U.S. Court of Appeals in Pacific Legal Foundation vs.

Department of Transportation, 593 F. 2d 1338 (D.C. Circuit 1979). William A. Boehly, "The Safety and Fuel Economy Performance of the NHTSA's Research and Safety Vehicles," Proceedings of the First International Automotive Fuel Economy Conference, 1979. D. E. Struble, "A Summary of the Minicars RSV Program," Proceedings of the Sixth International Technical Conference on Experimental Safety Vehicles, October 1976. G. J. Fabian and G. Frig, "Status Report on the Calspan/Chrysler Research Safety Vehicle," Proceedings of the Seventh International Technical Conference on Experimental Safety Vehicles, June 1979. "Occupant Protection Program, Progress Report No. 1," NHTSA, August 30, 1978. "Occupant Protection Program, Progress Report No. 2," NHTSA, April 1979, DOT/HS 804 418. "Seat Belt Performance of Manual and Automatic Systems Installed in the GM Chevrolet Chevette and the VW Rabbit," December 1979, DOT/HS 805 203. "Restraint Usage and Effectiveness in the National Crash Severity Study," September 1979, DOT/HS 805 151. "1979 Survey of Public Perceptions on Highway Safety," July 1979, DOT/HS 805 165. "An Examination of the Comfort and Convenience of 1979 Safety Belt Systems," January 1979, DOT/HS 8 01984. "Safety Belt Usage, Survey of Cars in the Traffic Population," Interim Report, December 1978, DOT/HS 7 01736. Andrew R. Hricko, "T. J. Hooper and the Air Bag," Federation of Insurance Counsel Quaterly, 27(30:2330243) (Spring 1977). Background Manual on the Occupant Restraint Issue, Insurance Institute for Highway Safety, June 1, 1978. Albert Benjamin Kelley, "Passive vs. Active - Life vs. Death," Automotive Engineering Congress, Society of Automotive Engineers, Warrendale, PA, February 24-28, 1975, SAE Paper #750391. National Motor Vehicle Safety Advisory Council, "Head and Neck Injury Seminar," Washington, March 1978, HS-803240. This report discusses methods for identification of head and neck injury mechanisms and injury tolerance levels and provides recommendations for future research. Brian O'Neill, "A Comparison of the Potential Reductions in Passenger Car Occupant Crash Deaths and Injuries from Manual Seat Belts, Automatic Seat Belts, and Air Bags," American Public Health Association Conference, October 21, 1980 (available from the Insurance Institute for Highway Safety, Watergate 600, Washington, D.C. 20037). Allan F. Williams, "Air Bags and Out-of-Position Children - A Survey," Accident Analysis and Prevention, 8(2) (June 1976). William Haddon, Jr., M.D., "Strategy in Preventive Medicine: Passive vs. Active Approaches to Reducing Human Wastage," The Journal of Trauma, 14(4) (1974).

54. Accident investigation report, in progress by National Center for Statistics and Analysis, NHTSA, on April 20, 1981, air cushion restraint system crash in Salt Lake City.
55. "Final Regulatory Impact Analysis on the Amendment to FMVSS No. 208, Occupant Crash Protection," NHTSA, April 1981.

56. Kunitaka Suzuki, presenter, statement by Toyota before the Committee on Energy and Commerce, U.S. House of Representatives, April 30, 1981.
57. Statement on behalf of John DeLorean at public hearings held by Secretary of Transportation Brock Adams in April 1977 on amendments to Federal Motor Vehicle Safety Standard No. 208 to require automatic crash protection for occupants.
58. Insurance Institute for Highway Safety, Status Report 16(8) (June 10, 1981), article on statement by the Automotive Occupant Protection Association, a group of air bag suppliers, stating the costs to consumers for air bags when manufactured in different volumes and allowing 20% profit to the auto manufacturer and dealer:

Air bags in 2 million cars	$185 per car
Air bags in 1 million cars	$240 per car
Air bags in 500,000 cars	$280 per car
Air bags in 100,000 cars	$500 per car
Air bags in 10,000 cars	$1100 per car

Various figures, including a statement by GM vice president Betsy Anker-Johnson at the April 30 hearing that the unit cost for 100,000 air bags would be $1100, were discussed by the manufacturers and suppliers at congressional hearings in April and May 1981; before the Subcommittee on Telecommunications, Consumer and Finance, Committee on Energy and Commerce of the U.S. House of Representatives, April 27 and 30, 1981; before the Subcommittee on Government Activities and Transportation, Committee on Government Operations, U.S. House of Representatives, May 13 and 14, 1981.

59. Statements by General Motors Corporation to NHTSA FMVSS No. 208 docket on Secretary Brock Adams' proposal for passive occupant restraints, May 1977.
60. Ralph Rockow, president of Talley Industries, Phoenix, statements before the Subcommittee on Telecommunications, Consumer Protection and Finance of the Committee on Energy and Commerce, U.S. House of Representatives, April 27, 1981; "Automatic Occupant Crash Protection, Progress Report No. 3," NHTSA, July 1980, p. 9; Small Car Safety in the 1980's, NHTSA, December 1980, p. 47.
61. Unpublished study by Pioneer Corporation for NHTSA.
62. R. C. Reed, Train Wrecks, Superior Publishing Co., Seattle, 1968, pp. 142-143.
63. Center for Policy Alternatives, Policy Choices, A Review Discussing Technology, Engineering, and Social Policy, Massachusetts Institute of Technology, Cambridge, MA, Spring 1981, p. 5.

APPENDIX A

NOTES ON ENVIRONMENTAL EFFECTS, LIFE SPAN,
AND REPAIRABILITY OF RESTRAINT SYSTEMS
AND ON CHILD RESTRAINTS

Environmental Cleanliness in Production, Operation, and Disposal

While questions are rarely raised about the environmental degradation from use, sunlight exposure, etc., of safety belts, concern has been expressed about the sodium azide chemical inflater used in air bags. This chemical, which has been used for some years as a sterilizer in agricultural production, is toxic but has not been shown to be carcinogenic. If mixed with other substances not used in air bag production, it could become explosive, but there is no likelihood of such a result in this application. It is far less dangerous than many chemicals commonly used in manufacturing, and appropriate production safeguards are readily available. When the air bag inflates, the sodium azide turns into harmless nitrogen gas which is heavily filtered through a series of wire mesh and paper filters. For disposal, the air bag need only be inflated to dispose of the chemical, and this can readily be achieved in the scrapping process by notifying the dismantler with a sign on the gas tank that is removed from the vehicle by the dismantler at the beginning of the scrapping process.

Proper planning for the handling of the inflater can mitigate any potential environmental hazards.

Life Span

Both belt and inflatable restraint systems have a life span equal to the life of the vehicle in which they are installed.

Belt systems have held up well for the average life of most automobiles. However, exposure to sunlight and the stess of a crash seriously deteriorate belt systems without any warning or notification to the owner, and virtually no effort has been made to inform owners of belt system limitations.

The inflatable restraint system is hermetically sealed in a thick steel container and protected from most environmental effects. These systems have continued to work effectively in vehicles at least 8 years old. There is no known reason why they would not operate effectively for the life of the car. Additionally, the inflatable restraint has a built-in electronic diagnostic system with a light indicator on the dashboard which measures its operationl capability each time the ignition key is engaged. The diagnostic system not only indicates whether there is some malfunction, but, through a coded flashing signal, will also reveal the source of the problem.

Repairability

Both belts and inflatable restraints can be readily repaired although inflatable restraints are more costly to repair. In both instances, only the inflater and bag would have to be replaced after a crash in which the car was not totaled. The cost of this repair would be covered by insurance.

Various technical issues have been raised during the past decade, sometimes by auto manufacturers and sometimes by a few members of Congress, to undermine public confidence in the air bag system. While this process has served the purpose of diverting and delaying decisions on installation of air bags, none of the adverse claims has survived scrutiny.

It was at first claimed that the noise made during inflation would divert the driver or harm an occupant's hearing. The noise of the crash far exceeds any noise from the air bag, and both come at a moment when the crash is already in progress.

It has been said that the sodium azide inflater could be harmful to workers or the environment when disposed. This chemical is far less hazardous than most others used in industrial production, and it is easier to dispose of then left-over gasoline or battery acid - you merely inflate the air bag and it disappears.

It has been argued that the inflater can come apart and cause a vehicle interior fire. The argument was based on the failure of a research-type inflater which was not hermetically sealed in order to make it cheaper and easier to reuse. The air bag inflater, as manufactured for sale to the public, is of a different design and could not experience this problem.

It has been said that an air bag could harm small children leaning against the dashboard at the moment a crash occurs. This is not a necessary result if the air bag system is properly designed. Following the intial problems General Motors experienced with its design, that have now been resolved, NHTSA worked with Michael Fitzpatrick who developed a computer program that can test various types of air bag designs (bag shape, height of location on the dashboard, width of opening, speed of inflation, etc.) against the dimensions and crash performance of a particular vehicle to make sure that the design fits the car. In short, the technology is available, has been tested, and has been waiting too long to be employed.

Child Restraints

The challenge of achieving significant restraint usage for children is enlarged by the absence of standard equipment child restraint systems sold with each car.

Usage of infant restraint systems is highest, averaged about 45% [A-1]. Toddler restraint usage is far lower, but increased from 10% in May 1980 to 17% in May 1981 [A-2]. The inflatable restraint, if manufactured, would protect children in the front seat. In the absence of this or any other designed-in child restraint system, further increases in child restraint usage will depend upon parental recognition of risk, advice and education from pediatricians and other medical sources, and restraint system use laws where enacted.

In 1977, the state of Tennessee enacted a child restraint use law which has increased usage from 9% to 29% [A-3]. In 1980, Rhode Island enacted a front seat child restraint use law. In 1981, West Virginia, Kansas, Michigan, New York, and North Carolina enacted laws, and the Rhode Island and Tennessee statutes were improved. In contrast to the potential enactment of adult restraint use laws, mandatory child passenger safety evokes sympathetic responses and avoids the philosophical debate about the right to take risks. It remains to be seen whether child restraint laws, combined with a healthy dose of medical guidance and delicate but firm police enforcement, can significantly enhance usage.

REFERENCES

1. B. M. Phillips, "Safety Belt Usage Among Drivers; Use of Child Restraint Devices," U.S. Department of Transportation, May 1980, DOT/HS 805 398.
2. B. M. Phillips, unpublished progress report to NHTSA, May 21, 1981.
3. R. L. Perry, K. W. Heathington, J. W. Philpot et al., "Impact of a Child Passenger Restraint Law and A Public Information and Education Program on Child Passenger Safety in Tennessee," U.S. Department of Transportation, October 1980, DOT/HS 805 640.

SUMMARY OF PANEL DISCUSSION AND COMMENTARY

Christoph Hohenemser

Hazard Assessment Group
Clark University
Worcester MA 01610

In summing up the lessons learned at this meeting, I would like to pass up further discussion of the symposium I chaired and comment instead on the overall theme of the conference: the comparison of "perceived" vs. "actual" risk.

With a couple of exceptions, the speakers at this meeting have adopted a definition of "actual" risk which equates it with scientifically calculated or experienced mortality. This was true for the auto risk numbers summarized by Hans Joksch [1], the ways of expressing smoking risks explained by Jeff Harris [2], the discussion of nuclear power risk presented by Leonard Hamilton [3], the variability of cancer risk described by Rebecca Gelman [4], the new criteria for risk acceptability put forward by Chauncey Starr [5], the very interesting approach to risk classification described by Norman Rasmussen [6], and several other papers. In all these cases "perceived" risk, if it was considered, was equated with various ways of judging or valuing mortality. The persistent use of expected mortality to define risk should not be surprising since it is but an expression of the dominant "count the bodies" school of risk analysis.

Yet I want to remind you that insofar as we have evidence, perceived risk is not equivalent to expected mortality, not even approximately. Rather, as our colleagues at Decision Research [7] and others [8] have shown, perceived risk is a "global" cognitive construct which in psychometric studies with lay subjects is associated with a wide range of qualities beyond mortality. To illustrate, I remind you in Table 1 of 18 risk qualities which in the work of Fischhoff, Lichtenstein, and Slovic [7] are shown to correlate with the level of perceived risk.

Table 1. Hazard Descriptors Studied by Decision Research in Psychometric Studies (see Ref. 7)

1. Voluntariness of risk
2. Immediacy of effect
3. Knowledge about risk
4. Knowledge to science
5. Control over risk (1)
6. Newness of the hazard
7. Chronic vs. catastrophic
8. Commonness vs. dread
9. Severity of consequences
10. Control over risk (2)
11. Number of people exposed
12. Equitability of exposure
13. Effect on future generations
14. Degree of personal exposure
15. Global catastrophic character
16. Degree of observability
17. Changing character of risk
18. Ease of reduction of risk

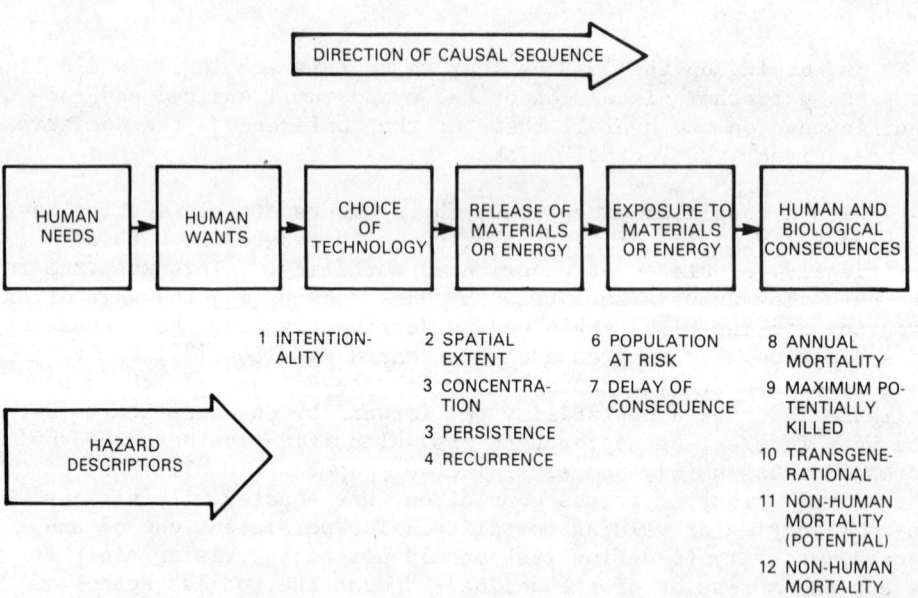

Fig. 1. Causal structure of technological hazards illustrated via a simplified causal sequence. Note the arrow defining the direction of the sequence from human needs and wants, via choice of technology, to human and biological consequences. Hazard descriptors used to define "hazardousness" in Table 2 are shown below the stage to which they apply.

It seems, therefore, that in order to avoid fundamental misunderstandings in risk discussions between experts and the lay public we need to find some way of accommodating the two divergent definitions of risk. Logically there appear to be two solutions. On the one hand, we risk experts can try to "educate" a sometimes recalcitrant public to accept our definition of risk as mortality. Wilson [9] and Cohen and Lee [10] have on several occasions taken this approach. Alternatively, we may try to broaden our own concept of risk using expressed public preferences as a guide. The efforts by Rasmussen et al. [6] and earlier by Starr [11] to classify risks by qualities such as voluntary/involuntary character are noteworthy steps in this direction, but do not in fact abandom the definition of risk as mortality.

With my colleagues at Clark University and Decision Research [12] I have recently been involved in an approach that, in effect, broadens the definition of risk and goes considerably beyond classification of mortality risk as described by Rasmussen et al. [6] at this conference. Because our approach may be of interest to some of you, I thought I would introduce you briefly to what we have done and what it implies.

We distinguish at the outset between "risk" and "hazard." We take "hazards" as threats to humans and what they value, and "hazardousness" as a description of such threats in terms of a causal sequence of events. Like everyone else we take "risks" as quantitative measures of hazard consequences which are conveniently expressed as mortality on injury probabilities. Thus we think of automobile usage as a hazard, and say that the lifetime auto fatality risk for the average American is 3%.

As shown in Fig. 1, we conceptualize the anatomy of hazards as a causal sequence of events that lead from human needs and wants, to choice of technology, to possible releases of materials and energy, to human exposure to eventual harmful consequences and health effects. Each stage along the causal sequence offers opportunities for modification of the flow of events. In its logic, the model is related to the partition of natural hazards into events and consequences [13], an approach now widely used in risk assessment [14, 15]. The model may also be thought of as a simplified fault tree with the branches missing. As such it is comparable to the methods used to analyze nuclear reactor safety [16, 17], to classify auto safety options [18], and to deal with a variety of consumer product hazards [19].

The focal point of the model is the stage "release," defined as the loss of control over flows of energy and materials. Such flows are essential to all living things, and this provides the basis for a powerful common language that applies to all hazards. For most technologies several hazard releases occur. For example,

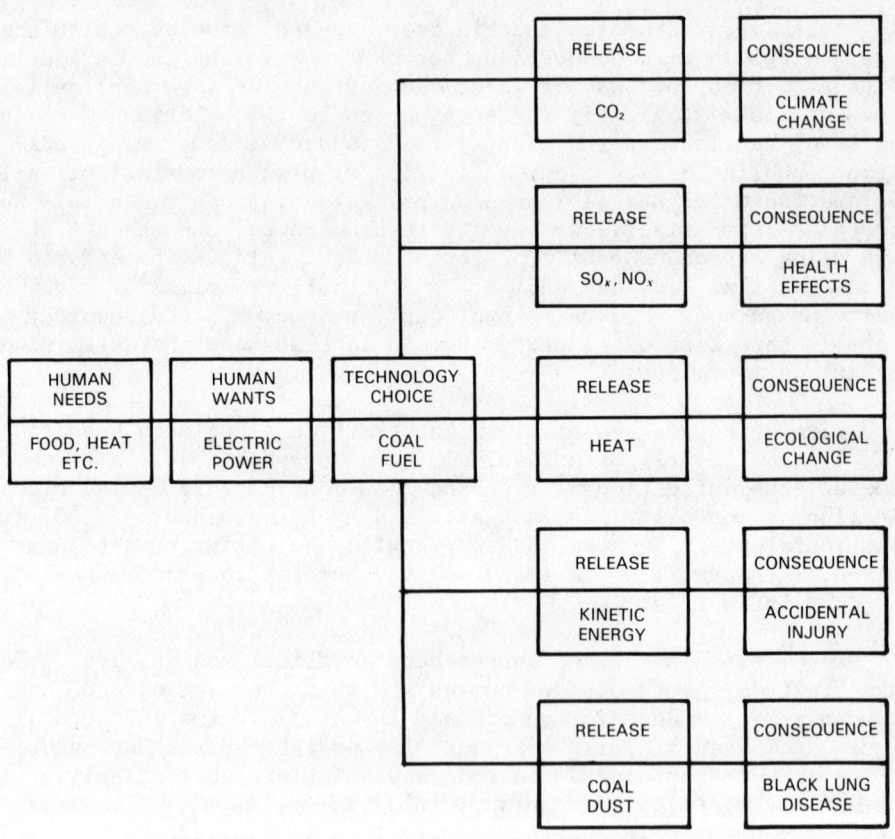

Fig. 2. Topology of technological hazards, illustrated for the case of coal fuelled electric power. Associated with this single technology are several distinct hazard sequences, each with their own release of energy and materials, and subsequent consequences.

the entire cycle of coal fuelled electric power, illustrated in Fig. 2, has a total of at least five separate hazard releases originating from a single choice of technology.

In defining hazardousness via causal chains, our principal ansatz was that the entire causal chain, not just the stage "consequences," is important in understanding hazards. In particular, it was our assumption that risk management is better thought of as hazard management, and that perceived risk is best renamed "perceived hazardousness."

Table 2. Hazard Dimensions and Scales

Name	Scale definitions
1. Intentionality	Measures the degree to which the technology is intended to harm. Scoring requires that a logical distinction is made on intended use. To obtain score, use the following scale.

    3 Not intended to harm living organisms (e.g., bicycles)

    6 Intended to harm non-human living organisms (e.g., pesticides)

    9 Intended to harm humans (e.g., handguns)

RELEASE DESCRIPTORS

2. Spatial extent — Measures the maximum spatial extent over which a single release exerts a significant impact. To obtain score, use the following scale.

Score	Lineal dim.	Geographical equiv.
1		inidividual
2	1-10 m	small group
3	10-100 m	large group
4	100-1000 m	neighborhood
5	1010 km	small region
6	10-100 km	region
7	100-1000 km	sub-continental
8	10^3-10^4 km	continental
9	>10^4 km	global

Table 2. Continued

Name	Scale definitions
3. Concentration	Measures the degree to which concentration of the released energy or material is above background. Two scales are required, one for energy, one for materials releases.

<u>Material releases</u>: We are interested in the ratio, R

$$R = \frac{\text{concentration averaged over release scale}}{\text{natural background concentration}}$$

To score, use the following scale

Score	Range of the ratio R
1	$R < 1$
2	$R \simeq 1$
3	$1 < R < 10$
4	$10 < R < 100$
5	$100 < R < 1000$
6	$1000 < R < 10{,}000$
7	$10{,}000 \leqslant R$

<u>Energy releases</u>: Limited to kinetic energy hazards which are a threat to humans because they exert a force or equivalently, accelerate the human body or a portion of it. Scored according to the number of "g's" involved.

Score	Range of acceleration	Qualitative
1	$g < 1$	Protected ordinary life
2	$g \simeq 1$	Ordinary life, small falls
3	$1 < g < 5$	Very few fatalities
4	$5 < g < 10$	A few unlucky fatalties
5	$10 < g < 20$	Significant fatalities
6	$20 < g < 40$	Protected individuals survive
7	$40 < g < 40$	Some protected individuals survive
8	$80 \leqslant g$	Rare survivors

Table 2. Continued

Name	Scale definitions
4. Persistence	Measures the time period over which the release remains a significant threat to humans. To obtain the score, use the following scale:

Score	Persistence time range
1	<1 minute
2	1-10 minutes
3	10-100 minutes
...	...
8	10^6-10^7 minutes
9	>10^7 minutes

5. Recurrence time	Measures the time period over which the minimum significant release recurs at least once within the U.S. To obtain the score for this variable, use the scale defined for "persistence" above.

EXPOSURE DESCRIPTORS

6. Population at risk	Measures the number of people in the U.S. exposed or potentially exposed to the hazard. To score, use the following scale:

Score	Number of people
1	0-10
2	10-100
...	...
9	10^8-10^9

7. Delay of consequences	Measures the delay time between exposure to the hazard release and the occurrence of significant consequences. To score, use the time scale defined for persistence.

(continued)

Table 2. Continued

Name	Scale definitions
Consequence Descriptors	
8. Annual mortality	Measures the average annual number of deaths in the U.S. due to the hazard in question. To score, use the scale defined for "population at risk" above.
9. Maximum potentially killed	Measures the maximum credible number of people that could be killed in a single release. To score use the scale defined for "population at risk" above.
10. Trans-generational effects	Measures the number of future generations which are at risk for the hazard in question. Scor- requires a logical distinction only.

Score	Definition condition
3	Hazard affects the exposed generation only
6	Hazard affects the children of the exposed generation, but not other future generations
7	Hazard affects several future generations

11. Maximum potential non-human	Measures the maximum potential non-human mortality as a result of the hazard. Scale is qualitative because of paucity of good quantitative data on species mortality. To score, use the following scale.

Score	Definition condition
3	No potential non-human mortality
6	Significant potential non-human mortality
9	Potential or experienced species extinction

Table 2. Continued

Name	Scale definitions
12. Experienced non-human mortality	Measures non-human mortality that has actually been experienced. Scale is qualitative, and similar to that for C_6.

Score	Defining condition
3	No experience non-human mortality
6	Significant experienced non-human mortality
9	Experienced species extinction

To describe hazardousness more concretely we sought descriptors that apply to the full hazard chain as expected mortality applies to risk. After some trial and error over a period of a year we came to 12 descriptors of hazards that are related to the stages of the hazard chain approximately as shown in Fig. 1. Our 12 descriptors are defined in terms of social, physical and biological scales, expressed numerically if possible in common units. Scoring hazards on our scales is in principle as scientific as scoring risks in terms of mortality, and has at this stage no connection to perception of lay people (though it does, of course, reflect our own perception of the problem). Because of the large range of relevant numerical values we chose logarithmic scales: i.e., scale increments of unity were defined to correspond to multiplicative factors of 10. In this sense our numerical scales are similar to other socio-physical scales in which events of human interest cover many orders of magnitude in physical intensity (viz., the decibel scale of sound intensity or the Richter scale of earthquake intensity). A complete definition of our 12 descriptors, including the scales to measure them, appears in Table 2.

At the outset we did not know whether our relatively crude scaling of hazards would be successful in drawing useful distinctions. To test this question, we prepared a representative list of hazards for scoring on the 12 scales defined in Table 2. Our initial base for hazard selection was the Clark collection of case studies and the hazard list employed in early risk perception work by Slovic, Lichtenstein, and Fischhoff [20]. The Clark collection included the case histories of technological concern prepared by Lawless [21], and a continuous inventory of medical and scientific literature over the last five years.

To score the hazards on our list we used the scientific literature to estimate mean values of the variables in question. Most hazards were scored by two or more individuals from our interdisciplinary group. Many cases were discussed by a larger group or referred to specialists in order to clarify the meaning of the available literature. After completion of scoring, one individual made a series of checks for inconsistencies, and in this way altered 20% of the scores by one scale point and a handful by 2 or 3 scale points. We therefore believe the reliability of our scoring to be ±1 scale points in most cases. The results of our hazard selection and scoring are summarized in Table 3.

Briefly stated, what we have in Table 3 is specification of hazardousness for 93 technological hazards. In contrast to mortality risk, which would require but one number, the hazardousness of a given case is expressed through a 12 variable profile.

To approach the problem of hazard perception, we asked two specific questions: 1) are lay people able to understand and judge

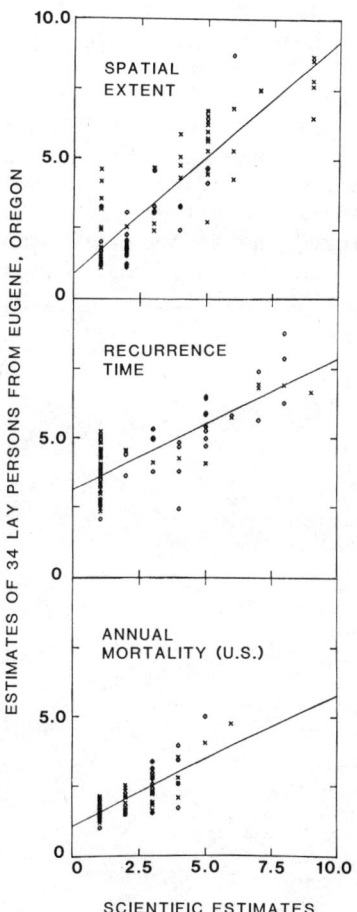

Fig. 3. Scatter plots with linear regression lines indicating the correlation between lay judgments and our estimates of hazard descriptor scales. The three cases are typical of the results obtained for the 11 cases for which both types of estimates are available. The plots illustrate the three principal features of these correlations: 1) a generally high degree of correspondence between the two types of judgments (for correlation coefficints see Table 4); 2) some deviations corresponding to a factor of 1000 (three scale points) from the regression line; 3) significant compression of the scale of lay judgments, indicated by a slope of less than unity.

Table 3. Hazards and Descriptor Scores. For each hazard in the following table, the 12 digit code represents the scores for the hazard on each of the 12 scales defined in Table 2. The order of the code is the same as the order of descriptors in Table 2: i.e., the first digit describes "intentionality," the second digit describes "spatial extent," and so forth

HAZARD	DESCRIPTOR CODE
ENERGY HAZARDS	
1. Appliances - fire	3-2534-93-32333
2. Appliances - shock	3-1512-91-31333
3. Auto - crashes	3-2611-91-51333
4. Aviation - commercial - crashes	3-4716-91-33333
5. Aviation - commercial - noise	3-5521-81-11333
6. Aviation - private - crashes	3-4713-91-42333
7. Aviation - SST noise	3-5634-71-11333
8. Bicycles - crashes	3-2411-81-31333
9. Bridges - collapse	3-3515-91-13333
10. Chainsaws - accidents	6-2411-71-11366
11. Coal mining - accidents	3-3425-63-33333
12. Dams - failure	3-5547-82-24396
13. Downhill skiing - falls	3-1312-61-21333
14. Dynamite blasts - accidents	3-3513-61-22333
15. Elevators - falls	3-2615-91-22333
16. Fireworks - accidents	3-2313-81-11333
17. Handguns - shootings	9-1614-91-41363
18. High construction - falls	3-2817-21-11333
19. High voltage wires - shock	3-3411-77-11333
20. LNG - explosions	3-5628-81-15363
21. Medical x-rays - radiation	3-2211-98-41933
22. Microwave ovens - radiation	3-2411-87-11333
23. Motorcycles - accidents	3-2611-71-41333
24. Motor vehicles - noise	3-3321-81-11333
25. Motor vehicles - racing crashes	3-2715-61-22333
26. Nuclear war - blast	9-6828-91-47396
27. Power mowers - accidents	3-2312-71-21333
28. Skateboards - falls	3-1311-71-31333
29. Skydiving - accidents	3-1815-41-21333
30. Skyscrapers - fire	3-4545-82-33333
31. Smoking - fires	3-1543-83-32333
32. Snowmobiles - collisions	3-2314-71-21333
33. Space vehicles - crashes	3-5838-91-14333
34. Tractors - accidents	3-2414-71-21333
35. Trains - crashes	3-3425-81-33333
36. Trampolines - falls	3-2415-71-11333
MATERIALS HAZARDS	
37. Alcohol - accidents	3-2531-91-41333
38. Alcohol - chronic effects	3-1541-88-51633
39. Antibiotics - bacterial resistance	6-1751-96-31366

Table 3. Continued

HAZARD	DESCRIPTOR CODE
40. Asbestos insulation - toxic effects	3-3651-58-31333
41. Asbestos spray - toxic effects	3-3351-88-11333
42. Aspirin - overdose	3-1741-95-31633
43. Auto - CO pollution	3-4431-94-21633
44. Auto - lead pollution	3-5591-97-21666
45. Cadmium - toxic effects	3-6491-78-21666
46. Caffeine - chronic effects	3-1551-96-11633
47. Coal burning - NO_x pollution	3-7551-96-31696
48. Coal burning - SO_2 pollution	3-7451-96-41396
49. Coal mining - black lung	3-3441-68-41333
50. Contraceptive IUDs - side effects	3-1771-66-21333
51. Contraceptive pills - side effects	3-1451-78-31633
52. Darvon - overdose	3-1751-75-41633
53. DDT - toxic effects	6-5783-88-12699
54. Deforestation - CO_2 release	6-9191-99-11396
55. DES - animal feed - human toxicity	3-1351-98-11633
56. Fertilizer - NO_x pollution	3-9361-98-11693
57. Fluorocarbons - ozone depletion	3-9781-98-11393
58. Fossil fuels - CO_2 release	3-9291-99-11393
59. Hair dyes - coal tar exposure	3-1721-88-11633
60. Hexachlorophene - toxic effects	6-1731-86-21366
61. Home pools - drowning	3-1324-82-31333
62. Laetrile - toxic effects	3-1551-55-11333
63. Lead paint - human toxicity	3-2571-77-31333
64. Mercury - toxic effects	3-5591-88-23666
65. Mirex pesticide - toxic effects	6-5782-68-12696
66. Nerve gas - accidents	9-5787-73-13666
67. Nerve gas - war use	9-7788-93-37696
68. Nitrite preservative - toxic effects	6-1171-98-11633
69. Nuclear reactor - radiation release	3-7698-96-16963
70. Nuclear tests - fallout	3-9197-98-33966
71. Nuclear war - radiation effects	9-9798-98-48996
72. Nuclear waste - radiation effects	3-6291-88-15963
73. Oil tankers - spills	3-6576-16-11366
74. PCBs - Toxic effects	3-6791-97-13666
75. Pesticides - human toxicity	6-5781-98-22699
76. PVC - human toxicity	3-4741-78-21633
77. Recombinant DNA - harmful release	3-9789-96-17993
78. Recreational boating - drowning	3-2325-82-41333
79. Rubber manufacture - toxic exposure	3-4791-58-31633
80. Saccharin - cancer	3-1741-88-11633
81. Smoking - chronic effects	3-1541-88-61633
82. SST - ozone depletion	3-9381-99-11393
83. Taconite mining - water pollution	3-6791-68-11366
84. Thalidomide - side effects	3-1745-15-11633
85. Trichloroethylene - toxic effects	3-4791-88-11333
86. Two,4,5-T herbicide - toxic effects	6-5782-78-12696
87. Underwater construction - accidents	3-3426-42-11333
88. Uranium mining - radiation	3-5491-68-22933
89. Vaccines - side effects	6-1451-85-21696
90. Valium - misuse	3-1751-86-31633
91. Warfarin - human toxicity	6-1761-85-11366
92. Water chlorination - toxic effects	6-5751-98-11366
93. Water fluoridation - toxic effects	3-5271-88-11633

Table 4. Correlation of Lay and Scientific Judgments of Hazard Descriptors

Hazard descriptor	Correlation coefficients		
	Energy hazards	Materials hazards	All hazards
TECHNOLOGY DESCRIPTOR			
1. Intentionality	0.95	0.84	0.89
RELEASE DESCRIPTORS			
2. Spatial extent	0.83	0.89	0.87
3. Concentration	N/A	N/A	N/A
4. Persistence	0.33	0.62	0.79
5. Recurrence time	0.85	0.73	0.80
EXPOSURE DESCRIPTORS			
6. Population at risk	0.77	0.73	0.74
7. Delay of consequences	0.88	0.92	0.96
CONSEQUENCE DESCRIPTORS			
8. Annual mortality	0.79	0.77	0.76
9. Maximum potentially killed	0.89	0.75	0.79
10. Transgenerational	0.34	0.56	0.65
11. Non-human mortality (pot.)	0.82	0.75	0.78
12. Non-human mortality (exp.)	0.63	0.73	0.71

Table 5. Correlation of Causal Structural Descriptors with Psychometrically Determined Values of "Perceived Risk" Across 81 Hazards

Hazard descriptor	Correlation coefficients (only r-values at greater than 0.95 confidence level are given)
TECHNOLOGY DESCRIPTOR	
1. Intentionality	0.28
RELEASE DESCRIPTORS	
2. Spatial extent	0.57
3. Concentration	—
4. Persistence	0.42
5. Recurrence time	—
EXPOSURE DESCRIPTORS	
6. Population at risk	0.42
7. Delay of consequences	0.30
CONSEQUENCE DESCRIPTORS	
8. Annual mortality	—
9. Maximum potential killed	0.53
10. Transgenerational	0.43
11. Non-human mortality (pot.)	0.53
12. Non-human mortality (exp.)	0.30
FACTORS	
1. Intentional/biocidal	0.32
2. Persistence/delay	0.41
3. Catastrophic	0.32
4. Mortality	—
5. Global/diffuse	0.30
Variance explained = Σr^2	0.50

our hazard descriptors, and 2) do our hazard descriptors reflect the issues that ordinary people worry about? While we can not offer a definitive answer to these questions, we can report on the results of a pilot study with a group of 34 college educated people in Eugene, Oregon.

To test lay understanding, we created definitions and simple scoring instructions for the causal descriptors of hazard, and with these asked our subjects to score a subset of our hazard sample. After an initial trial, "concentration of release" was judged too difficult for lay people to understand. For similar reasons, 14 hazards were omitted from our list. The subjects then scored 81 hazards on 11 scales, using only our instructions, their general knowledge, reasoning and intuition.

The results demonstrate reasonably high correlations between the scores derived from the scientific literature and the 34 lay subjects. As shown in Table 4, correlation coefficients ranged from a low of 0.65 to a high of 0.96 for the 11 dimensions scored. At the same time, as illustrated in Fig. 3, there are many deviations from the regression line which translate into discrepancies of 1000 or more in physical dimensions, suggesting that there is strong bias towards undervaluing or overvaluing specific dimensions for certain hazards. Thus, while ordinary people can do reasonably well in scoring our hazards, and in this sense certainly understand the dimensions of hazardousness, there are serious distortions that deserve further analysis.

To test whether our causal structure descriptors capture our subjects' overall concern with hazard, we collected judgments of "perceived risk," a global risk measure whose variability has been explained by a broad range of risk qualities in previous psychometric studies [7, 20]. The correlation of perceived risk to our hazard scores is illustrated in Table 5, top, and shows that modest positive correlation coefficients, between 0.30 to 0.57, exist in 9 of 12 cases.

Perhaps the most interesting aspect of these results is the fact that among the 12 descriptors of hazards, annual mortality is one of three which shows no significant correlation with perceived risk. Hence the variable we risk analysts chose most frequently to represent risk does not appear to be a strong factor in the judgment of our subjects.

In an independent investigation, we determined that our 12 descriptors of causal structure explain about 50% of the variance in perceived risk for our sample of 34 young Oregonians. Our conclusion was based on a two step process: first we derived five orthogonal dimensions of hazardousness from our 12 descriptors, using the method of factor analysis; then we obtained the correlation between

Table 6. Factor Structure of Combined Energy and Materials Hazards, 93 Cases

No.	Factor Name	Variance explained	Hazard descriptors Name	Factor loading
1.	Intentionality	0.32	Non-human species mortality (experienced)	0.87
			Non-human species mortality (potential)	0.79
			Intentional design of technology	0.81
2.	Persistence/delay	0.19	Persistence of release	0.81
			Delay of consequences	0.85
			Transgenerational effects	0.84
3.	Catastrophic potential	0.11	Rarity of occurrence	0.91
			Maximum potential killed	0.89
				0.89
4.	Mortality	0.10	Annual mortality	0.85
5.	Global diffuseness	0.09	Population at risk	0.73
			Concentration above backg.	-0.173

these five dimensions and perceived risk, as shown in Table 5, bottom. The definition of the five factors in terms of the original 12 descriptors is given in Table 6, and will be discussed in further detail in a forthcoming publication [12].

We conclude, therefore, that while our hazard descriptors are reasonably well understood by our pilot sample of lay persons, they explain (in a correlational sense) only about half of the sample's overall concern with hazard, which we assume is expressed by the variable "perceived risk."

Hence, if our analysis is substantiated by investigations with more representative samples of lay persons, it follows that "perceived risk" is not to be understood as a qualitative distortion of "actual risk," defined as mortality; rather, "perceived risk" reflects qualities of hazard that may be objectively defined, that the public understands, and to which risk analysts have probably paid too little attention.

If some kind of multi-variate conceptualization of hazardousness is accepted by the risk analysis community (and here our particular formulation is only a tentative first step), then it is a short next step to ask: how should each component of hazardousness be weighted? This, I believe, is a largely normative question for which we risk analysts have no special qualification. We may still wish to argue that mortality is the most important component of hazardousness, but we must be prepared to have others argue that CATASTROPHIC POTENTIAL (Factor 3), PERSISTENCE/DELAY (Factor 2) or some other dimension of hazard is more important.

I expect that the views of the public will change through such a discussion, with the result that the relation between various hazard dimensions and perceived risk (Table 5) is altered. Ths is, in my view, desirable and consistent with the goals of a democratic society. In fact, I envision future application of a multi-variate approach to hazard in which specialists use their expertise to scale the descriptors (as we have done in Table 3), and the policy process determines the appropriate weighting.

ACKNOWLEDGMENTS

In making these remarks I have drawn freely on the results of a broader project on hazard management and perception that has been jointly undertaken by the Clark University Hazard Assessment Group and Decision Research in the past five years. Some of my interpretative remarks here may not be shared by all members of this collaborative, though I would expect broad agreement on the main points. Senior members of the collaborative are Baruch Fischhoff, Sarah Lichtenstein, and Paul Slovic at Decision Research, and Roger Kasperson, Robert Kates and I at Clark University. The work of both groups has been supported for the past two years by N.S.F. grant PRA 79-11934.

REFERENCES

1. H. Joksch, this conference.
2. J. Harris, this conference.
3. L. Hamilton, this conference.
4. R. Gelman, this conference.
5. C. Starr, this conference.
6. N. Rasmussen, this conference.

7. P. Slovic, B. Fischhoff, and S. Lichtenstein, In: Societal Risk Assessment: How Safe is Safe Enough?, R. C. Schwing and W. A. Albers, Jr., eds., Plenum Press (1980).
8. Workshop on Perceived Risk, December 11-13, 1980, Eugene, Oregon, P. Slovic, ed., Decision Research (1981).
9. R. Wilson, Technology Review, 81, 40 (1979).
10. B. Cohen and I. Lee, Health Physics, 36, 707 (1978).
11. C. Starr, Science, 165, 1232 (1969).
12. A fuller statement of the collaborative work of the Clark and Decision Research groups is forthcoming under the title "The Nature of Technological Hazard" (submitted for publication, July, 1981).
13. I. Burton, R. W. Kates, and G. F. White, The Environment as Hazard, Oxford, New York (1978).
14. W. D. Rowe, The Anatomy of Risk, Wiley, New York (1977).
15. R. W. Kates, Risk Assessment of Environmental Hazard, Wiley, Chichester (1978).
16. U.S. Nuclear Regulatory Commission, Reactor Safety Study, NUREG 75/014 (1975).
17. H. W. Lewis, Risk Assessment Review Group Report to the U.S. Nuclear Regulatory Commission, NUREG 75/014; N.R.C., Washington (1978).
18. W. Haddon, Jr., Technology Review, 77, 52 (1975).
19. T. Bick and R. E. Kasperson, Environment, 20, No. 8, 30 (1978).
20. P. Slovic, B. Fischhoff, and S. Lichtenstein, Environment, 21, No. 3, 14 (1979).
21. E. W. Lawless, Technology and Social Shock, Rutgers, New Brunswick, N. J. (1977).

THE ASSESSMENT OF NUCLEAR RISKS:

SOME EXPERIENCES FROM THE SWEDISH ENERGY COMMISSION

> Bengt Hansson
>
> AB EPISTEME Scientific Consultants
> Glimmervägen 8
> S-223 78 Lund
> Sweden

BACKGROUND DATA ON THE PRODUCTION AND CONSUMPTION
OF ELECTRIC ENERGY

Production Structure

For a long time, electricity was almost equivalent to hydroelectric power in Sweden. Very little power from other sources was produced before World War II. The rivers of northern Sweden are rich sources of electric power and probably a major factor behind the industrial upswing in Sweden during the first half of this century. The government agency for electricity production, Statens Vattenfallsverk, founded in 1909, usually translates its own name as State Power Board (or Administration), but the literal meaning is State Waterfall Agency.

Thermal power was of little importance until the beginning of the sixties, when several large condenser stations were put into operation. Before then, back-pressure constituted a large proportion of the thermal power produced.

The first commercial nuclear power station was put into operation in the late sixties, but not until a decade later did nuclear power contribute substantially to the total production figures.

Table 1 gives the yearly average production figures in TWh for 5-year periods from 1921 to 1980.

Table 1. Power Production in Sweden

Period	Hydro	Non-Nuclear Thermal	Nuclear	Net Imported	Domestic Use
1921/25	3.0	—	—	—	3.0
1926/30	4.6	—	—	—	4.6
1931/35	5.7	—	—	—	5.7
1936/40	7.4	0.8	—	—	8.2
1941/45	10.8	0.4	—	—	11.2
1946/50	14.1	1.1	—	-0.1	15.2
1951/55	20.6	1.6	—	-0.1	22.1
1956/60	28.0	2.6	—	-0.4	29.5
1961/65	40.6	2.2	—	-0.6	42.2
1966/70	45.4	10.5	—	1.6	57.5
1971/75	56.1	14.7	3.5	1.5	75.8
1976/80	57.1	13.5	21.4	0.2	92.2

Comments

Comment 1. Hydro power production increased exponentially at a yearly rate of about 6% until the mid-sixties, when it tended to level off. Technically, there are still a number of rivers and waterfalls to be exploited, but the hydro power program has been halted for environmental reasons.

Comment 2. There was a sharp increase in thermal power production in the late-sixties and early-seventies, followed by a slight decrease. The increase would be even more striking if the fairly steady back-pressure production were subtracted. The increase was, of course, intended to compensate for the drop in the rate of increase of hydro power production, but rising oil prices prevented further growth of a thermal program.

Comment 3. Nuclear power came into the picture suddenly and recently. The increase in demand for the last 5-year period was met in full by nuclear production.

Comment 4. Import and export figures are small. They refer to the other Nordic countries and reflect temporary imbalances in the production rather than conscious trading efforts. Basically, domestic production has to be planned to cover domestic demand.

Comment 5. The rate of increase in the total consumption figures shows a tendency to decrease. The average yearly increases for the different 5-year periods range from 4.0% (from 1971-1975, to

Table 2. Average Installed Power Capacity in Sweden

Period	Hydroelectric GW	%	Non-Nuclear Thermal GW	%	Nuclear GW	%
1946/50	3.21	50	0.96	13	—	—
1951/55	4.75	50	1.29	14	—	—
1956/60	6.97	46	1.93	15	—	—
1961/65	9.71	48	2.98	8	—	—
1966/70	11.67	44	4.29	28	—	—
1971/75	13.41	48	7.85	21	1.07	37
1976/79	15.06	43	9.73	16	4.41	52

1976-1980) to 8.9% (from 21-25 to 26-30). The tendency is more pronounced in the last few yearly figures: the increase from 1978 to 1979 was 2.0% and from 1979 to 1980, 1.2%.

Table 2 shows the average installed capacity in GW for 5-year periods (or, in one case, 4-year period) from 1946 to 1979 and its relation to actual production, measured by the percentage which actual production makes up of the theoretical maximal production (full effect for 8760 hours per year).

Comments

Comment 1. The degree of utilization of hydro and nuclear installations is fairly high and even. The limiting factor for nuclear installations is purely technical (yearly revision, malfunctions), while the situation for hydro is more complicated and to a high degree dependent on the current content of the reservoirs. The inflow is concentrated during the spring months when the mountain snow melts, and we must be prepared should the spring floods for the next few years turn out to be smaller than expected. Given a basic nuclear production, the hydro stations will also be used to regulate daily variations.

Comment 2. Thermal installations are utilized to a much lesser degree. There was a marked dip in the early sixties, followed by a rise to historically high levels. From 1956-1960 to 1971-1975 the installed capacity increased at an average yearly rate of 10%. The need for more thermal electricity in the late sixties was thus well anticipated. The growth rate decreased markedly after 1973, and very little capacity has been added since 1976. (See also the following section on institutional structure, especially the paragraph on planning.)

Relationship to Other Forms of Energy

Most uses to which electricity is put are such that no immediate large-scale substitution is possible. The most notable exception is house heating. Usually, oil is used for this purpose, but electrical heating accounts for some 12-13 TWh of the electricity consumption. The future of electrical heating is a strategic issue of some importance. The electricity required for increased use of electrical heating (or, for that part, continued use, if other types of consumption increase) has to be produced either in nuclear plants or in oil- or coal-based condenser plants (until so-called alternative energy sources have been sufficiently developed). The first alternative is economically preferable but politically impossible (at present); the second is economically rather indifferent - it is a matter of where to burn the oil or the coal. A law against electrical house heating (which has in fact been proposed in some quarters) would eliminate one important factor pressing for more nuclear power.

BACKGROUND DATA ON INSTITUTIONAL STRUCTURE

Production, Distribution, and Planning

There are four categories of electricity producers in Sweden: Statens Vattenfallsverk (the State Power Board), the semi-publicly owned company Sydsvenska Kraft AB, power stations owned by and serving particular cities and local communities, and a small number of private companies.

Statens Vattenfallsverk, commonly known as Vattenfall, is wholly state owned. It is the largest single producer of electric power, producing a good 50% of the hydro power output, and the sole owner of the three-aggregate PWR nuclear plant at Ringhals and 74.5% owner of the one-aggregate BWR plant at Forsmark. Because of its dominating position in the market, Vattenfall's price policy strongly affects the whole industry.

Sydsvenska Kraft AB, commonly known as Sydkdraft, is a joint stock company listed on the Stockholm stock exchange in which six large local communities in southern Sweden own 54.5% through an ad hoc consortium. Boliden AB, a major mining company in northern Sweden listed on the Stockholm stock exchange, owns 17%. A number of smaller local communities own a couple of percent, and the rest is owned by a large number of individual investors. The six local communities in the majority consortium buy their electrical power from Sydkraft and distribute it through their own facilities, which makes the economy of Sydkraft interwined with those of their owners. Yet, the stock is briskly traded at the stock exchange and has some speculative appeal. This is a unique ownership construction in Sweden.

Sydkraft produces 4-5 TWh hydro power per year and is the sole owner of the two BWR aggregates at Barsebäck. The company also owns 35% of the two-aggregate BWR plant at Oskarsham and operates a number of thermal power plants.

Cities and local communities often operate combined heat and power stations and in some cases own small minority shares of the nuclear stations in Forsmark and Oskarsham.

Some 20 of the companies listed on the Stockholm stock exchange together produce hydroelectrical power in excess of 25 TWh per year (including Sydkraft). However, many of these companies are primarily in the engineering or forest industry and use the power for their own internal demands. Some forest companies also produce sizable quantities of thermal electricity through back-pressure.

The local distribution of electric power is usually effected through local communal companies in urban areas and by Vattenfall or Sykdraft in rural areas.

Practically all electrical power stations are interconnected through the main power line system, a network of 400 kV lines. This system is wholly owned by Vattenfall. Since not every producer can serve every consumer independently, a high degree of cooperation is necessary, and the control of the main power line system becomes strategically important. For example, a local producer can only sell occasional surplus power through the main system and at conditions acceptable to its owner. Another important strategic aspect is the bidding process by which it is determined which stations should be added to the system when the load increases. The combined role of Vattenfall as a major producer and a dominant distributor gives it a very strong position.

In such an integrated system the planning has to be highly co-ordinated. Traditionally, planning has been well advanced and new capacity has been installed to meet expected increases in demand. The attitude of the producers has been to promote the use of electrical power by keeping prices low. The prices have increased lately but are still low in an international perspective, a typical rate in southern Sweden is 4 US cents per kWh for ordinary households and 3 US cents per kWh for those using electrical heating.

Licensing Procedure

In reality, there are two different licensing procedures in Sweden. The first one is formally a matter for the local planning authorities who grant building permits for all sorts of buildings and hence also for power plants. The question to be tried there is: Is it permissible to erect this particular type of building, for this particular purpose, at this particular site? The matter is

thus purpose- and site-specific but not at all concerned with what
is inside the shell, so to speak. Formally, the authorities are
not required to make any risk considerations (except for the constructional safety of the building as such). Really, risk enters
the picture to the extent it is felt to be important on general political grounds in that community and nationally.

A refusal of the communal planning authority to grant a permit
cannot be appealed. This is known as the "communal veto." The
granting of a permit, however, can be appealed to the national
government, which then makes the final decision. So what is needed
is in fact the joint approval of local and national authorities.

The second licensing procedure is supposed to take place only
after the first one is completed. The site and the general permissibility have then been approved, and an application is made to
Statens Kärnkraftsinspektion (the State Nuclear Power Inspectorate),
commonly known as SKI. Several other state agencies are involved
in the licensing, notably Statens Strålskyddsinstitut (the State
Radiation Protection Institute), but they all give their permits or
express their opinions through SKI, which grants the license in relation to the applicant.

There are no formal requirements on the application given by
law, but SKI determines the format and content from case to case.
A custom has developed, closely modeled on the U.S. system, and that
calls for preliminary safety report before construction is started
and a final safety report before the plant is taken into operation.
The demands on these reports have become much more exacting during
the last decade, and they are now required to contain very detailed
descriptions of all the safety systems, although no attempts are
made to include quantifications of probabilities for various types
of accidents like those aimed at by WASH-1400.

Such quantifications are now, however, being prepared for two
of the older aggregates, in operation since 1972 and 1975 respectively. These are done in cooperation with SKI but are not required
by them for permission to continue operation, although they are
probably done in anticipation of such requirements in the future.
These studies include only what happens inside the reactor station,
and do not include population distribution, meteorological conditions, or other factors determining the consequences. (This is in
accordance with the presumption that siting problems are settled
before SKI enters the picture.)

The Role of Risk Assessment Studies

There is thus no formal role for risk assessments in the Swedish decision process. Whatever role they play is performed through
their impact on opinion formation in the public, in the political

parties, in the media, etc. Two general conclusions follow from
this: 1) Although the impact of risk assessment studies has indeed
been great in the last half-decade, there is little inertia in the
system and the role of these studies may change drastically and suddently if public interest becomes focused on some aspect such as the
economics; 2) Not only the actual content but also the mode of presentation, the form of publication, media coverage, and similar factors can be assumed to be important. Therefore, the role of risk
assessment studies in Sweden is tightly coupled to general mechanisms of opinion formation and opinion diffusion. A study of this
role is underway but cannot be presented here. However, one previous study, that of relating attitudes towards nuclear power to
other factors, should be briefly mentioned. It does not directly
deal with the effects of risk assessment studies, but it gives a
hint of what to look for.

The study, "Kärnkraftsomröstningen i kommunerna" ("The Nuclear
Power Referendum in the Local Communites") by Leif Johansson concerns the Swedish referendum on nuclear power in 1980. It is an
ecological exmination of covariation between referendum results and
a number of characteristic features of local communities and is
therefore not directly applicable at the level of individuals. By
far the strongest covariation found was that with party sympathy
(measured by the number of votes cast for the various parties in
the election of 1979.) All the major political parties officially
supported one of the three alternatives in the referendum, and the
agreement in number of votes is remarkable. All other factors showed
much less correlation. Of some interest is the fact that the pronuclear attitude seemed to correlate positively with the proximity to
actual nuclear installations. This is not a spurious effect due to
an accidental coincidence with the party effect, but is found in an
analysis of the differences between party figures and referendum
figures. It is also true that communities with uranium deposits
were more pronuclear than those without. No explanation will be attempted here.

The role of the press is difficult to ascertain because different political parties are very unevenly represented in the press.
Particularly, the antinuclear parties (the Center Party and the
Communists) have few newspapers supporting them, but in those communities where the dominating newspaper was antinuclear, an increased
difference between referendum vote and party vote (in antinuclear
direction) can be observed.

An implication of this study is that the opinion forming process
within the political parties would be of prime interest.

A SAMPLE STUDY: THE REPORT OF THE ENERGY COMMISSION'S COMMITTEE ON SAFETY AND ENVIRONMENTAL EFFECTS

Ad Hoc Committees in Swedish Energy Politics

Besides its lack of formal structure, the licensing procedure loses in importance also because the number of nuclear aggregates to be licensed is so small and the political interest in nuclear power so intense that the number of new laws governing nuclear power tends to keep an even pace with the number of aggregates, which means that the main problem for a prospective nuclear plant builder is not to obtain the license from SKI but to stay clear of all political obstacles during the time of planning and construction.

The usual thing to do when a problem of the nuclear type is to be solved is to appoint an ad hoc committee, usually made up of representatives from the four or five major political parties and from other major organizations with an interest in the matter, such as state agencies, trade unions, employers' associations, trade associations, etc. This committee then publishes a report, usually including proposed new laws, which is circulated for comments to a large number of organizations, both public and private. A bill to the parliament is then prepared on the basis of this material.

Two properties of this procedure are of special interest in the present connection: 1) Unity is sought and usually achieved in those parts of the matter where no real difference of opinion exists, and the extent and nature of real differences is chiseled out rather precisely; 2) The detailed policy formation in the parties is often entrusted to the committee members, who therefore achieve the status of party experts on the matter at hand. It then becomes especially interesting to find out about the impact of risk assessment studies on such committee members.

Several such committees have been appointed to solve energy questions in the seventies. The first one was Närförläggingsutredningen (The Committee on Close Siting), appointed in 1970. Its report was published in 1974 and contained the recommendation that nuclear heat plants should be built in metropolitan areas, although extremely close siting (5 km or less from the metropolitan centre) should be avoided until further experience had been gained.

The Aka-committee of 1973 dealt with the problems of spent fuel and radioactive wastes. It recommended containment in vitreous or ceramic materials and deposition in the primary rocks of Sweden as the best method of solving the waste problem. It also recommended that planning for a Swedish reprocessing plant be started immediately.

(The KBS study, Kärnbränslesäkerhet (Nuclear Fuel Safety), was, however, not by a government committee but was a private initiative by the nuclear industry.)

NUCLEAR RISK: EXPERIENCES FROM THE SWEDISH COMMISSION

There were also two reports in 1979-1980 about the implications for the Swedish nuclear program of the Three Mile Island accident.

But the largest effort was no doubt the Energy Commission of 1976. The commission itself was composed of various representatives in the usual way, but they appointed five subcommittees (or "groups of experts") with more real expertise. Each of these subcommittees then contracted a large number of special studies with various groups of scientists which were then summed up by the subcommittees in their reports to the Commission. The report by the subcommittee on safety and environment will be presented here. (The other four subcommittees dealt with energy supply, energy conservation, means of control over energy flows, and research and development, respectively.)

The Content and Organization of the Energy Commission Report

The report from the subcommittee to the Commission contains some 1400 pages divided into 22 chapters, being mostly a systematization and a critical summary of the content of the underlying research reports.

The report's 12 chapters focus on the different energy sources: coal, oil, natural gas, peat, shales, nuclear power, hydro power, wind energy, solar energy, biomass, geothermic energy, and others. Two chapters are devoted to special energy carriers, electricity and alcohols. The remaining chapters include a summary and discussions of some special topics, such as health effects of energy conservation (!), wastes, risks for catastrophic accidents, and the relation between nuclear power in Sweden and proliferation of nuclear weapons.

An important aim of the report was to provide background data for a rational comparison of different sources of energy. The chapters on the different energy sources are therefore written in a uniform format, and an attempt is made to discuss such areas as waste problems and to compare the risk for catastrophic accidents for all energy sources.

A major problem in all comparative studies is where the boundaries are to be put. The following principles were adopted: effects within and without Sweden and direct and long-term effects were to be treated equally; the energy system was to include everything from extraction at the source to final use; human health effects were to be given first priority and quantified whenever possible, with second priority given to effects on plants and animals, climatological effects, and landscape transformations.

Of special interest in the present connection are the chapters on oil, coal, nuclear power, hydro power, electricity, and the summary.

The chapters on oil and coal are fairly parallel. The fuel cycle is split up into extraction, transport, storage, refining, and combustion, where combustion is differentiated according to its various purposes. Different changes to the enviroment are identified at each step, and so are possible health effects to humans. The different types of emissions are quantified in a detailed way, but they are not transformed into quantitative estimates of health effects (such as expected number of deaths). Occupational risks are estimated on the basis of available statistics, but their relevance and accuracy are acknowledged as sometimes doubtful. So are, e.g., the numbers for strip and shaft mining of coal, which are not always separated. Estimates from different sources for the number of deaths per TWh in land-based oil extraction vary by a factor of 40. It is not always clear whether the figures include risks connected to prospecting, etc.

The nuclear cycle is split up in the usual way into mining, conversion and enrichment, manufacture of fuel elements, plant operation, transport of radioactive material, reprocessing, waste handling and deposit, and demolition of nuclear facilities. For each of these steps, the process is described, possible risk factors identified, radiation doses to the public estimated, and occupational risks estimated. Proper data bases are often not available and estimations have to be made by analogy or on the basis of few data.

The chapter on hydro power is short and deals mainly with environmental effects. The only health effects worth mentioning are accidents during the construction phase.

The chapter on electricity deals with its distribution and use. There exist reliable statistics for accidents in this area, and they are added to the risk spectre of electrical energy.

The summary tries to bring some order into the vast collection of facts and figures. Clearly, the uncertainties in the quantification of most of the risks involved are much too great to permit a meaningful all-embracing numerical index to be constructed, not to speak of the conceptual difficulties in comparing, e.g., deaths to limitations in outdoor recreation possibilities. In its report, the subcommittee proceeds as follows.

First, it goes through the different types of health and environmental effects and determines what limitations to set to different types of energy production. It is estimated, e.g., that the sulfur content of fossil fuels limits their use for the year 1990 to a quantity corresponding to 220 TWh, given certain standards. Rough estimates are many times sufficient to remove a problem from the agenda: If modern electrofilters are used, coal corresponding to 1200 TWh can be combusted before the resulting cadmium fallout amounts to 1% of what emanates from fertilizers and atmospheric fallout.

Second, with this in mind, the subcommittee goes through the effects once again and points out to what extent these effects affect the choice of energy sources and transformation processes.

Two Comments on the Usefulness of the Energy Commission Report

It is clearly a great achievement to bring together such an amount of data and to arrange them in a form that permits comparison, at least to a degree. The following remarks are not intended to belittle this, but only to point out how difficult it is to paint a complete picture of the risk panorama which can be directly useful in decision making.

(1) My first remark concerns the putting of boundaries to different energy systems and to the energy question as a whole. It has two parts: 1) To achieve comparability between different systems and 2) to find a reasonable line of demarcation between energy and society at large. The study discussed included accidents connected with the use of electricity, i.e., these involving ranges, lights, and TV sets. Does comparability then require that the study also include accidents connected with the use of petroleum products, such as motor accidents, deaths resulting from slipping on a patch of oil on the floor, or the like (which they do not)? Or, regardless of comparability, should such accidents be accounted to energy use or to something else? We can surmise that the place of the ultimate boundary of the energy system will be important to the comparison of different energy sources, since energy is so strongly intertwined with our daily lives.

The subcommittee has certainly been aware of such boundary problems and tried to solve them, but the problem is whether there exists at all any correct or objective solution. Ultimately, it seems to be a matter of opinion where the right boundary is to be placed, unless we want to make risk analyses of complete societal systems include all their minute details.

(2) The second problem concerns the fact that risks are pinned down to different energy sources, not to alternatives, in a decision situation. An alternative may be made uo of a certain mix of energy sources, but even if the risk figures for all the energy sources are completely known and accurate, these would not suffice because changes in the overall system may induce changes in the risk figures for individual sources. Increased use of coal may necessitate comparatively more shaft mining, thereby increasing the average risk of coal energy; a different price of electricity in relation to other kinds of energy may radically change the structure of some industries and thereby deeply affect the pattern of environmental

impacts; but risk figures are not known and accurate, and the actual task of putting such figures on specific alternative energy policies would be beset by a vast number of definitional and conceptual problems, interpretations of data, and assessments of their quality, and would require as large an effort as the writing of the subcommittee's report itself, and it would probably require as much technical expertese. There is still a long way to go if we want to appraise energy policies in an objective way - if that is at all possible.

NUCLEAR POWER PLANT: WEST GERMAN MANAGEMENT

OF RISK — A PROBLEM ANALYSIS

> H. Paschen, G. Bechmann, and G. Frederichs
>
> Kernforschungszentrum Karlsruhe
> Abteilung für Angewandte Systemanalyse
> 7500 Karlsruhe, Postfach 3640
> Federal Republic of Germany

THE CURRENT SITUATION OF NUCLEAR ENERGY
IN THE FEDERAL REPUBLIC OF GERMANY:
REASONS FOR STAGNATION

Introduction

The expansion of nuclear energy use in the Federal Republic of Germany has almost come to a standstill. In 1980, not only did no new nuclear power plant go into operation, but not one single construction permit was granted. And only one order was placed in the entire country (A letter of intent for Block 2 of the nuclear power plant ISAR was given to the Kraftwerk Union, a subsidiary of Siemens). Every one of the nuclear power plants presently under construction is several years behind its scheduled completion date (11 blocks with a little under 13,000 MWe) [1]. Naturally, the resulting cost overruns, from inflation, interest, etc., are considerable. Licensing delays, experts say, are holding up an estimated $15 billion of investments for nuclear power plants. Safety measures and quality demanded by German licensing authorities make German nuclear power plants much more expensive than those of other countries with which Germany is competing in the nuclear export business. Nevertheless, some feel that in the post-Three-Mile Island era, these expensive safety and quality control features will pay off. It is still too early to tell.

Economic arguments have been diminishing in importance as the growth in electricity consumption has slowed down as a result of low economic growth, saturation tendencies in the use of household appliances, and successful conservation efforts. Furthermore, the argument of direct oil substitution by nuclear energy in electricity production is not of particular importance for the Federal Republic of Germany (in contrast to France, e.g.,), because the portion of electric power generated by oil-fired power plants is very small - just a few percent. A host of other economic arguments, such as the substitution of gas in electricity production, indirect oil substitution, lack of base-load generating capacity, and cost advantages of nuclear-generated electricity, do not appear to take hold and are, in part, controversial.

The major reason for the stagnation of the German nuclear industry is seen by many in the continuous tightening of safety requirements. Ordering a new nuclear power plant, they argue, has become a largely incalcuable investment risk for the utility that places the order. Moreover, the existing uncertainty regarding the further development of the backend of the fuel cycle - reprocessing, interim storage, waste disposal - will certainly not help speed up the development of nuclear energy use within the next few years, in the light of court decisions which declare concrete measures regarding the further treatment of spent fuel elements to be an integral part of the damage-preventing measures required by law.

Certainly, the protests and demonstrations, hearing interventions, and citizens' action suits against nuclear energy, especially in connection with siting decisions, have played a decisive role, both directly and indirectly, in bringing about the situation described in the beginning of this paper. Public opinion polls carried out in the Federal Republic of Germany show, on an average, that there are two opponents of nuclear energy for every three proponents, with many respondents undecided.

In sum, a detailed analysis reveals a whole complex of developments and factors that reciprocally influence one other. Uncertainly, bewilderment among business people and scientists compounded by contradictory judgments and reactions by politicians, responsible authorities, and the courts, are the result.

THE TREATMENT OF RISK IN LEGAL AND OTHER REGULATIONS
ON THE PEACEFUL UTILIZATION OF NUCLEAR ENERGY
AND IN THE LICENSING PROCEDURE FOR NUCLEAR POWER PLANTS

The risk management system* in the nuclear area is related to the above described developments in many ways; it influences these developments and it reacts to them. In the following sections, a

---

*The risk managment system is to be understood as the totality of the societal groups and instutions which are concerned with the planning, organization, and implementation of policies to control potential adverse consequences to man, his environment, and his well-being.

few issues that have turned out to be crucial in the current nuclear energy discussion, as well as for the present condition and future development of nuclear risk management in Germany, will be analyzed. The issues include:

1) the requirement of the German Atomic Energy Act that the "existing scientific knowledge and technology," i.e., the state of the art in science and technology, be the criterion by which adequacy of prevention of damages and risks is to be judged; the difficulties resulting from the above requirement; and the suggestions for overcoming these difficulties that are currently being debated;

2) the role of special risk studies, parliamentary commissions, and hearings of experts; and

3) the role of the public in the risk management system.

THE LEGAL BASIS FOR THE LICENSING PROCEDURE, INTERNAL ADMINISTRATIVE REGULATIONS, AND TECHNICAL RULES [2]

Nuclear facilities in the Federal Republic of Germany have to comply with a special federal law, the Law on the Peaceful Utilization of Nuclear Energy and the Protection from Its Dangers (Atomic Energy Act, 1960). According to the Atomic Energy Act, a license to build, operate, or modify a nuclear facility may be granted only if (among others) "every necessary precaution has been taken in the light of existing scientific knowledge and technology to prevent damage resulting from the construction and operation of the installation" [3].

This is the central clause in the Atomic Energy Act regarding the treatment of damage, danger, and risk. The demand for the taking into account of "existing scientific knowledge and technology" serves to adapt legal regulations as closely as possible to scientific knowledge and technology" serves to adapt legal regulations as closely as possible to scientific and technical progress.

Statutory ordinances, serving to define more closely with regard to certain aspects the goals and measures for safety laid down to Section 7 of the Atomic Energy Act, are located on a second stage in the system of regulations for the licensing of nuclear facilities. Among them are the Radiation Protection Ordinance (1976) and the Ordinance on Procedure for Licensing Nuclear Facilities (1977).

The third stage consists of the internal administrative regulations which aim at specifying more concretely the state of existing scientific knowledge and technology. These include: 1) the safety criteria for nuclear power plants, issued by the Federal Ministry of the Interior (Bundesministerium des Innern/BMI), and 2) the recommendations by the Commission on Reactor Safety (Reaktorsicherheitskommission/RSK) and the Commission on Radiation Protection (Strahlenschutzkommission/SSK), both advisory committees to the BMI, on issues of reactor safety and radiation protection, and the RSK guidelines on standards of technical safety.

Technical rules from a number of areas are a fourth stage of specification. Examples are: 1) the set of rules of the Committee on Nuclear Technology (Kerntechnischer Ausschuss/KTA, situated at the BMI), 2) DIN norms, and 3) VDI guidelines.

It should be noted that the adminsistrative regulations and technical rules are not legal norms and are thus not legally binding for the courts. Problems arising from this fact, with regard to determining the adequacy of measures for the protection against dangers and the prevention of risks, will be dealt with later.

THE STRUCTURE AND COURSE OF THE LICENSING PROCEDURE FOR NUCLEAR POWER PLANTS AND OTHER NUCLEAR FACILITIES

As would be expected in view of the importance of the decisions involved, the licensing procedure in the Federal Republic of Germany is very complicated; there is a very stringent control of every nuclear utilization. A great number of government authories and institutions participate in the process (Fig. 1). Characteristic features of the licensing procedure are the following [5]:

Participation of the Federal State (Bund), the Constituent States (Länder), and Local Governments. The licensing authority is a state ministry (e.g., the Ministry of Economics or the Ministry for Social Affairs). The state is subject to instructions from and control by the Federal Ministry of the Interior (Federal Mandate Adminstration). The BMI is intensively involved in each and every licensing procedure! The communities affected are consulted in the process by state ministries.

Multiple Participation of Experts. Experts participate 1) on behalf of the licensing authority, as a rule TOVs (Association for Technical Monitoring) and the Association for Reactor Safety (Gesellschaft fur Reaktorsicherheit/GRS) for special issues such as plane crashes and sabotage; 2) on behalf of the Federal Ministry of the Interior (RSK and SSK, independent advisory bodies addressing recommendations to the BMI and not to the licensing authority. These recommendations provide the basis for the Ministry of the Interior's instructions to the licensing authorities in the Länder, or constituent states, in regard to standards of safety. The GRS, too, provides advice for the BMI). One may speak of a twofold expert review procedure. The responsibility for the examination of licensing prerequisites and for the orderly course of the procedure is, however, in the hands of the appropriate state authorities; 3) on behalf of courts, if an action is brought against a partial construction permit; and 4) on behalf of the public involved.

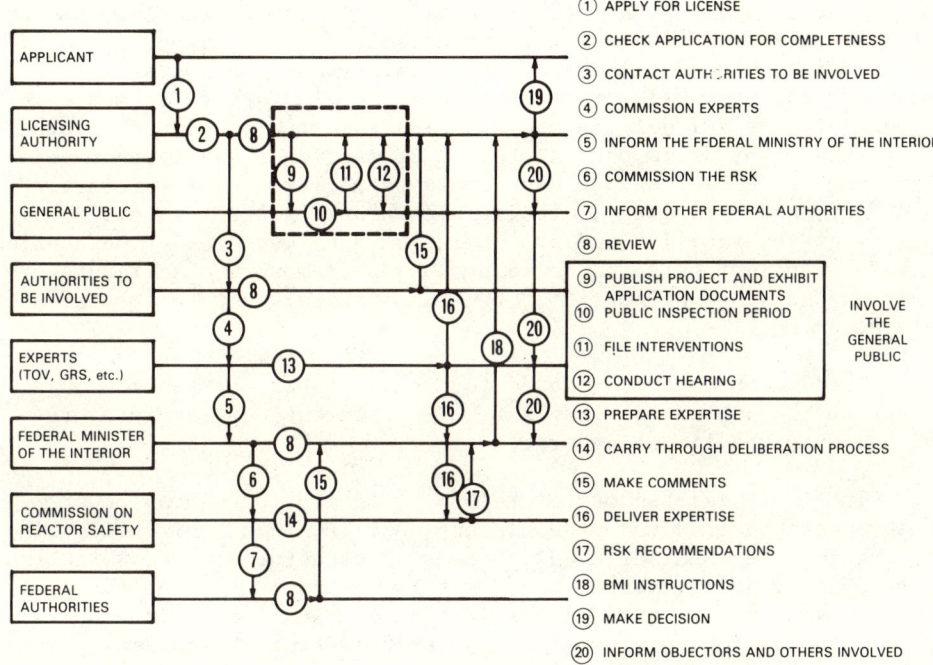

Fig. 1. Schedule for the Nuclear Licensing Procedure.

Permits. The licensing of the construction and operation of nuclear power plants involves a whole series of permits for part-construction and part-operation (the number can become very large).

Participation of the General Public. During the review procedure, the request of the applicant for a license is to be made public. The application, the safety report, and a brief description of the plant to be constructed are to be exhibited publicly. Interventions, which may be filed in the course of a two-month period during which the documents are exhibited, are dealt with in the course of a hearing. Most importantly, in the Federal Republic of Germany, each citizen whose individual rights are directly affected, has the right to appeal to the administrative courts against decisions issued by the administrative authorities.

Risk Considerations and Safety Requirements in the Licensing Procedure

The applicant for a license, usually a large energy utility or a group of utilities, is required to provide certain documents with the application (in accordance with Section 3 of the Ordinance on

Procedure for Licensing Nuclear Facilities). Among these documents is a so-called safety report which describes, among others, the impacts and dangers arising from the facility and its operations and explains the preventive measures provided. The other documents include: 1) special details on measures against disturbance or other interference by third parties, e.g., precaution against sabotage; 2) a list of measures intended to prevent pollution of water, air, and soil; and 3) a list including all information relevant to the safety of the facility and its operation, the measures to cope effectively with incidents and accidents, and a framework plan for the required control of all parts of the facility relevant to its safe operation.

The documents are prepared for the most part by the manufacturer of the nuclear power plant. They are subject to expert review during the licensing procedure.

The first assessments of technical safety for German power plants used the maximum credible accident (MCA) as their point of reference. The MCA, defined as the sudden rupture of the largest tube conducting reactor coolant, set the standard for the safety requirements with regard to the design and operation of the nuclear power plant. Today, a greatly expanded spectrum of accidents is taken into account when setting the standards for the design of the systems.

The deterministic concept of safety applied in the nuclear licensing procedure requires proof by simulation of previously defined accident sequences that the consequences of these accidents will be brought under control. Uncertainties in the analyses are taken into account by means of conservative assumptions. Defining the range of accidents to be considered is largely a matter of convention. Because of their minimal probability, accident sequences where safety systems largely or completely fail are neglected for the purposes of the licensing procedure. However, these are precisely the most interesting cases for genuine risk studies such as the Rasmussen Study or the German Risk Study [6, 7].

The Controversy on Section 7 Subsection 2 No. 3 Atomic Energy Act

Dispute on the licensing of nuclear power plants in the courts has increasingly concentrated on issues of protection from dangers and prevention of risks. The legal basis for the controversy is Section 7 Subsection 2 of the Atomic Energy Act (AtG). This regulation does not grant the applicant a legal claim to the issuance of a license, even if the conditions for receiving a license, as specified by the Atomic Energy Act, are met. The licensing authority is bound by law to refuse the license if one of the conditions is not fulfilled; it still has a scope of discretion to grant or deny the license after due consideration if all the conditions are fulfilled.

This legal construction is intended to enable the licensing authority to carefully consider everything necessary for the licensing decision. Key here is that the concept "existing scientific knowledge and technology," which, according to Section 7 Subsection 2 No. 3 Atomic Energy Act, is used to determine the necessary precautions and is an indeterminate legal term without administrative scope for discretion. Therefore, the precautions conforming to the latest standards of scientific knowledge and technology may, under existing law, be fully examined by the courts. It is thus required of the courts that they determine the current state of reactor research and technology and participate in the controversy between scientists and engineers on what is to be judged technically necessary, adequate, appropriate, and avoidable. Since neither laws nor ordinances provide sufficient normative evaluation criteria for the courts, each court must, more or less, produce its own criteria to assess the necessity of precautionary measures. This is the reason for the heterogeneity of court rulings and their partially contradictory results. For example, the administrative court in Freiburg refused to allow the construction of a nuclear power plant at Wyhl because a separate burst protection was not provided for (referring to Section 7 Subsection 2 No. 3 Atomic Energy Act); the administrative court in Würzburg, however, refused to impose such a requirement. The latitude for interpretation of Section 7 is illustrated by a court ruling demanding that the necessary protection against dangers include waste-disposal measures; as long as this problem has not been solved, permission to construct nuclear power plants should, according to this ruling, be refused.

The Federal Constitutional Court (BVerfG = Bundesverfassungsgericht) has commented on Section 7 of the Atomic Energy Act in two rulings, Kalkar and Mülheim-Kärlich [8]. The BVerfG came to the conclusion that Section 7 Subsection 2 No. 3 not only requires protection against dangers but also precautions against risks. It proceeds from the assumption that the legislative body, in the Atomic Energy Act, required all kinds of damage, danger, and risk resulting from the operation of the plant to be taken into consideration. Especially with the reference to the latest state of the art of science and technology, the Atomic Energy Act normatively commits the executive to the principle of best possible protection against dangers and prevention of risks. However, the court accepts the notion that some residual risk has to be put up with, which means that licenses may be granted even if the probability of future damage cannot be excluded with absolute certainty. On the other hand, the BVerfG has also established that the legislation has, with the principles incorporated in Section 1 No. 2 and in Section 7 Subsection 2 of the Atomic Energy Act, created a standard by which licenses may be granted only if existing scientific knowledge and technology practically rule out the occurrence of such events. Uncertainty beyond this threshold of common sense is, according to the BVerfG, attributable to the limits of human intellectual power and perception and is to be regarded as inevitable, and thus as a burden to be accepted and to be borne by the public as a whole. At the same time, the court has raised standards for precautionary measures. In order to conform to existing scientific knowledge and technology, the court says, "those precautionary measures against damage must be adopted that are deemed necessary according to the latest scientific knowledge. If they are not as yet technically feasible, the license may not be granted. The appropriate measures of precaution are thus not determined by technical feasibility."

This provides the courts with a broad scope for review. To aid them, the courts must commission experts. There is a danger that the courts practically have to turn into specialized bodies on scientific and technological matters. Controversial administrative decisions may thus be replaced by equally controversial court rulings. This has already led to the charge that political decisions in the energy field are being altered by judges without political answerability and without sufficient specialized knowledge.

Suggestions for Limiting Judicial Review Powers

Several suggestions have been made by scientists and members of the judiciary to overcome the present unsatisfactory situation of overburdened courts and increasing legal uncertainty. The aim of the suggestions is to limit judicial review powers. There appear to be three main strategies to achieve this [9].

Introducing Administrative Discretion. The introduction of adminstrative scope for discretion is one suggestion. This means that the administrative authorities would gain a certain degree of discretionary power for the application of indeterminate legal terms, especially in relation to regulations on technical safety. The courts then would no longer have the power to examine the substance of decisions, but would merely have the task of examining whether the discretionary powers have been exceeded. A second suggestion aims in a similar direction. It concerns the adoption of the so-called American model of administrative justice. Under this model, the main task of the courts is to examine whether the administrative ruling was reached by means of fair procedure.

Objections to both suggestions are raised on the basis of Article 19 Section 4 of the German Constitution (Grundgesetz) which guarantees complete legal protection when basic rights are threatened. However, there appears to be a certain tendency in the BVerfG's rulings to establish, to a certain degree, the idea of "protection of basic rights by procedure" in the Federal Republic of Germany.

Giving Legal Competence to Experts. It has been suggested to upgrade representative and independent expert advisory commissions by giving them the powers to determine legally binding norms for technical safety, thus enabling them to specify indeterminate legal terms. Strong objections to this kind of suggestion are being raised on constitutional grounds, especially with regard to the democratic legitimacy of such bodies.

Giving Legal Status to Technical Standards. At present, the simplest and constitutionally least problematic method of overcoming the dilemma described above would appear to be to convert existing technical standards into legal norms, i.e., to incorporate certain industrial technical norms, such as DIN norms, VDI guidelines, and the set of rules of the Minister of the Interior's Nuclear Technology Committee, as norms into safety legislation. As a consequence, both the administrative authorities and the courts would be bound by them. One precondition would be, however, that affected interests be guaranteed the right to participate on an equal footing in the process of developing the technical norms.

Another suggestion involves the increased application of the instrument of ordinances in order to specify general legal clauses and to relieve the courts. On the one hand, as opposed to administrative regulations which merely bind administrative authorities, ordinances bind both administrative authorities and the courts. On the other hand, ordinances may go into greater detail and be altered more easily than laws.

Two points are of importance when considering the above and similar suggestions: 1) They must not infringe upon the right to legal protection guaranteed under Article 19 Section 4 of the Constitution; 2) The priority of the protection of life, health and property over the purpose of the promotion of nuclear energy must be preserved without reservation.

Concluding Remarks

The various "actors" in the nuclear risk management system react differently to the problems and suggestions for their solution which were discussed earlier. Understandably, the executive branch, especially the Interior Ministry, the administrative agencies, and the legislative branch, are more guarded than industry. Concerning the criteria for preventive safety measures, industry has explicitly demanded that the rules of the Committee on Nuclear Technology (KTA) and other sets of technical regulations be considered as the binding state of the art of science and technology and that the specific requirements derived from them be accepted, not merely as minimally necessary but also as sufficient conditions for licensing. Additionally, industry is demanding that such sets of technical rules be fixed for a definite time period in order to make the expert advisory and licensing procedures more predictable, i.e., that the rules be amended only periodically to take account of scientific and technological advances [10].

Nevertheless, there is wide industry and government agreement that the licensing procedures should be accelerated. This issue was even touched upon in a recent government declaration by Chancellor Schmidt. There is, however, less agreement about the ways to reach the goal of a streamlined licensing procedure. Just how delicate the whole matter is can be illustrated by reference to a few examples from the set of industry licensing-reform demands:

- drastic reduction in the number of required partial construction licenses;

- procedural simplification by a systematic standardization of nuclear power plant design, and

- reciprocal recognition of expert evaluations of similarly designed installations by the different authorities.

In sum, one can say that the nuclear risk management system is in flux; the result is difficult to foresee.

SPECIAL RISK STUDIES, PARLIAMENTARY COMMISSIONS, AND HEARINGS OF EXPERTS

Increasing efforts have been made recently in the executive and legislative branches of government to support or carry out special projects aimed at further clarifying the problems of nuclear risks. To what extent are the types of systems that are now in operation or under construction or in the planning stage safe? Can and must they be made safer? The most important projects of this type are:

- generic studies of the risks of different reactor types and of alternative waste management and disposal systems,

- the so-called Gorleben Hearing on the German Nuclear Reprocessing and Waste Disposal Center that was to be built in Lower Saxony, and

- the Federal Parliament's Special Inquiring Commission on Germany's future nuclear energy policy.

A brief overview follows of these special projects which have influenced nuclear risk management in the Federal Republic of Germany and will influence its future development.*

---

*An important activity that must be mentioned in this context is a forthcoming program of the Federal Ministry for Research and Technology (BMFT) to promote risk and safety research. Among others, this program aims at 1) identifying weak points in the safety con-

The German Risk Study for Nuclear Power Plants/Phase A [11]

The German Risk Study for Nuclear Power Plants/Phase A was carried out between 1976 and 1979 by the GRS (Association for Reactor Safety), as the main contractor, and several other institutions. It was initiated and supported by the Federal Ministry for Research and Technology. The aim of the study was to assess the collective risk caused by potential reactor accidents. In order to make a comparison of results possible, the methods and assumptions of the German study follow largely those of the U.S. Reactor Safety Study (Wash-1400). The nuclear power plant Biblis B, which has a typical pressurized water reactor of German make, was chosen as the reference facility. To determine the risk, 19 sites with a total of 25 reactors were considered in the analyses. A general result of the study is that risks caused by accidents in nuclear power plants do not differ significantly between the two countries.

Work on Phase B of the German Risk Study has begun. There is participation by nuclear critics. Phase B will be characterized, among others, by further methodological refinement and the in-depth treatment of such problems as common mode failures and human failure.

To the authors' knowledge, no really systematic attempt has been made so far to perform an overall evaluation of the German Risk Study as to its usefulness, e.g., its influence on regulatory policy. Many groups and individuals have made isolated comments on the usefulness of the study, however. Most of these comments fall within one of the following categories [12]:

Technical Arguments. There is general agreement that the German Risk Study has been and will be very useful in helping to (among others): 1) detect weak points of the technical systems, 2) assess the effects of certain safety systems, 3) minimize the frequency of severe accidents by optimizing the plant design with respect to accident prevention, 4) effectively mitigate the consequences of severe accidents, and optimize emergency procedures.

Arguments Concerning the Study's Influence Upon Licensing Procedures. There are several comments to the effect that the results of the study should influence, and are to some extent already influencing, regulatory policy and practice. Deutsches Atomforum

---

cepts of technical facilities and systems by means of new methods, especially of quantitative risk analysis, and 2) contributing to a better understanding of risk perception and evaluation by the public, thus providing impulses for improving the management of risk.

(German Atomic Forum) has demanded that probabilistic analyses like the German Risk Study be used in the licensing procedures to help in assessing the necessity of additional technical safety measures, e.g., measures to control hypothetical accidents (such as a core catcher or underground construction).

Arguments Concerning the Study's Influence on the Public Acceptance of Nuclear Energy. There is general scepticism as to the possibility of directly influencing public opinion in favor of nuclear energy by means of probabilistic analyses like the German Risk Study or Wash-1400. Some point out that carrying out such studies may cause people to become even more anxious about nuclear energy because they may, for the first time, really become aware of the catastrophic potential of the accidents analyzed in the study (ignoring or not understanding the meaning of their extremely low probabilities). There has been no attempt so far to assess the real impact of the German Risk Study upon public opinion about nuclear energy. Anyway, there is reason to suspect that the study has been largely ignored by the general public.

The Federal Parliament's Special Inquiry Commission on Germany's Future Nuclear Energy Policy (Enquete-Kommission "Zukünftige Kernenergie-Politik" des Deutschen Bundestages) [13]

The Special Inquiry Commission was set up early in 1979 by the 8th Bundestag and presented the results of its work in June 1980. It was composed of parliamentarians and experts, some of the latter known to oppose nuclear energy. (More recently, the 9th German Parliament decided to institute another commission on Germany's future nuclear energy policy to continue the work of the first commission.)

The Inquiry Commission discussed, and made recommendations for, a very broad range of topics, including, e.g., energy conservation and the use of renewable energy sources. Problems of reactor safety and nuclear risk management were also dealt with. One of the most important recommendations of the Commission in this area is to carry out a risk-oriented analysis for the SNR 300, the German prototype of a sodium-cooled breeder reactor currently under construction at Kalkar. With regard to the subject to this paper, it is interesting to note the following remarks made in the Commission's report in connection with this SNR 300 risk study:

- On the one hand, the projected risk study is not to hamper the ongoing licensing procedure for the SNR 300. On the other hand, the Commission states that it expects the study to suggest risk-reducing modifications of the SNR 300 concept in case weak points in the concept should be detected.

- The Commission raises the point of political responsibility in connection with the implementation of the breeder technology. It states that, at this point, not all of its members are convinced that they can take this responsibility. The question is, the Commission says, to what point risk-reducing efforts must be carried to make such a decision possible. The Commission believes that a political evaluation of the SNR 300 should not be made without supplementing the technical safety analyses by a risk-oriented analysis of the kind proposed.

The Federal Ministry for Research and Technology has commissioned the GRS to carry out the SNR 300 risk-oriented analysis as the main contractor. Opponents to nuclear energy will participate in this project.

Safety Study for High-Temperature-Reactor Concepts [14]

The Safety Study for High-Temperature-Reactor Concepts, which follows the example of the U.S. Accident Initiation and Progression Analysis (AIPA) study and adapts it to German conditions, has just been completed. It was carried out jointly by the Kernforschungsanlage Jülich and the GRS and was supported by the Federal Ministry of the Interior. The study does not contain analyses of the consequences of accidents.

Comparative Study of Alternative Waste Management and Disposal Systems

In 1979, a political decision was taken by federal and state authorities to evaluate and compare two alternative waste management and disposal systems, i.e., the classical system based on reprocessing and disposal of high-level waste and the direct disposal of spent fuel. In fulfillment of that decision, on R & D effort with a financial volume of about DM 60 Mio was started, with almost equal amounts for hardware development and technology assessment activities. The comparative risk (accidental and routine) of the two alternatives will be the centerpiece of the study. The political decision on which alternative to choose will be made in the mid-eighties, and will be based on the results of this study. The study is financed by the Federal Ministry for Research and Technology and coordinated by the Kernforschungszentrum Karlsruhe.

The Gorleben Hearing [15]

From March 28 through April 3, 1979, the state government of Lower Saxony organized, parallel to the formal licensing procedure, an expert symposium to discuss the technical safety problems of the projected Integrated Nuclear Reprocessing and Waste Disposal Center that was to be built in Gorleben, Lower Saxony. Critics and advocates from Germany and abroad were invited. On May 16, 1979, the Minister President of Lower Saxony made a twofold declaration concerning the conclusions drawn from the symposium by his government: 1) As far as considerations of technical safety are concerned, the project is to be judged practicable. The integrated concept cannot, however, be realized politically; therefore reprocessing is to be taken out of the concept. Since then, the project has not been pursued in its original form.

THE ROLE OF THE PUBLIC IN THE RISK MANAGEMENT SYSTEM

The Public Discussion About Nuclear Energy in the Federal Republic of Germany

At the beginning of the seventies, orders for nuclear power plants increased. In almost all cases, the decisions on the siting of these plants led to protests by the local population. The events in connection with the planned site with Wyhl in 1974 and 1975 became prototypical for the course of the conflict: 1) mobilization of a strong opposition among the rural population; 2) effective support by national environmental organizations; 3) demonstrations of solidarity by the most diverse groups; 4) extraordinary actions of resistance, such as occupation of sites, mass demonstrations, and collective complaints with several thousands of signatures; and 5) transfer of decisions to the courts.

The example of Wyhl showed that the resistance against nuclear energy could reach a momentum which nobody had expected, at least on the occasion of siting decisions. The events in Brokdorf near Hamburg in 1976 and in Grohnde near Hannover in 1977 were a repetition and escalation of what had happened before. Though the number and extent of demonstrations have declined since 1977, a mass demonstration was staged again in 1981 in Brokdorf, with occasional violent clashes between the police and demonstrators.

Due to the attention which these protests attracted, not only in the Federal Republic of Germany but worldwide, parties and parliament could no longer remain aloof from the nuclear energy controversy. The public discussion on nuclear energy, which so far had been led nearly exclusively in the extraparliamentary arena by such organizations as environmental groups, church-supported institutions, Volkshochschulen (a sort of community college), and, above all, the mass media, entered into a new phase marked by special party conventions on nuclear energy policy held by the three major political parties.

As in other countries, the nuclear energy controversy is investigated scientifically in the Federal Republic of Germany. The numerous opinion polls conducted showed, on the average, a relation of promoters to opponents of 3:2, nelgecting fluctuations over time and between the surveys.

The substance of the major arguments in the broad public discussion in the Federal Republic of Germany is similar to that in other countries. In the scientific literature on the problem of public acceptance, three main types of issues are distinguished: 1) specific issues relating to the risks of particular nuclear facilities and processes and of specific sites; 2) more general issues related to nuclear energy, such as waste management, proliferation, and problems for the constitutional state (Rechtsstattsproblematik); and 3) very broad and comprehensive issues relating to the direction of technological, economic, and industrial development.

The Influence of the Public in the Risk Management System

Formally, three ways of public participation in nuclear licensing procedures are provided for: 1) interventions by third parties against the application for a license and oral deliberation of the objections in the nonpublic hearing between the licensing authority, those appealing, and the applicant; 2) the private law or constitutional complaint in the case of a violation of private interests or constitutional considerations; and 3) the complaint before an administrative court after a (partial) construction or operating permit has been granted by the licensing authority.

Such forms of public participation, constructed along constitutional principles, are tailored to take into account isolated or small numbers of protests. In the face of such broad opposition as developed in the nuclear energy field during the seventies, they proved, however, more or less inadequate. Ample use was made of the right to intervene. In some cases, the appeals had clearly demonstrable intentions (mass interventions). The hearings often took a turbulent course because the opposing opinions of the participants could not be bridged by procedural means. The goal of conflict settlement was not achieved. Rather, the hearings themselves provoked new protests. Complaints before the administrative courts against licenses granted were (and still are) filled regularly.

Contrary to a widely held opinion, the operation or construction of nuclear power plants is at present hampered only slightly by court decisions [16]. The considerable delays in licensing procedures which can be observed can, therefore, probably be attributed mainly to problems in the relationship between applicants and manufacturers on the one hand and licensing authorities on the other. The hypothesis which can be derived from this is that the influence of the public through protesting and filing complaints becomes effective primarily indirectly; politicians and administrative authorities, as also part of the judiciary, are sensitive to the far-reaching political implications of the nuclear energy issue.

The increasing delays in licensing are presently the object of sharp controversy. One major point of dispute is the extent to which the public should participate. The endeavor, above all, of industry to prevent at least a further extension of citizen participation is primarily fed from the fear that the opportunities to intervene in the hearing procedures could, in the future, multiply in the wake of technical modifications of the facilities under construction. The concrete background for this fear is a ruling by the federal constitutional court involving the Mülheim-Kärlich nuclear power station, which stipulates that § 4 of the Ordinance on Procedure for Licensing Nuclear Facilities, containing important provisions concerning the participation of third parties in licensing procedures in the case of modifications of the facilities under construction, be interpreted more in favor of the citizens than before.

The political debate on the problem of citizen participation in the nuclear risk management system is again in full swing due to this court ruling; its impacts upon the prospects of nuclear energy utilization in the Federal Republic of Germany may be considerable.

THE ROLE OF TECHNICAL RISK FOR PUBLIC ACCEPTANCE [17]

The problems of nuclear risk management in the Federal Republic of Germany discussed in the previous sections suggest that it would be worthwhile to investigate in what way and to what extent nuclear technical risks and their reduction to a minimum level influence the acceptance of nuclear energy by the public. The following critical remarks can be made.

The central topic in the nuclear energy controversy is its risk to health and the environment. This suggests the thesis that the problem of acceptance is a reaction to the potential catastrophic consequences of highly improbable, but, nevertheless, possible nuclear accidents. This thesis leads to contradictions, however. There are other risks which should (but do not) provoke the same reactions, such as, until recently, the stationing of nuclear weapons on the territory of the Federal Republic of Germany. On the other hand, problems of acceptance with similar manifestations of citizens' protests are encountered in other fields of policy, as in education, town planning, and housing policy, and in the transport sector, where a risk like that of nuclear energy is not involved. Such considerations make a modification of the thesis necessary.

Within the context of this modified thesis, the issue of risk assumes a symbolic function in the nuclear controversy. Risk is the thematic focus which has become the permanent issue in the course of the controversy; it is the common topic which is the precondition for any public discussion, without, however, necessarily comprising all the dimensions which are at the root of the controversy. Such a theme does not crystallize by chance but is dependent on the extent to which it follows for the articulation of opposing opinions and can be used for defining the respective individual standpoints.

Both parties use the conflict about nuclear energy for supporting their respective lines of argumentation on the issue of risk.

On the part of the proponents of nuclear energy, risk considerations are used to point out how safe nuclear energy is as compared to other technologies, and safety research represents a field of governmental promotion to which a certain legitimating value can be attributed.

The objections to nuclear energy, to the extent that they refer to risks, are based on the irrefutable premise that catastrophic accidents are possible in nuclear facilities. Their impacts on health and the environment are described, and the conditions are shown which might lead to such accidents. Each such presentation conveys the impression of a threat which, irrespective of probability considerations, gives rise to a general fealing of concern. If probabilities are considered, the argument is used that even in cases of extremely low probability the accident might, nevertheless, happen tomorrow.

These arguments are given even greater weight when, in addition to the risk due to technical failure, deliberate human actions are mentioned - such as sabotage by states, terrorist organizations or mentally deranged individuals - as possible triggers of catastrophes. The thesis that the precautionary measures necessary to protect nuclear installations and operations might even lead to police state-like circumstances has met with great resonance. The dimensions of possible accidents, as well as the time periods for which radioactive waste is left to future generations, exceed normal imagination and can easily be used, therefore, to argue against the acceptability of such a technology.

The arguments concerning the risks of nuclear energy mentioned so far can evoke fear and give rise to a general feeling of concern. Accordingly, the theory is widely held that resistance against nuclear energy is an irrational reaction of fear. But however valid one might think this thesis is, it does not seem to suffice to assess the political importance of the nuclear controversy. Nuclear opponents do not merely paint the dark picture of possible catastrophes but endeavor to found their arguments on detailed information, to point out the relationship of their arguments with more general political objectives, and to operationalize them for the political debate. Thus, attempts can be made to show where nulcear energy policy contradicts other political goals, and the spectrum of nuclear risks can be extended to problems in the field of foreign policy, aid policies to developing countries, and to economic, environmental, and defense policies.

The arguments concerning the risk of nuclear energy put forward in the press and in books, as well as in hearings and in courts of law, force experts and politicians to take a stand. Usually, these arguments concern the experience that problems of safety have so far always been solved, the high state of the art of safety technology, and the existence of comparable or much higher risks in other fields of technology and daily life. In all, no major progress has been made in this dialogue for years. Both sides repeat their arguments in a more or less stereotypical manner.

This obvious talking past each other confirms the above-mentioned thesis that risk primarily has the function of symbolizing the opposing positions in a controversy whose causes are more complex than can be preceived under the narrow perspective of safety and risk. If this thesis proves correct, a consensus on the issue of risk will not be reached as long as the complex causes and underlying motives of the conflict remain effective.

REFERENCES

1. Neue Kernkraftwerke in der Bundesrepublik Deutschland 1981, atomwirtschaft-atomtechnik, 257-271 (April 1981).
2. Stellungnahme des Bundesministeriums des Innern zu § 7 Atomgesetz (AtG), Umwelt Nr. 75, pp. 21-23, March 21, 1980.
3. Bekanntmachung der Neufassung des Gesetzes über die friedliche Verwendung der Kernenergie und den Schutz gegen ihre Gefahren (Atomgesetz), Bundesgesetzblatt, Teil I, Nr. 131, pp. 3053-3072, October, 31, 1976.
4. Bundesminister für Forschung und Technologie (ed.), Zur friedlichen Nutzung der Kernenergie, Bonn, 1977, p. 379.
5. H. A. Ritter, Genehmigungsverfahren für kerntechnische Anlagen, Kernthemen, November 1978 (edited by Deutsches Atomforum e.V.).
6. A. Birkhofer, Risikoanalysen unter besonderer Berücksichtigung der Kerntechnik, paper held at the Seminar on Legal and Technical Aspects of Risk Evaluation in Baden-Baden, July 1979, supported by the Federal Ministry for Research and Technology (manuscript).
7. F. W. Heuser, Die Risikoanalyse in der Kerntechnik, Aussagafähigkeit und Grenzen ihrer Anwendung, paper held at the Seminar on Risk and Safety Research in Romrod/Alsfeld, September 1980, supported by the Federal Ministry for Research and Technology (manuscript).
8. Amtliche Sammlung der Entscheidungen des Bundesverfassungsgerichts, (49):128 et seq. (Kalkar). Neue Juristische Wochenzeitschrift, 1980, pp. 763 et seq. (Mülheim-Kärlich).
9. B. Bender, Das technische Risiko als Rechtsproblem, paper held at the Seminar on Risk and Safety Research in Romrod/Alsfeld, September 1980, supported by the Federal Ministry for Research and Technology (manuscript).
10. Deutsches Atomforum e.V., Probleme des atomrechtlichen Genehmigungsverfahrens, Analysen 6/81.
11. Gesellschaft für Reaktorsicherheit, Deutsche Risikostudie Kernkraftwerke, eine Untersuchung zu dem durch Störfalle in Kernfraftwerken verursachten Risiko, Hauptband Bonn, 1980.

12. A. Birkhofer, The Expanding Role of Quantitative Risk Analysis in Germany, paper held at the IAEO International Conference on Current Nuclear Power Plant Safety Issues, Stockholm, October 1980 (manuscript).
13. Deutscher Bundestage 8. Wahlperiode, Bericht der Enquete-Kommission "Zukünftige Kernenergiepolitik" über den Stand der Arbeiten und die Ergebnisse, Drucksache 8/4341, June 27, 1980.
14. Kernforschungsanlage Jülich/Geselschaft fur Reaktosicherheit, Sicherheitsstudie für HTR-Konzepte unter deutschen Standortbedingungen, Hauptband zur Phase Ib, 1981.
15. Kerntechnische Gesellschaft e.V. (ed.), Auf dem Wege zu einer deutschen Wiederaufarbeitungsanlage, Die Diskussion seit 1979, Bonn, 1981.
16. Kernergie und Umwelt, Informationsdienst der Zeitschrfit atomwirtschaft-atomtechnik (March 1980).
17. G. Frederichs, Ursachen und Entwicklungstendenzen der Opposition gegen die Kernenergie, Zeitschrift für Umweltpolitik, September 1980, S. 681-705.

COPING WITH THE RISKS OF NUCLEAR POWER PLANTS

IN THE UNITED KINGDOM*

> Timothy O'Riordan
>
> School of Environmental Sciences
> University of East Anglia
> Norwich, U. K.

RISK MANAGEMENT AND NATIONAL ENERGY POLICY AND POLITICS

Risk management is only one element in any national energy policy. There is a possibility that too much emphasis is currently being placed on nuclear-related safety issues at the expense of wider political considerations relevant to a national energy strategy. Of equal, and possibly of greater significance, are the following factors:

1) Institutional inertia and political expectations have been built into existing commitments to a particular mix of energy supply options. In Britain, the coal industry supplies about 80% of the current electricity output, and the electricity supply industry in turn takes about 70% of U. K. coal production. Some 225,000 miners are thus dependent on the U. K. electricity industry, and it would be politically suicidal to substantially alter that commitment in a hurry. Already the National Union of Mineworkers has forced the present government to alter declared policy of running down the old, uneconomic pits in favor of a £400 million investment in existing and new coal mines [1].

Equally powerful is the tenacity of the nation's nuclear industry to maintain its commercial viability, to protect its highly skilled employees, and to boost its morale as a leader in advanced

---

*This research has been funded by the Swedish Energy R & D Commission and is sponsored by the Beijer Institute in Stockholm. The author is very grateful to these two institutions for the support he has received.

Table 1. Revised U.K. Electricity Demand Forecasts*

Estimate adopted	Average cold spell demand (gw)	Year to which estimate refers
March 1970	54.0	1975-76
March 1973	56.5	1979-80
Marh 1975	54.0	1981-82
March 1977	51.5	1983-84
October 1979	52.0	1986-87
February 1980	48.5	1986-87

*Source: Select Committee on Energy, HC Paper 114-i, HMSO, London, p. 20.

technology. Some 2,500 key nuclear engineering jobs and as many as 75,000 jobs in associated engineering and construction industries are at stake should the nuclear program be significantly run down. In addition, the national taxpayer and electricity consumer have invested at least £3,000 million in R & D and construction costs in the advanced gas cooled reactor (AGR) program over the past 20 years [2]. Currently, the investment in the proposed pressurized water reactor (PWR) is of the order of £30 million, though a 1,300 mw plant is expected to cost at least £1,500 million in 1982 prices. Possible public expenditure on this scale is not readily abandoned when major construction consortia stand to lose hundreds of millions of pounds should proposed nuclear schemes be delayed or abandoned.

2) Security of national energy supplies is probably the single most important variable in determining a future energy strategy. It is a complex policy objective which can be achieved in a number of ways:

a) By boosting indigenous sources of energy. In the United Kingdom, this means North Sea oil and gas, plus coal, at the expense of a dependence on middle eastern crude or imported liquified gas.

b) By encouraging greater energy conservation through an aggressive pricing policy, economic support for conservation technologies, and public information campaigns. Present U. K. government policy is to make all the nationalized energy industries (coal, oil/gas, electricity) self-financing and subject to explicit borrowing limits. Large investments in new sources of supply can prove embarassing as demand falls and prices have to be increased to finance debt. This is currently a matter of major concern to the

Table 2. Estimated Net Effective Cost of New Stations*
(£ per kw per year, March 1979 prices)

	AGR	PWR	Coal
Basic net effective cost	−26	−40	+18
Derating nuclear by 20%	+16	+12	—
Delay nuclear by 6 years	+33	+33	—
Delay coal by 2 years	—	—	+4
Average annual availability −8% pts nuclear	+12	+12	—
Nuclear fuel +0.13 p/kwh	+ 7	+ 7	—
Revised mean net effective cost (NEC)	+42	+24	+22

*Source: Select Committee on Energy, HC Paper 114-i, HMSO, London, p. 30.
Notes: Net economic cost is simply a version of net present value. The CEGB claims that negative values indicate that installation of new nuclear plant is a net benefit because it represents energy cost savings, i.e., savings on the total cost of providing electricity. NEC is a reflection of systems cost savings, not just of the cost of supplying power from a particular station.

Critics have pointed to the sensitivity of net effective costs to a variety of probable circumstances. the Monopolies and Mergers Commission has recently recalculated these figures in favor of maintaining old coal field stations over new nuclear power plants until the mid-eighties [5]. The Commission notes that when all these assumptions are taken into account, the NEC of a typical AGR would fall from −£8/kw to +£20, but for a new coal-fired plant it would fall from +£22 to +£25/kw. The net avoidance cost for an old coal-fired plant would rise from +£10/kw to −£1 to +£4/kw. The Commission also concludes that the two recent AGR estimates by the CEGB were made on strategic, not economic, grounds [6].

British electricity supply industry, which is faced with significant reductions in demand over its optimistic forecasts of the early seventies (Table 1).

c) By investing R & D into renewable energy resources and into nuclear fusion power. Currently the Department of Energy is spending about 10% of its total R & D budget on renewable energy resources, while the Central Electricity Generating Board (CEGB) is designing a number of large experimental windmills. While nuclear fusion is technically still a long way off, about £100 million has been invested in R & D to date.

3) For cost effectiveness, the present U. K. government policy is to reduce public expenditure and keep price rises to a minimum. Thus it encourages its electricity supply boards to produce electricity at lowest unit cost. The major reason for investments in new coal developments in eastern England is the low price of coal (£30 per ton from the Belvoir scheme as opposed to £60-70 per ton from old pits in Wales and central Scotland) [3]. And the major reason for its advocacy of the PWR is its alleged low unit cost of supply compared not only with coal-fired developments, but also with the AGR (Table 2). Recent calculations, however, suggest that an American designed PWR built to British safety standards will cost at least 15% more than an AGR [4].

It should be noted here that risk reduction may impair cost effectiveness. The British nuclear industry takes the view that the safety components of both the AGR and the PWR can be designed into new plants at relatively modest costs. For the PWR, this has been put at about £50 million for a plant costing at least £1,500 million when constructed. Nuclear critics would argue that this figure is preposterously low [7].

It is also worth noting that the cost of estimating nuclear plant risks and correcting for these is internationally shared by the nuclear industry. The results of U.S. risk assessments and Swedish, West German, and French analyses are all readily available to the industry. While the British industry will and must conduct its own risk analyses of its nuclear plants, the job is assisted by these international transfers of information, and the R & D costs should be proportionately reduced. Whether this sharing of data has its own "locking in" effect to the extent that risk appraisals become unquestioned after a time is anybody's guess. In Britain, the presence of an aggressive antinuclear lobby and the inevitability of exhaustive public inquiries before any new nuclear plant is licensed may reduce this danger, although this cannot be guaranteed.

4) Unlike the cases in the United States and France, the British political parties have adopted specific stances on nuclear power, notably as to the choice of reactor. In a country where official secrecy abounds and investigative journalism is restricted, it is impossible to obtain a clear picture of the party political infighting vis-á-vis nuclear power. One source of information is the carefully planted leak, another is the recollections of politicians formerly associated with nuclear power decisions. Whatever the case, it is worth emphasizing that both major British political parties support a continuing program of nuclear power (the liberals being more ambivalent but generally less nuclear-minded, while the social democrats have not yet announced a policy). The Conservative Party is the more nuclear-minded of the two major parties partly because it would like to reduce the economic stranglehold of the National Union of Mineworkers, which is closely affiliated with the Labor

Party, and generally because it believes in keeping its options open [8]. The political issues with the choice of reactor are not so much a function of risk, but of cost and the support of British industry and technology.

The Labor Party advocates the AGR on the grounds that it is now of proven design and that it should become more cost-effective as the prototype stage passes into more regular commercial production. It also believes that the AGR will maintain jobs in the declining British engineering industry especially in the recession-hit regions of north-east England and west central Scotland (traditionally Labor strongholds). The Conservative Party supports the PWR at least as an experimental venture, and, in this stance, it has the support of the General Electric Company, the largest engineering group in the nuclear business (though its main commitment is to turbine construction), the Atomic Energy Authority (UKAEA), the CEGB, and, allegedly, senior officials in the Department of Energy [9]. This is a powerful alliance that will not readily be thwarted.

The party political battle over reactor choice is an extremely relevant matter in any discussion of risk management. To begin with, it almost held up the report of the Parliamentary Select Committee on Energy which analysed the government's proposed nuclear power program [10]. The Committee divided repeatedly (but not purely on party lines) on the relative merits of PWR vs AGR and narrowly voted in favor of clearance) [11]. The Committee also concluded that a second PWR should not be contemplated for at least another five years, thus challenging the government's intention of constructing about one reactor per year over the period 1982-2000 [12].

Another sign of disarray within the British nuclear industry vis-á-vis its choice of thermal reactor is the resignation of the dynamic chairman of Britain's nuclear design and construction public agency, the National Nuclear Corporation (NNC), allegedly because he fell afoul of the struggle within the nuclear component industry by channeling investment into AGR components [13]. The indication here is that the nuclear construction industry, with enormous financial commitments at stake on each reactor, is deeply divided over the future of the British reactor. While safety considerations play an important role here, when feelings run high and so much is at stake, it is possible that ambiguities over risk analyses may be drawn into this wider conflict and be blown out of proportion.

This very lengthy introduction has emphasized that risk appraisal of energy supply options intermingles with political and economic considerations that may be of much greater significance when determining a nuclear energy strategy. The intermingling of risk appraisal with matters relating to pricing policy, the viability

Table 3. Fuel Consumption in U.K. Electricity Generating Stations 1977* (million tons coal equivalent)

Coal and coke	78.2	69.8%
Oil	17.7	15.8%
Nuclear	12.6	11.3%
Gas	1.7	1.6%
Hydro	1.7	1.5%
Total	111.9	

*Source: Department of Energy, Energy Policy: A Consultative Document, HMSO, London, 1978, p. 47.

of an important industrial sector in a faltering economy, precedent, and political expectations, render of dubious value any analysis of the risk component in isolation of an energy strategy.

ELECTRICITY SUPPLY OPTIONS IN THE UNITED KINGDOM

As already indicated, the electricity supply industry, which provides about 33% of the current national energy requirement, relies very heavily on a legacy of coal fired plants (Table 3).

The supply industry is run by three Boards, the North of Scotland Electricity Board (NSEB), the South of Scotland Electricity Board (SSEB) and the CEGB. Though financially and managerially autonomous, the three boards connect their supplies through a national grid and have close working relations. The SSEB recently committed itself to a new AGR plant at Torness in eastern Scotland and is now so oversupplied with baseload power that it is not keen to take on any PWR plant in the forseeable future. It has also publicaly expressed its satisfaction with the safety aspects of the AGR and its doubts over the risks associated with the PWR [14].

The CEGB is presently committed to maintaining its program of 9 of the 11 first-generation magnox stations (the other two are in Scotland, Fig. 1), most of which are approaching the end of their design lives and some of which are showing signs of wear and tear, but all of which are presumed "safe." Friends of the Earth, the major British antinuclear group, has expressed its dissatisfaction over magnox safety, but in rather muted terms. It is basically frustrated, as safety details and operational histories of each plant are simply unavailable [15]. Of some interest is the admission by representatives of the industry to the author that the safety analyses of the old magnox stations, built in the mid-fifties and early sixties, are below the standards required for such analyses

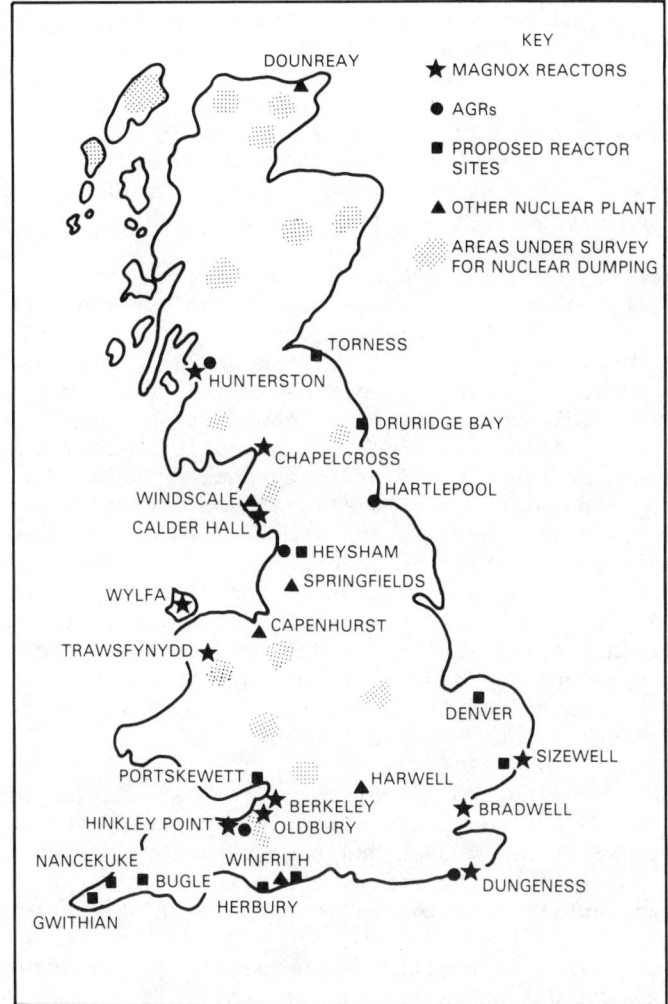

Fig. 1. The location of nuclear power plants in Britain.

today. The CEGB has invested in five AGR plants since the mid-sixties, only two of which are currently in operation. The others have suffered from design faults and construction delays that have been the source of considerable public criticism. All U. K. AGR plants have experienced cost overruns of at least 200% over the rate of inflation, while the cumulative time delay on the five plants is 23 years [16].

On December 18, 1979, the Energy Secretary announced a new nuclear program based upon estimates by the electricity supply industry

and endorsed by the government of at least one new nuclear power
station per year over the decade from 1982, amounting to the installation of 15,000 mw over 10 years [17]. This opened the way for the
CEGB to propose investment in one "trial" PWR proposed for Sizewell
in Suffolk (where there is already a successful magnox reactor), and,
pending its safety clearance, the CEGB may invest in more PWRs as
the need arises. The safety clearance will be reviewed in a public
inquiry which is scheduled for late summer, 1982.

The background to the proposed PWR plant is interesting to
record, for it emphasizes the points made earlier about the international collaboration of nuclear expertise. The plant is to be
based on the nuclear steam supply system (NSSS) design as approved
by the U.S. Nuclear Regulatory Commission and which is being constructed at the Calloway plant in Fulton, Missouri, and at Wolf
Creek in Kansas. This reference design and its safety case are the
result of the combined efforts of four teams - the NNC, Bechtel,
Westinghouse, and the CEGB. Recently it was announced that in order
to speed up the design process and reduce total costs (which were
becoming alarmingly high), the government had established a task
force consisting of representatives of the CEGB and the NNC, under
the chairmanship of Dr. Walter Marshall, chairman of the U. K.
Atomic Energy Authority (UKAEA), and a self-confessed advocate of
the PWR, to speed up the design process and to buy most of the NSSS
direct from the United States under the terms of the standardized
Nuclear Unit Power Plant Systems Agreement [18].

The novelty of the PWR in Britain, its association with the
Three Mile Island accident, and scientific concern over the integrity of the pressure vessel and the emergency core cooling system all combine to make the safety analysis of this PWR the most
exhaustive and public that the nation will ever have experienced.

But equally relevant will be the so-called need issue, namely,
whether an additional baseload reactor really is required in the
light of existing commitments to the coal industry and in view of
falling demand forecasts. Two recent reports by the Parliamentary
Select Committee and the Monopolies and Mergers Commission cast
serious doubt on the CEGB's case for the cost-effectiveness of the
PWR and its argument that it would be cheaper to run down obsolete
coal fired plants in favor of a major new nuclear investment. This
is a controversial matter which will not be resolved in the short
run, but the detailed economic arguments contained in these two reports are certainly not going to advance the case for the PWR.
Nevertheless, the political and institutional investment in a PWR,
and to nuclear power in general, is such that the PWR may well be
passed as both safe and necessary unless the Labor Party is returned
to power in 1984, in which case the project may well end up as an
AGR.

THE NATURE OF NUCLEAR POWER RISK MANAGEMENT IN BRITAIN

Regulation of risk in the United Kingdom takes place within a set of conventions which have a long and cherished tradition. These conventions include: 1) confidentiality and selective consultation, 2) cooperation, 3) professional integrity, and 4) gradualism.

Confidentiality and Selective Consultation

Under the Official Secrets Act, 1911, all governmental information is regarded as confidential until it is officially authorized to be made public. The bias lies with secrecy, not with disclosure. Legally it is an offence for any governmental official to disclose any information without permission, though in practice consultation among parties known to be helpful and sympathetic (though not necessarily uncritical), does take place with and without official sanction.

Cooperation

Cooperation is one of the principal hallmarks of British regulatory practice, based as it is on mutual respect born of trust and professional competence. Given the containment of confidentiality, the assumption holds that the qualities of honesty and frankness, together with a professional ethos of competence and noncorruptability, combine to maintain high standards outside formal and continual public scrutiny and accountability. In practice, the regulatory agencies work closely with their clients and frequently regard their role more as advisors then policemen. Their aim is not just to understand the problems facing their clients when attempting to meet certain safety standards, but to help them design plant and safety procedures that will meet the required standards. Hence, in Britain, self-regulation is more important than externally imposed regulation.

Professional Integrity

Safety regulation operates through a kind of club of people who are not only highly trained but who genuinely believe in professional competence and high standards. At times this conviction can be so overwhelming that any criticism from an "outsider," a fellow professional, or an informed layperson may be regraded almost ipso facto as erroneous and possibly irrelevant. The major critics of nuclear safety in Britain have to be people of highly reputable scientific standing in the community. One must note here that the dictates of official secrecy preclude scientists employed within the official nuclear regulatory and research organizations from speaking or writing in public. Their disagreements can only be channeled through carefully plants leaks. Studies of "mistakes" in the industry are undertaken by only quasi-independent analysts [19].

Also relevant to this point is the practice of establishing professional task forces, study groups, or advisory committees to review the work of safety design and regulation. These bodies are either set up to look at a particularly thorny issue, such as pressure vessel integrity, or they are appointed to advise the regulatory agencies on various aspects of their work. Their job is to sift all the relevant evidence and to report confidentially to their respective masters. Generally their meetings are not minuted, and their deliberations are kept secret even from other members of the regulatory and research organizations. In general, the membership of these bodies is drawn from scientists and public figures with well-established public reputations. Whether these groups are genuinely independent and freely able to criticize remains a matter for speculation. The "containment" element to all such deliberations provides almost an impenetrable barrier to the outside investigator.

Gradualism

Institutional policymaking and decisionmaking are no different in Britain than elsewhere. Decisions are usually made incrementally, bearing in mind the momentum and institutional inertia of the organizations involved and their commitment to previous decisions and practices. In the context of risk analysis, this means a tendency to add only marginally to established techniques so that any novel methodology is often accreted onto tried practice. There is nothing inherently wrong in this; indeed, there is much to commend it, but it can mean that major criticisms of a whole philosophy of risk analysis are not fully taken on board at a suitable stage. This will particularly be the case where such criticisms emanate from sources not regarded as professionally competent.

This problem emerges when incalculable probabilities such as operator error or multiple, yet independently caused, faults are grafted onto the well established fault tree analysis. As Dr. John Collier remarked in a detailed defense of that methodology, "Obviously no assessment can deal with the unforeseen; it can merely ensure that its likelihood is acceptably low" [20]. The problem is to define by whose judgment that risk of incalcuable probability is made and by whom that judgment is believed to be acceptably low. The gradualist approach to risk assessment may not be the most suitable to cope with this kind of problem.

THE STRUCTURE OF NUCLEAR RISK MANAGEMENT IN BRITAIN

In Britain, the management of nuclear related risk takes place through a number of organizations, all of which are interconnected. Figure 2 displays these relationships.

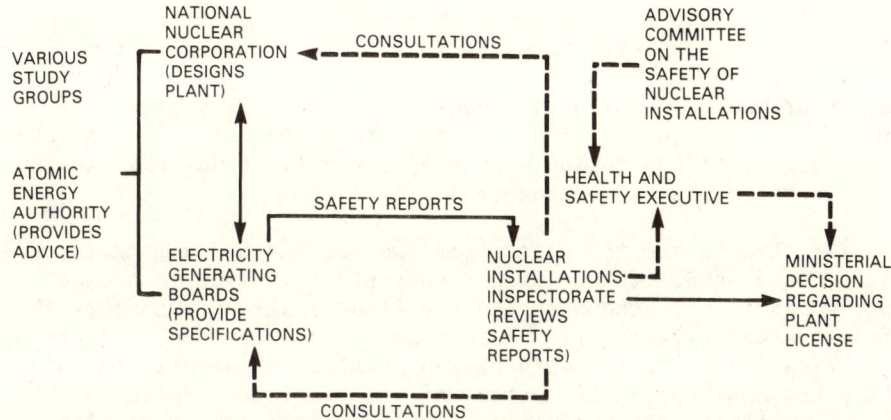

Fig. 2. The relationship between the principal agencies involved with nuclear power plant safety in the U. K.

The UKAEA is responsible for undertaking research into various aspects of nuclear power, and through its Safety and Reliability Directorate, it advises the nuclear industry in risk appraisals themselves and risk appraisal methodologies. The UKAEA is primarily an R & D agency where much of its work is completed in advance of a commercially designed plant. At present, the UKAEA needs to provide relatively little advice on the safety aspects of either the AGR or the PWR in terms of detailed investigation, as much of this work has either been done or is available through international data transfer. For example, the results of U.S. rulemaking hearings are available to the U. K. nuclear industry, and one Bechtel and two Westinghous employees are currently seconded to the NNC to assist in the PWR design. Currently, about 5% of AEA's resources are devoted to R & D on the PWR.

Nevertheless, in response to criticism of PWR pressure vessel reliability first presented by a former government chief scientist and distinguished metallurgist, Sir Alan Cottrell, to a Parliamentary Select Committee in 1974 [21], the UKAEA appointed a study group under its current chairman (then its vice-chairman), Dr. Walter Marshall. This group, composed allegedly of the best minds in the country with a knowledge of this problem, concluded that the pressure vessel could be made to be acceptably safe from catastrophic failure, at least at the start of service, "provided that the vessel when manufactured is in accordance with the design intentions, that it is made from the specified materials, that extensive inspection, quality control and testing is [sic] carried out, and that service conditions comply with those defined by the vessel specifications... . In addition... all the associated and auxiliary systems and structures perform as specified" [22].

This matter will be dealt with again below, the point here being to emphasize that the UKAEA is seen as an authoritative advisory souce for matters of technical complexity, including the quality of risk assessment methodologies. In this capacity it serves the whole nuclear industry in Britain and provides assistance, upon request, to the nation's principal regulatory agency, the Nuclear Installations Inspectorate (NII).

The clients are the electricity generating boards, namely the SSEB and the CEGB. (NSHEB is not currently involved in nuclear power generation.) These boards are financially and managerially autonomous and largely free of government interference, except when major financial debt is contemplated. These boards are also ultimately responsible for the safety of the completed nuclear plant, so it is in their interest to see that adequate safety studies are completed and that quality assurance of component ordering and manufacturing is properly followed through. Thus the CEGB not only works closely with the design teams in the construction consortia to the extent of sharing documents throughout the safety review and design process, but it also employs its own safety and design engineers who report to the board's director of health and safety, who, in turn, reports directly to the chairman of the board. Indeed, it is now evident that the CEGB, not the NCC, is taking the lead over the safety aspects of the PWR design.

In both the client and vendor organizations in Britain, the health and safety divisions are independent of line management, and senior officials have free access to all divisional heads and, most important, to the chief executives and chairmen. This suggests that health and safety matters can readily be channeled down to the very lowest levels of management without impediment or distortion. It does, of course, also mean that potentially controversial interpretations of health and safety are contained within a relatively narrow managerial spectrum, with no "fail safe" device to allow contrary points of view to be aired at top management level. This is clearly a matter for speculation, for it is all but impossible to prove that such a situation might arise. Nevertheless, it is an accusation that is sometimes made against the nuclear industry in various countries by the ever suspicious antinuclear lobby. In Britain, the answer to this issue is that of professional commitment and self-regulation.

The vendor is NNC, a jointly owned public corporation in which the UKAEA has 35% of the shares, a consortium of nuclear component manufacturers has another 35%, and the General Electric Company which owns the remaining 30%. The NNC also has its own safety and policy group (Fig. 3) which reports directly to the chief executive, though each line department is responsible for its own safety aspects. The actual draft safety documents for the proposed PWR at Sizewell will be prepared by the safety and licensing

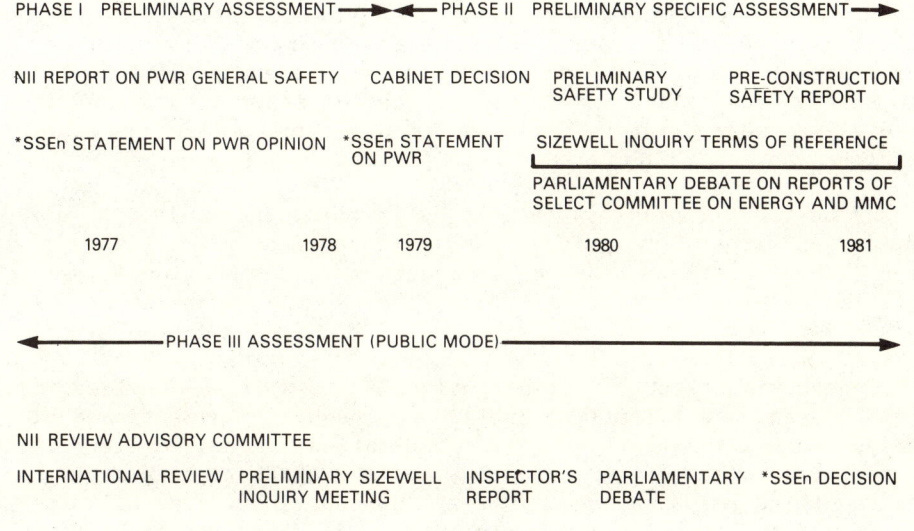

Fig. 3. Pathway to a decision in the U. K. PWR proposal. (*SSEn = Secretary of State for Energy.)

group of the PWR team, but, as Fig. 2 indicates, this will be done in close consultation with both the CEGB and the NII. The NNC is particularly proud of its quality assurance procedures which are really managerial devices to ensure that all components arrive on time and in order and are fully checked via quality control procedures. This is the operational heart of the British nuclear risk reduction business. Theoretical calculations are of no use if the design and construction are faulty. The NNC regards quality assurance as both an executive and policy-related activity that is dictated largely on the grounds of safety [23].

The regulator is NII, a branch of the Health and Safety Executive, which is responsible for all industrially related health and safety in Britain. The NII is composed of engineers and scientists with considerable experience in the nuclear industry (About 80 of the 110 employees have actually worked in the industry), but whose particular expertise is to see safety "in the whole," i.e., as a complete phenomenon not as a series of interacting parts. The NII is required to be independent of the industry, and its officials go to great lengths to avoid any suspicion of collusion or involvement. Thus the NII plant inspector does not sit on the plant safety committee, nor does he have direct relations with the plant safety officer. The Inspectorate is also responsible for advising the Secretary of State for Energy (or the Scottish and Welsh secretaries for plants built in Scotland and Wales as to the validity of all the safety documentatin and must give the Minister its consent before any nuclear plant is licensed. (In practice, the industry regulates and the NII oversees this process.)

Thus the NII plays a key role as reviewer and advisor on all aspects of safety both prior to and after nuclear plant construction. However, in the post-construction phase, the client's self-policing role over operational safety is of crucial practical importance.

The current issue of great public interest is the degree of true independence actually exercised by the NII. As mentioned already, the NII is a party to all consultations among the electricity boards, the construction consortia, and the UKAEA throughout the period of the prepearation of safety documentation. The NII also has issued safety assessment principles [15] for all classes of nuclear reactors. These provide general guides, although the NII has separately endorsed the principle that a PWR can be designed so that it is acceptably safe. Critics feel that the NII was too precipitious in its analysis and may have prejudged the Sizewell PWR case. But, to be fair to the NII, all it has done is to indicate that a PWR can be made safe. It still has to be convinced that a British version of a Westinghouse design will be safe. A point worth noting here is that the NII guidelines are not rigidly defined with set criteria, but are based on the "best practicable means" principle, a well-known concept in British regulatory circles.

While the NII is very close to the nuclear industry in Britain, it steadfastly maintains that it is professionally independent of it. What does worry a number of people is that the NII is currently suffering from a decline in recruitment and, more serious, a loss of experienced officers who are either retiring or are tempted into better paid and more interesting jobs in the industry. In addition, the government has not abandoned its intention of moving the Inspectorate to Bootle on Merseyside in 1985, a prospect that does not inspire any enthusiasm among middle-level management. The Select Committee on Energy was so worried about this proposed move that it recommended that the government think again about this proposed office relation and that it provide better pay so staff can continue and experienced officers can be tempted to remain. The Committee went to some lengths to demand that the NII remain "respected, independent and effective and be seen to be so" [24]. The Committee also urged that the NII be provided with sufficient expertise to enable it to "reexamine all data and recalculate all calculations" on risk estimation and that there be no doubt in the public mind that this would be done.

The loss of morale among the NII staff is a serious matter, for it seems that only professional integrity will enable the Inspectorate to retain its standards and independence. Yet, the public needs more reassurance, and the Select Committee was right in arguing for a considerably strengthened and obviously more independent Inspectorate. While the Inspectorate claims that under present conditions it will be able to complete the safety review of the PWR satisfactorily, it has expressed reservations about its ability to check post-construction plant safety, especially with regard to worker health protection.

The NII is advised by an independent scientific committee, the Advisory Committee on the Safety of Nuclear Installations. This committee does not report pubicly and is not privy to all the relevant documentation on health and safety matters. For example, it is alleged on the basis of reliable evidence that the committee has not seen all the minutes and correspondence associated with the work of the Marshall Study Group on pressure vessel integrity, which are of crucial significance when determining the safety of a PWR design. Yet, its job is to review all materials sent to it, sift the key arguments, and report to the Chief Nuclear Inspector. How well the committee does this is anybody's guess, as few people know if it is adequately informed, let alone how it reports. Yet, on the really controversial risk issues, its comments are supposed to be enormously significant for the officially sanctioned safety position.

THE STRENGTHS OF THE BRITISH APPROACH

The British make no secret of the fact that their cooperative, flexible approach, coupled with professional integrity and high standards of plant construction, component manufacture, and operator training, produces a reliable safety management system. In evidence before the Commons Select Committee on Energy, the Nuclear Installations Inspectorate reinforced this point.

> One major difference between the nuclear regulatory policy of the U. K. and that of some other countries, notably the USA, is in the extent to which the U. K. regulatory regime places the responsibility for safety squarely upon the licensee... . The U. K. view, exemplified by the NII's Safety Assessment Principles is that undue detail in formal regulatory requirements is not only time consuming for the regulator but also tends to be counterproductive and tends to reduce the responsibility of the licensee and industry for safety. This can lead to the belief that so long as the formal requirement is met, no more needs to be done to ensure safety [25] (emphasis added).

The key to British nuclear policy lies in flexibility and self-regulation, backed up by a variety of expert analyses and duplication of safety assessments by a number of participating agencies.

Strict standards are not laid down; instead, emphasis is being placed on guidelines, principles, and the concepts of reasonableness and practicability. Here is where the consultative and advisory role of the NII and various safety research and operational divisions is claimed to be so effective. The designer, NNC, has to produce safety cases from first principles which include all practicable faults and fault sequences and which will stand up to systematic examination by the CEGB and the NII.

As regards these principles, the CEGB has announced that its acceptable risk envelope is a design such that the total predicted frequency of all accidents that could give rise to uncontrolled releases of radioactivity should be less than around 10^{-6} per reactor year. For any single accident sequence, the design guide requires a predicted accident frequency of 10^{-7} per reactor year, though such estimates can only be best guesses and have a spurious air of accuracy about them. Accident frequency is defined as the product of the initiating fault frequency and the probability of failure to control the reactor. As the CEGB's Direct of Health and Safety put it,

> The assessors (NII) will judge that a reactor has an acceptably low risk on the basis of qualitative assessment, supported by numerical analysis, and taking into account various factors not readily amenable to numerical analysis, such as safety administrative procedures, operator recovery and damage control arrangements. Overall acceptance of a design is not based on the achievements of a particular numerical standard [26].

To achieve acceptable levels of safety, the industry believes not only in redundancy (usually of four systems) but also in diversity to account for common mode failure. In both cases, the least effective system must be capable of being able to perform effectively on its own. Naturally, the analysis will only be as good as the assumptions upon which it is based, but another strength of the British approach lies in the continuous nature of the risk appraisal process from the very earliest stages of plant design right through to decommissioning.

For a major new plant there are four stages of risk analysis:

1) <u>The preliminary safety case</u>: This is prepared at a very early stage and is based on a reference design. The objective is to outline the safety case in general terms and to pinpoint key issues and possible limitations.

2) <u>The preconstruction safety report</u>: This is the major safety analysis upon which the reactor license will be deciced. It makes the case for the engineering feasibility of the specific re-

actor design and includes information on quality assurance and detailed specification of the whole plant. Any engineered safety features (e.g., fault and transient analyses) must be specified in detail. This document is prepared by the NNC and CEGB, but is formally presented to the NII by the CEGB. The NII must give its approval to this document before the plant can be licensed. That approval and the CEGB report are made public and subjected to cross-examination at a public inquiry.

3) The plant completion report: This must be available before commissioning takes place, and requires that a series of specific tests be undertaken and that the results of each "as far as possible" be fully documented. The results are reviewed by a plant completion committee chaired by the station superintendent and serviced by the plant safety officer, both CEGB employees. This committee contains representatives of the NNC as well as operational managers of the CEGB, but it does not include an official of the NII.

4) The integrated commission report: This follows the actual startup of the station and is prepared by the station commission committee, which must ensure that all plant personnel are suitably trained. Again, the NII is not represented. This report is basically an operation and maintenance manual providing a detailed case history of each item in the plant.

Each of these stages requires a lot of paper work and internal review. At each stage the safety analysis is carefully checked for consistency, and any new developments in research or methodological appraisal are incorporated where relevant. The industry believes that this continuing review, led jointly by the client and the vendor, is of crucial significance in maintaining plant reliability.

THE WEAKNESSES OF THE BRITISH APPROACH

The principal weakness of the British approach to nuclear risk managment lies in the lack of genuinely independent scrutiny of all relevant documents at the formative stage of their preparation. Public debate of a plant safety case or a reactor design is normally confined to the public inquiry, but the forthcoming Sizewell PWR inquiry will be the first opportunity for a major public investigation of a nuclear reactor case, despite the fact that 11 magnox stations and 7 AGRs are either operational or under construction. The first four magnox reactors, built in the mid-fifties, did not even go to public inquiry. Subsequent stations involved public inquiries, but these were largely confined to planning and amenity issues. The safety case was hardly ever presented, let alone cross-examined.

The electricity generating boards and the NNC will not give way on the issue of full disclosure of their safety analyses. Their po-

sition is supported by the AEA, the NII, and the Department of the
Environment on the grounds that to publish all the relevant documentation would be too costly and that their contents would be "misinterpreted" or "taken out of context" by laypeople. The argument
continues that this would or could fuel general public disquiet
about nuclear power and would lead to the industry representatives
spending enormous amounts of time explaining the minutiae of highly
complex technical information. In addition, there is a strong belief that technical matters such as risk analysis should be left to
the experts, and that when politicians and the public become involved, the issues get distorted and there is a danger of obfuscation
and delay. This point was made quite specifically by Dr. Marshall,
now chairman of the UKAEA, in evidence before the Select Committee
on Energy.

> I do not want a Minister to decide whether a defect that is
> half an inch in diameter is safe or unsafe. I do not want to
> have the Minister stand up on the floor of the House and try
> to explain which way his decision went and why... . I want to
> be sure that this is done by technocrats. I also want to make
> sure that their arguments and their thinking are also exposed
> to the public as much as we can so as to carry the public and
> Parliament with us... [27] (emphasis added).

Nevertheless, in view of continuing public apprehension over
nuclear power kindled from time to time by reports of radioactive
leaks or missing fuel rods which are usually not reported when they
occur and which tend to reinforce in the public mind that the self-regulation by the industry may not be as foolproof as it might appear, the British nuclear establishment is increasingly recognizing
that greater public disclosure and discussion would be desirable.
But this will be done cautiously, with careful monitoring of all
possible political repercussions constantly undertaken. As Dr.
Marshall commented,

> The Nuclear Installations Inspector should be the technical
> and dispassionate judge... . In the past this has been done
> in a dispassionate and responsible way but outside the limelight of public opinion... . I have come to realize that, in
> this matter, we must ensure not only that justice is done but
> that justice is also seen to be done and, therefore, we must
> find a method of exposing this technical process to the public
> in some appropriate way [28].

More specifically, the weakness in the British approach lies
not only in the inadequate disclosure of technical information and
the reviews that various key documents receive via the process of
selective consultation but also in the manner in which technical
judgments are transmitted through the civil service hierarchy for
political decision. A number of authoritative commentors have

noted that the British civil service is not always adequately competent to review technical evidence, that there are few fail-safe checks to ensure that civil service advice is properly thought through, and that once the senior civil service hierarchy is convinced of a course of action, it is very difficult for a minister to shake it off [29]. Brian Sedgemore, once a parliamentary private secretary to Tony Benn when he was Secretary of State for Energy, is convinced that there is a PWR conspiracy amongst Department of Energy officials and sections of the nuclear industry. He is equally convinced that any flaws in a safety analysis will be conveniently laundered out before public disclosure takes place [30]. These are powerful accusations, but, coming from an inside, admittedly, biased, source, they cannot be ignored.

To put these points into perspective, consider the PWR pressure vessel reliability issue. As already mentioned, in 1974 Sir Alan Cottrell voiced concern over the metallurgical properties of steel to withstand the enormous pressures without cracking precipitously and catastrophically. In 1976, the Marshall Study Group reported that, while this matter had to be viewed with concern, the technology of manufacture and inspection was adequate to produce a reliable vessel and that cracks of less than 100 mm would not result in catastrophic failure. This report was widely circulated among the profession for comment, but the summation of these comments has never been made public. It should be noted that the Marshall Study Group qualified its findings on the grounds that all auxiliary systems would work to specification (a point still to be confirmed). The Group suggested that a failure in the emergency core cooling system could result in meltdown with serious consequences. Meanwhile, in 1978, the Plate Inspection Steering Committee (PISC) an international advisory body to the OECD involved in inspection technologies, reported that cracks of up to 25 mm could not be reliably identified in more than 50% of inspections. The Marshall Group would probably comment that this was not too serious, as cracks of this size were not likely to result in catastrophic failure. But the PISC report also found that there was only a 50% chance of detecting crack group defects of 200 mm.

In 1979, Sir Alan Cottrell wrote directly to the Prime Minister that three particular design features of the PWR gave him concern, namely, vaporization of the coolant, cracking in the steel pressure circuit and difficulties of repair and maintenance. With regard to the question of cracks in the pressure vessel, Sir Alan wrote,

> I feel confident that if the Marshall recommendations were to be rigorously applied no dangerous cracks would escape detection. But this is only part of the problem. There would still remain the difficulty of deciding what to do if dangerous cracks became apparent after the reactor had come into operation and its interior was no longer accessible to workmen... in such circumstances one would become faced with a choice... either to

order the shutdown of the reactor after a very short life... or to risk operating the reactor knowing that it is no longer assuredly safe. I beg you not to enter upon a course which might eventually force such a decision upon you or your successors [31].

This letter was leaked to *The Guardian* newspaper, which subsequently published a prolonged and lively correspondence between a former nuclear engineer in Framatome (the French equivalent to the AEA), whose credentials were repeatedly questioned, and Dr. Marshall, who defended the report of his study group [32]. Yet, as mentioned, criticisms of that report have not been published. But on February 14, 1980, *The Guardian* published an article about a critique of that report allegedly prepared by the NII, but in fact prepared by the Institute of Professional Civil Servants, the union representing part of the scientifiec civil service, some of whose members work in the UKAEA. That critique stated that the Marshall Study Group failed to document all relevant pressure vessel failure rates, that it concentrated on both thin-walled and thick-walled vessesl (when the thin-walled variety may leak before break) and that the critical crack size could be as low as 12 mm (about an order of magnitude lower than that which the Marshall Study Group felt was serious).

The Commons Select Committee on Energy, well-briefed by highly competent advisers, was aware of all this correspondence and cross-examined both Dr. Marshall and Sir Alan Cottrell on their views. Dr. Marshall stuck to his Study Group report and remained convinced that not only could the crack inspection technology using ultrasonic techniques be improved to the required standards (something that would take four to six man-years of AEA time he felt), but that crack repair, even under radioactive conditions, was possible and was "not as frightening as (the Committee) have interpreted Sir Alan's remarks" [33].

For his part, Sir Alan maintained that a crack size of 25 mm "was not far from the estimated critical crack size for a PWR in fault conditions" and that "to ensure safety... an improved system of ultrasonic inspection will be needed." When quizzed as to whather he thought this probelm was soluble, Sir Alan replied, "No. I would say that the trask of running a safe PWR system is achievable but with great difficulty. I think that to achieve it puts a strain on human qualities" [34].

The Marshall Study Group is apparently still in existence and the lead work on pressure vessel reliability is presumably still with the UKAEA, but no further public statements have appeared on this vital matter. In all probability, full public disclosure will occur only prior to the onset of the Sizewell PWR public inquiry when there may be too little time for a truly independent appraisal of all the relevant documentation.

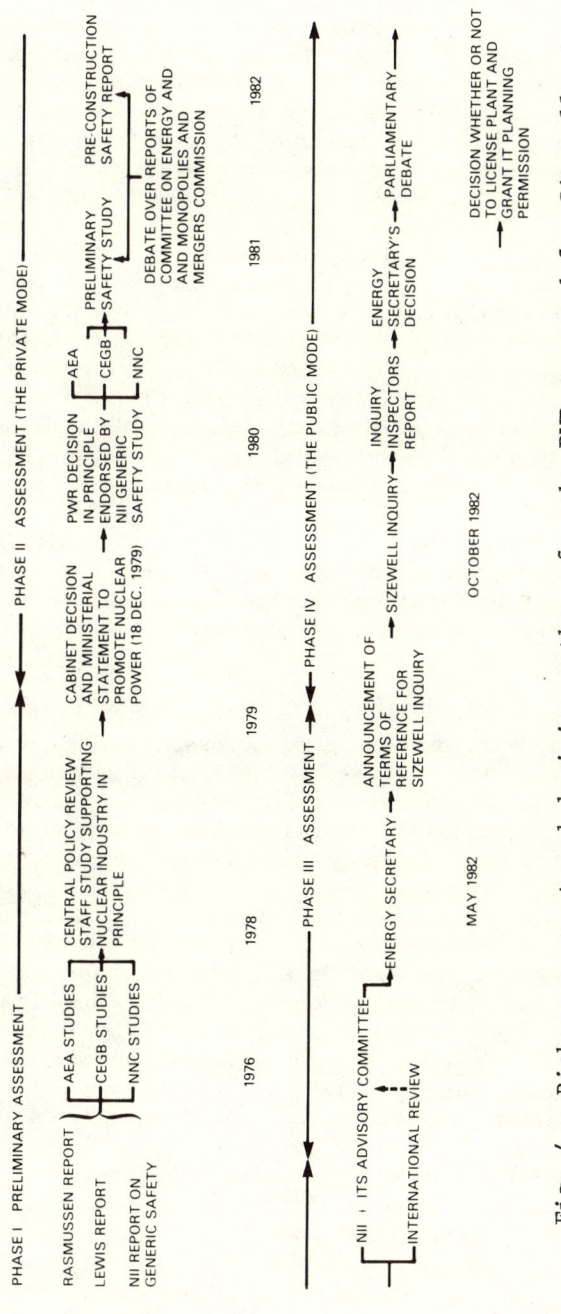

Fig. 4. Risk assessment and decision pathway for the PWR proposed for Sizewell.

Meanwhile, the government apparently is convinced by Dr. Marshall's views that any problems with the reliability of the pressure vessel "can be surmounted" [35]. The implication here is that such problems have not yet been technically overcome. Only time will tell whether this assurance is justified. Much will depend on the quality of the cross-examination on this matter at the forthcoming public inquiry. The quality of cross-examination is also a matter of concern, since the antinuclear groups may be ill-equipped with money and expertise to conduct an adequate case in this highly complex area.

This leads to the third major flaw in the British approach, namely, the overdependence on the public inquiry as the forum to review the safety analysis of a major nuclear plant. The vexed question of the purpose of the public inquiry over major technnological processes involving various kinds of risk has been much debated in Britain over the past 18 months [36]. Its importance as a device for full review of nuclear risk should be clear from Fig. 4 which depicts the decision pathway for the Sizewell PWR station. From this discussion the following points can be made:

1) The government is committed to a major public inquiry for the Sizewell proposal, but has not yet determined its terms of reference. In all probability, these inquiries will permit discussion of safety matters but will exclude debate on the policy relevance of the nuclear option in general.

2) The precise format of the inquiry cannot be determined in advance. Each inquiry of this kind is essentially novel. The government also has made it clear that it is the very flexibility of the inquiry system that is one of its strengths. However, the inquiry will be site specific; it will not be a two-stage process reviewing the merits of PWRs in general before turning to the local impact of Sizewell. Nevertheless, the nuclear industry, the NII, and, presumably, the government all want this inquiry to determine the safety case for the PWR once and for all, so that any further PWR inquiry would concentrate more on site-specific details than on the general safety case. This is probably wishful thinking.

3) The major antinuclear organizations regard this particular inquiry as a significant test case for the future political viability of the public inquiry as an extraparliamentary reviewing device. They are suspicious of the inquiry, accusing governments of using it as a window dressing exercise while the real political commitments are already made. Any indication that the government favors the PWR in advice of the inquiry is pounced upon, yet a cabinet leak suggests that this may be the case [37].

4) As part of their suspicions about the inquiry, the antinuclear groups want to see public money being made available to enable

them to take their case to the inquiry unimpeded by lack of resources. This is a vital issue, for, without some external funding, the opposition case may well be limited especially with regard to the safety review. One possibility is that some funds will be released by the ratepayers of Suffolk, the county in which the plant may be sited. But it now appears that the Council will employ a consultant, formerly associated with the nuclear industry, simply to review both the CEGB and NII safety analyses. How thoroughly and how independently this job is done remains to be seen.

5) The inquiry will be held under the auspices of the Department of Energy and will be chaired by an experienced inspector who is a planning lawyer and who will be advised by two expert assessors. In general, the British concede that the trained legal mind, deploying the techniques of cross-examination and natural justice (basically, the right of any objector to be heard and to examine and be examined), is best suited to deal with complex sociotechnical issues of this kind, though this is by no means a unanimous view [38]. It is also assumed that all documentation will be available to all objectors in sufficient time prior to the inquiry (at least four months) to allow the respective cases to be heard. If this happens, it will be the first time that such an arrangement has been guaranteed [39].

CONCLUSIONS

The main conclusions of this paper can be summarized thus:

1) In Britian, and probably in other nations, one must be careful to place energy-related risk management in the context of the wider economic and political aspects of energy policy making. In some instances, risk analyses become tools for political and economic arguments that are only partially related to the risk evidence. Thus practitioners of risk analyses must be extra careful to stick to the technological facts and predictions. When forced to make judgments of a qualitative nature, they should be sure that such opinions clearly relate to these facts and predictions. This is especially important in the case of nuclear-related risk assessments, as these tend to attract a lot of public interest and arouse controversy.

2) With respect to nuclear power in Britain, the safety aspects of nuclear reactors are beginning to take second place to economic aspects (especially investment policies, cost overruns, poor records of performance, and erroneous forecasting methodologies). This is particularly the case with the AGR, whose safety features are rarely questioned (with one notable exception [40]), though it may not be the case with the PWR.

3) The novelty of the PWR and doubts about two-phase flow, its coolant system, and the reliability of the pressure vessel and associated circuitry may give rise to public controversy over its safety. But any such controversy will be also fueled by nonsafety considerations particularly relating to the non-British element of its component construction.

4) The strength of the British approach lies in the high standards of the practitioners, the close working relations of all those involved in safety analysis, the flexibility of the safety guidelines, the self-policing element of the construction industry and the generating boards, and the dynamic nature and continual assessment of the whole safety case from plant design to decommission. It is worth repeating that the whole of the industry knows that it is under investigation and that it is more concerned than the antinuclear lobby that an accident, no matter how minor, must not occur.

5) The weakness of the British approach lies in its excessive commitment to secrecy, the lack of proper and open independent analysis at various stages of the preparation of the safety case, the nondisclosure of reviews of agency safety analysis when undertaken by expert internal advisory bodies, and the faults of the public inquiry as a device for analyzing in depth and complete fairness all the relevant documentation. It should be noted that the British nuclear establishment is steadily becoming aware of these failings and is slowly and cautiously opening up its whole safety review process. However, more needs to be done in this respect.

6) Because professional expertise counts for so much in Britain, a relatively small number of doubters, if they have the right credentials and public standing, can have a disproportionate influence on the risk debate. This equally applies to those who can give reassurance over areas of doubt regarding risk estimates. The Commons Select Committee on Energy was biased towards favoring the PWR safety case as a result of a detailed Westinghouse briefing, but remained sufficiently cautious following evidence from Sir Alan Cottrell to fuel further doubt in the minds of the anti-PWR lobby [41].

7) This indicates that what may be a matter for public concern with respect to nuclear plant risk management may not be the same as the risk elements that concern the professionals. The British professionals feel that the "big" controversies over the PWR can be handled, for they will be studied in depth, but that many "smaller" and less dramatic features of plant design and operation are more likely to go wrong.

NOTES

1. The British coal industry is suffering from the effects of recession. Despite a planned output of 135 million tons a year for 1985, the industry needs only to supply 121 million tons to meet the projected national need. At present there is much overcapacity and low productivity. Because of this, the previous Labor government forced the Central Electricity Generating Board to invest two to three years early in Drax B, a 660 mw coal-fired plant in Yorkshire, largely to absorb excess coal. For a detailed critique of U. K. coal forecasts, see C. Robinson and E. Marshall, What Future for British Coal?, Hobart Paper No. 89, Institute for Economic Affairs, London, 1981.
2. For a good analysis of the costs of the AGR program, see D. Burn, Nuclear Power and the Energy Crisis, London, 1979, and P. D. Henderson, "Two British Errors; Their Probable Size and Possible Consequences," Oxford Economic Papers, 1977, pp. 159-205.
3. These figures hinge on productivity per worker in new pits being twice as high as in the old pits and the wage rates not rising more than 30% in real terms to 2000. See Department of Energy, Energy Technologies for the United Kingdom, Energy Paper No. 39, HMSO, London, 1979, p. 152.
4. This assertion comes only from newspaper reports; it has not been officially acknowledged. See D. Fishlock, "Sizing Up the Cost of Britain's New PWR Project," The Financial Times, June 2, 1981, p. 10, and J. Huxley, "Nuclear Industry Going Critical," The Sunday Times, June 28, 1981, p. 41.
5. Monopolies and Mergers Commission, The Central Electricity Generating Board, HMSO, London, 1981, p. 106.
6. Ibid., p. 108. This conclusion was endorsed by the CEGB itself which confessed that the ordering of two additional AGRs (instead of one) kept skilled labor employed and reduced costs per plant by as much as 20%.
7. See, for example, the evidence of Charles Komanoff to the Select Committee on Energy, H. C. Paper 397-v, HMSO, London, 1980. Komanoff estimates that nuclear stations built today would be 55% more expensive in real terms than those built in 1979, largely due to the requirements for meeting safety regulations.
8. In its reply to a Select Committee report on nuclear power (see note 10 below), the Department of Energy, speaking for the government, stated, "A supply of energy from one source is subject to interruptions from time to time, so that it is prudent to establish as wide a range of options as possible," Department of Energy, Nuclear Power Cmnd. 8317, HMSO, London, 1981, p. 3.

9. For a political analysis of the British approach to nuclear reactor choice, see R. Williams, The Nuclear Power Decisions, Croom Helm, London, 1980, and B. Sedgemore, The Secret Constitution, Hodder and Stoughton, London, 1980.
10. The House of Commons Select Committee on Energy, 1980-81, The Government's Statement on the New Nuclear Power Program, H. C. Paper 114-i with appendices, HMSO, London.
11. For a discussion of the battles within the Select Committee, see P. Rogers, "Cool Gas, Hot Air and a 15 Year Fallout," The Sunday Times, February 22, 1981.
12. This intention was announced by the Secretry of State for Energy in a Commons Statement on December 18, 1979.
13. See N. Hirst, "Bitter Struggle Led to Nuclear Chief Relinquishing Post," The Times, May 22, 1981, p. 17.
14. In response to a question for the Commons Select Committee on Energy, Mr. Berridge, Chairman of the SSEB, stated,

> "There is a wide range of views in this country as to what is the best reactor system... we hold the view that it is by no means clear that the pressurized water reactor is the best reactor system... It would be a very long effort to assess a completely new system, something that perhaps none of us has heard of, but to assess another AGR is something we can do perfectly well."

Commons Select Committee on Energy, 1979-80, H.C. Paper 397-xi, pp. 357, 341.
15. See M. Flood, The Big Risk: Nuclear Power, Friends of the Earth, Poland Street, London, 1980, p. 6.
16. These data are available from the Monopolies and Mergers Commission, op. cit., pp. 271-275.
17. The Energy Secretary hedged his bets slightly over the precise requirements, noting, "The precise level of future ordering will depend upon the development of electricity demand and the performance of the industry, but (the government) considers this a reasonable prospect against which the nuclear and power plant industries can plan." In the white paper by the Department of Energy (op. cit.), the government noted that "it would keep its strategy under review and does not propose to authorize specific new nuclear power station orders until it is fully satisfied that each is justified," p. 5.
18. See D. Fairhall, "Sizewell Nuclear Power Station Design Changed," The Guardian, August 6, 1981, p. 15.
19. See, for example, the report of the Health and Safety Executive, Windscale: The Management of Safety, HMSO, London, 1980. This was conducted by the Nuclear Installations Inspectorate on behalf of the Executive.
20. J. Collier, The Guardian, November 8, 1979, p. 20.
21. A. Cottrell, Report of the Commons Select Committee on Science and Technology, Choice of a Thermal Reactor System, H.C. 73, HMSO, London, 1974, Annex III, p. 124.

22. W. Marshall (Chmn.), An Assessment of the Integrity of PWR Pressure Vessels, UKAEA, Risley, Cheshire, 1976, pp. 140-141.
23. Quality assurance differs from quality control. The former is a system of managerial supervision of design work, while the latter simply ensures that a designed part is manufactured precisely to specification. Quality assurance aims to ensure that a design is sound; quality control seeks to ensure that sound design is put into practice.
24. Commons Select Committee on Energy, H.C. Paper 114-i, HMSO, London, p. 76.
25. Evidence of the Nuclear Installations Inspectorate to the Commons Select Committee on Energy, 1979-80, H.C. Paper 397-vi, HMSO, London, p. 193.
26. Part of the section which follows is drawn from R. Matthews, "Using Quantitative Analysis in Britian," Nuclear Engineering International, 25:45-47 (November, 1980).
27. In evidence before the Commons Select Committee on Energy, 1979-80, H.C. Paper 397-ii, HMSO, London, p. 98.
28. W. Marshall, The Guardian, February 21, 1980, p. 11.
29. See in particular, P. D. Henderson (op. cit., note 2), R. Williams (op. cit., note 5), and B. Sedgemore (op. cit., note 5). For a more general critique of the British civil service, see P. Kellner and N. Crowther-Hunt, The Ruling Class, McDonald, London, 1980, and K. Middlemas, Politics in Industrial Society, Deutsch, London, 1980.
30. Sedgemore, op. cit.
31. Letter from Sir Alan Cottrell to Mrs. Thatcher leaked to The Guardian, February 14, 1980, p. 1.
32. The French nuclear engineer was S. Etemad who published articles in The Guardian on October 25, 1979, p. 20; January 17, 1980, p. 13; February 7, 1980, p. 13; and May 22, 1980, p. 13. He also gave evidence before the Select Committee on Energy, 1979-80, H.C. Paper 397-iii, pp. 125-132. While his accusations were grave, his credibility was repeatedly called into question, and the Select Committee ignored his evidence in its final report.
33. In evidence before the Commons Select Committee on Energy, 1979-80, H.C. Paper 397-ii, HMSO, London, p. 102.
34. In evidence before the Commons Select Committee on Energy, 1979-80, H.C. Paper 397-iii, HMSO, London, p. 117.
35. Department of Energy op. cit., p. 20.
36. See in particular, The Outer Circle Policy Unit, The Big Public Inquiry, 4 Cambridge Circus, Regent's Park, London, 1980, and The Two and Country Planning Association, Planning and Plutonium, 17 Carlton House Terrace, London, 1979.
37. Minutes of the Cabinet Energy Subcommittee, leaked to The Guardian on December 6, 1979, suggested that the Cabinet maintain a "low profile" over the PWR issue, presumably so as not to excite antinuclear suspicion.

38. See especially, The Outer Circle Policy Unit report (op. cit., note 6).
39. This remains to be seen. In its reply to the Select Committee report, the government noted that "extensive information and documentation" should be available to all parties [op. cit., p. 22] "in time for study" prior to the inquiry. Some feel that this reassurance falls short of the recommendations of the Select Committee on Energy that "the maximum amount of information and documentation should be made available well in advance" of the inquiry (op. cit., p. 88).
40. Political Ecology Research Group, Safety Aspects of the Advanced Gas Cooled Reactors, Commons Select Committee on Energy, H.C. Paper 114-iv, HMSO, London, 1981, pp. 1281-1315. For a more favorable view of the safety aspects of the AGR, see A. Cottrell, How Safe is Nuclear Energy?, Heineman, London, 1981.
41. See Friends of the Earth, The Pressurized Water Reactor: a Critique of the Government's Nuclear Power Program, London, 1981.

RISK ASSESSMENT FOLLOWING CRISIS IN THE UNITED STATES:

THE KEMENY COMMISSION

>Roger E. Kasperson and Arnold L. Gray
>
>Center for Technology, Environment, and Development
>Clark University
>Worcester, Massachusetts 01610

U.S. Management of Risk: An Example

The past decade has witnessed a confluence of two important developments in energy policy-making: the recognition of the transition role of oil in the mix of energy resources and the rapid growth of interest in and use of risk assessment. On the one hand, it is clear that the United States and other nations must reduce their dependence upon imported (especially) and domestic oil supplies. This has been the common message of all the major energy reports. On the other hand, there is evident public concern over the risks, particularly those catastrophic in nature, presented by various energy sources. Opposition to the expansion of nuclear power has been most notable, but energy planners and utility executives ponder the long-term responses likely to other power sources such as the new coal-burning plants.

Much of the increased scientific attention to energy risks has occurred in the form of major probabilistic assessments of risk, both for individual energy facilities and for comparisons among energy sources. Beginning with the U.S. Nuclear Regulatory Commission's Reactor Safety Study (WASH 1400) of 1975, major risk studies have emerged as a central ingredient in energy policy deliberations in the United States and other countries. 1) Event tree and fault tree analyses are accepted methodologies, 2) the Lewis Report has called for increased use of quantitative risk analysis by the regulatory agencies, 3) there is evident congressional interest in comparative risk assessment, and 4) energy utilities and consulting firms are acquiring increased capability in such assessment skills.

This embrace of "big" risk assessments has preceded, however, thoughtful examination of the contributions of such studies. Sweden has been perhaps the largest per-capita generator of such assessments Accordingly, the Swedish Energy Research and Development Commission in 1980 engaged the Beijer Institute to conduct a cross national study inquiring into the impacts of such assessments.

The paper which follows examines one special type of risk assessment: the Kemeny Commission post-mortem on the 1979 accident at the Three Mile Island nuclear plant [1]. As an assessment, it is more concerned with risk management institutions and practices than with the quantification of risk. Also, the linkage of the assessment with a major crisis undoubtedly puts this study at the higher end of likelihood of regulatory and policy change. The intent of this particular paper is to provide some empirical evidence from which to assess the more generic impacts of major energy risk assessments.

THE STRUCTURE OF NUCLEAR RISK MANAGEMENT

To facilitate comparisons with other national energy management systems, some pertinent features of nuclear risk management in the United States are noted. The most fundamental are those embedded in American political culture.

The basic tenets of American political culture are rooted in the inherent heterogeneity of American demography and in the Madisonian theory of democracy. In the Federalist, Paper No. 10, Madison noted the existence of factions which emanated from the nature of individuals - the differences of interest associated with differences in property, the attachment to various leaders, and the fallibility of human reason itself. Madison's central goal - which has become an article of faith in the American political credo - is the achievement of a nontyrannical republic. To achieve it, the United States has erected a complicated network of constitutional checks and balances: the separate constituencies for electing President, senators, and representatives; the presidential veto power; a bicameral congress; presidential control over appointments and senatorial confirmation; and federalism itself. Other checks and balances have been added - judicial review, decentralized political power, the Senate filibuster, and indeed almost every organizational mechanism to provide an external check on any identifiable group of political leaders [2].

Central to this political culture has been the assumption of inherent conflict among the diversity of interests. The Constitution provides means for limiting and channeling conflict, not avoiding it. It is not surprising, then, that regulation in the United States is strongly adversarial in nature and is yet another arena where social conflicts are played out. Just as the constitutional fathers feared the power of central government, so the populace at large has distrusted bureaucracy. The expert in American culture has always been suspect by a populace more convinced in the pragmatic good sense of the individual American. In this sense, President Reagan taps deep-seated aspects of American political culture with his program to deregulate economic and social activity and to "rescue" private initiative from the bureaucrats.

The management of nuclear power risks owes most of its characteristics to these fundamentals of American political culture. Nuclear risk management in the United States is spread over a bewildering array of institutions and interests. Notable among them are the Environmental Protection Agency, the Department of Energy, the Nuclear Regulatory Commission, the Department of Transportation, state public utility commissions, state environmental quality agencies, local zoning and planning authorities, a complex system of some 44 different public and private utility companies, some two dozen congressional committees, and, of course, the courts. It is an institutional situation in which the actors tend to be independent and without obligation to respect the jurisdiction or mission of the others, a fact which ensures immobilization of decisions when conflict occurs, uncertainty if and when final decisions in fact will be made, and numerous pressure points to register opposition. Since legislators owe their primary allegiance to constituents and party discipline is weak, political parties do not, as in many other political systems, overcome opposition in policy areas characterized by social conflict.

In this management system, the utilities and regulators share the major responsibility for protecting the public health and safety from nuclear power plants. Yet, because of the adversarial nature of the regulatory process, ready communication and informality between regulator and regulatee is limited by law. Decisions by informal negotiation are replaced by a complex system of rules and regulations, usually with quantitative specifications. While protecting from the "capture" of the regulator by industry (a criticism frequently noted for the former Atomic Energy Commission) the expanded regulatory system provoked the Kemeny Commission to note that"... once regulations become as voluminous and complex as those regulations now in place, they can serve as a negative factor in nuclear safety" [1]. Utilities orient their programs to meeting standards rather than ensuring safety.

Meanwhile, the diffuse policy and regulatory system articulates conflict at a thousand points. Herein is the inherent dilemma: whereas a viable energy system in the 1980s demands central planning and institutional resiliency to conflict, the basic tilt of the American political system is to prevent action where minority factions prevail.

Table 1. Major Post Mortems on the Accident at Three Mile Island

Study	Date	Extent	Comments
Kemeny Commission	October 1979	201 pages	The President's Commission makes 43 recommendations in 7 general areas. Echoing other studies, the report charges the NRC with "complacent mindset" and recommends reorganization. Also, the report cites "severe mental stress" as the overriding health effect of the accident. Special emphasis is given to the human factor in nuclear safety, the need to make the licensee responsible, the siting of plants in remote locations, and the establishment of an oversight committee on reactor safety.
Rogovin Report	January 1980	2 volumes (volume 1, 183 pages, volume 2, 1272 pages)	Labeling the TMI accident a management (not a hardware) problem, the report stresses the importance of an awareness of risk and acknowledgement of and confrontation of "what if" questions. The 12 recommendations - which range from remote siting, to step licensing, to possible prohibition of the same company's both generating and selling electricity - are supported by three substantial volumes of detailed technical and analytical documentation.
Pennsylvania Governor's Commission	Feburary 1980	211 pages	In an attempt to ascertain regional health, environmental, and economic impacts, the Commission concludes that physical health effects were short-lived and economic impacts were probably temporary. Citing the lack of federal/state coordination in emergency planning, the report found the state's plan adequate for the TMI accident but doubtful for an accident requiring quicker action. Moreover, the Commission judged the state ill-prepared to cope with a health emergency.
Hart Subcommittee (U.S. Senate)	June 1980 (report)	423 pages	The report, which comprises a history of the TMI plant and has several introductory chapters for the general reader, a glossary, etc., focuses on the first 24 hours of the accident and on the state of recovery and clean-up one year later. The 13 staff studies assess in depth the adequacy in post-TMI programs to address the shortcomings identified in the initial report. Taken together, these two volumes offer a comprehensive review of the facts and implications of the TMI accident.
	July 1980 (staff studies)	546 pages	

Study	Date	Extent	Comments
Nuclear Safety Analysis Center	July 1979 October 1979	380 pages + appendices; supplement: additional appendices	Widely acknowledged as the best of the technical analyses of the accident, the NSAC Report identifies some 24 contributing factors and points out that had almost any one of these factors been slightly different, there would have been no TMI damage. The report cautions against trying to correct everything.
NRC Short-Term Lessions Learned	July 1979	101 pages	Geared to correct promptly the specific and narrow defiencies that produced TMI, the report identifies and sets time tables for 23 short-term requirements in 12 broad areas (9 in design and analysis, 3 in operations).
NRC Office of Inspection and Enforcement	August 1979	850 pages	Designed to establish the facts and to evaluate licensee performance as a basis for corrective or enforcement action, the investigation emphasized mindset and operator error. The report provides no evaluation of agency or NRC actions, no analysis of the NRC regulatory process.
Moffett Subcommittee (U.S. House of Representatives)	August 1979	105 pages	The report makes five major recommendations to the NRC: 1) admit the possibility of serious accidents, 2) upgrade emergency planning and implement improvements respectively, 3) abandon LPZ concept, 4) review and upgrade state and local emergency plans, 5) review emergency response capability and require that such capability be established during the licensing process.
NRC Long-Term Lessons Learned	October 1979	38 pages	Unlike its TMI-specific short-term counterpart, the report confronts safety questions of a fundamental nature and sets general goals for improvements in design, operation, and the regulatory process. A categorization of 25 recommendations (in 13 general areas) gives priority to operational safety and emphasizes the role of personnel in responding to and/or preventing accidents. Moreover, the report gives more attention to mitigation, as opposed to prevention, of nuclear accidents.

THE KEMENY COMMISSION

The accident at Three Mile Island nuclear power plant on March 28, 1979 was by common consensus the worst to occur in the history of commercial nuclear power generation in the United States. It is not surprising, therefore, that the accident provoked a series of assessments [4-9] of its meaning as to the safety of nuclear power. Prominent among these was the President's Commission on the Accident at Three Mile Island, popularly known as the Kemeny Commission.

Appointed two weeks after the accident, the 12-member commission was carefully balanced to reflect a diversity of viewpoints. The chairman was John Kemeny, President of Dartmouth College and a one-time Manhattan Project researcher. Only two Commission members - Thomas Pigford and Theodore Taylor - could be considered nuclear power experts. The remaining eight members included representatives from industry, labor unions, and universities, the President of the National Audubon Society, a former deputy secretary of the Army, and a mother of six who lived across the river from the nuclear plant at Three Mile Island. The charge to the Commission was:

> ...to conduct a comprehensive study and investigation of the recent accident involving the nuclear power facility on Three Mile Island in Pennsylvania. The Commission's study and investigation shall include:
>
> a) a technical assessment of the events and their causes; this assessment shall include, but shall not be limited to, an evaluation of the actual and potential impact of the events on the public health and safety and on the health and safety of workers;
>
> b) an analysis of the role of the managing utility;
>
> c) an assessment of the emergency preparedness and response of the Nuclear Regulatory Commission and other federal, state, and local authorities;
>
> d) an evaluation of the Nuclear Regulatory Commission's licensing, inspections, operation, and enforcement procedures as applied to this facility;
>
> e) an assessment of how the public's right to information concerning the events at TMI was served and of the steps which should be taken during similar emergencies to provide the public with accurate, comprehensible, and timely information; and
>
> f) appropriate recommendations based upon the Commission's findings.

The Commission labored for six months, eventually taking some 150 formal depositions, interviewing hundreds of individuals, hearing testimony under oath from numerous witnesses, and collecting sufficient material to fill 300 feet of library shelf space. Its final report [1], issued in October of 1979, received undoubtedly more media coverage and congressional attention than any other document on nuclear power safety.

As noted above, the report is a special type of risk assessment. Unlike the Reactor Safety Study (WASH 1400) or the Risk Assessment Review Group Report (The Lewis Report), which relied heavily upon expert assessment dealing with the quantitative probabilistic assessment of risk, the Kemeny Commission inquired into the larger issues of nuclear risk management as indicated by a particular accident. Because of the significance of the crisis event and the direct responsibility of the Commission to the President, the report had a unique opportunity to contribute to the shaping of nuclear safety policy in the United States.

A Note on Methodological Issues

Assessing the impacts of the Kemeny Commission Report of nuclear safety policy requires the isolation of the report from the numerous othe risk assessments (summarized in Table 1) conducted after the accident, from the accident itself, and the 10 congressional subcommittees which had held hearings [10-15] on the subject by the first anniversary of the TMI accident. This cannot be done. In fact, the Nuclear Regulatory Commission quite explicitly and systematically integrated the various report findings in order to fashion a coordinated response. In addition, a number of safety problems were quite evident in the accident itself, and it is quite futile to determine which source stimulated a particular response.

Within these constraints, however, there are some opportunities. A substantial part of the industry and governmental response occurred well in advance of the issuance of the Kemeny Report some seven months after the accident, most of which would have presumably occurred even in the absence of the Commission. Also, several post-accident evaluations and congressional inquiries appeared prior to the Kemeny Report and thus provide a benchmark from which to assess the particular contributions of the Kemeny Report. Finally, there is not a complete overlap in these reports so that some of the individual findings and recommendations of the Kemeny Report can be distinguished and assessed as to impacts.

This analysis proceeds by identifying 12 key areas of recommendation from among the 43 specific recommendations made by the Commission. For each key recommendation, the major societal responses are noted, major unresolved issues specified, and our overall assessment of the response provided (Table 2).

Table 2. Societal Response to Key Kemeny Commission Recommendations (as of March 1981)

Recommendation	Response
Restructure/improve NRC (A1)	President does not accept Kemeny reorganization recommendations. Congress retains collegial structure of NRC with strengthened powers of chairman. Chairman designated as spokesman in emergencies. Assessment: Basic problems of the Commission referred to report remain unresolved and restructuring is not achieved, but some improvement in emergency response and regulation of operating reactor capabilities. In September 1980, the Nuclear Safety Oversight Committee finds evidence of a "business as usual mindset in NRC."
Improve ACRS (A3)	NRC opposes any mandatory response to ACRS recommendations. On February 11, 1980, ACRS charges NRC "largely ignores" its input on Kemeny Commission responses. Assessment: Little substantial action undertaken to improve ACRS. It is unlikely that the ACRS can and/or will influence major changes within the NRC.
Establish New Oversight Committee (A2)	Executive Order establishes Nuclear Safety Committee on March 18, 1980. Committee issues three letter reports to the President on NRC action plan, radiological consequences of nuclear accidents, and emergency response planning. Assessment: Committee has provided limited but useful function. Future is unclear.
Upgrade Reactor Operator and Supervisor Training	Nuclear Safety Analysis Center establishes computerized communication system connected to all utilities on operating incidents. NRC proposes upgrading in formal education: senior reactor operators, 60 college credits in engineering; shift supervisors, a BS degree in engineering. Utilities improve training in emergency events. No change proposed in formal education of reactor operators. Memphis State University inaugurates new training program in cooperation with utilities. Severity of licensing examinations increased; failure rate rises from 5 to 30%. NRC declines to accredit training programs.

Recommendation	Response
	Assessment: Upgrading becoming evident though requirements still lag behind those in Europe.
Increase Safety Emphasis in Licensing (A10)	NRC reorganizes licensing staff to correct weaknesses in licensing process. Increased attention to operator training, utility management, emergency planning, reactor design features, and evaluation of plant operating experience. NRC decides against Office of Hearing Counsel. 1981 licensing plan threatens to reduce role of intervenors but promises to speed up licensing. Assessment: Actions to date fill a number of gaps in safety coverage, but the degree of substantial improvement unclear. NRC licensing of Sequoyah plant questions commitment to safety. Reduced role of intervenors may weaken safety emphasis.
Improve Safety Inspection and Enforcement (A11)	NRC establishes resident inspectors at power plants, requires annual evaluation of licenses, improves reporting requirements. A new NRC Office for Analysis and Evaluation of Operational Data established (prior to Kemeny Report) in July 1979. Fines for utilities increased. Bingham Amendment calls for "systematic evaluations, of all operating nuclear power plants, a possible 5-8 year effort which has evoked opposition. Assessment: Although too early to tell, indications are of substantial improvement in inspection and regulation of operating reactors. But position of top leadership of NRC during Reagan Administration will be important. Bingham Amendment will require significant new NRC resources.

(continued)

Table 2. Continued

Recommendation	Response
Improve Technical Assessment and Equipment (D1-D3)	Utilities initiate improvements in control room design and instrumentation. Assessment: Substantial improvements implemented or ongoing in improved instrumentation, equipment, and monitoring.
Initiate New Reactor Risk Assessments (D4-5, D7, E1)	NRC reorients risk assessment research program with new attention to higher probability events, accident mitigation, and human factors. Retrospective iodine release study of TMI accident suggest possible past overestimate of consequences by factor of 10. Utilities establish improved monitoring and dissemination system of operating incidents. NRC establishes Division of Human Factors and initiates effort to define level of acceptable risk. Epidemiological studies of effects of low level radiation initiated. EPA recommends against 10-fold reduction in occupational standard. Probabilistic risk assessments initated by utilities at eight power plants. Radiation Policy Council established in Executive Branch. Assessment: Significant changes instituted to give new priority to TMI-like events, to human factors, and accident mitigation. Individual plant risk assessments should improve safety performance and enlarge accident response capability.
Improve Industry Attitudes and Performance (B1-B3, B5)	Industry establishes two new institutions: Institute for Nuclear Power Operations (INPO) with power plant evaluation and training as primary functions and Nuclear Safety Analysis Center (NSAC) with analysis of operating experience and other technical assessment its primary activities. International cooperation with NSAC makes world experience data base a possibility. Assessment: Substantial industry response: new institutions are important safety vehicles. Still unresolved are prevailing attitudes and assurance of high level of overall technical competence in individual utility management structure

Recommendation	Response
More Remote Siting of Nuclear Power Plants (A6)	NRC proposes (NUREG 0625) upper limits on population densities around plants and making siting criteria distinct from engineered safeguards. Estimates suggest 49 of 84 currently operating plants would fail to meet criteria. Strong industry opposition. Assessment: Proposal currently mired in controversy; no change to date, but no new plants presently being ordered in any event. Since no retrospective application of criteria, limited safety impact on 100-150 GWe nuclear system.
Improve Emergency Response and Mitigation (A7-A8, E3-E5, F1-F3, G1-G4)	NRC issues new rule on emergency response plans, extending 5-mile zone to 10-mile and 50-mile radii. All operating reactors required to have emergency plans approved by April 1981. NRC installs a crisis management communications link of all power plants to NRC headquarters. New rule mandates that state be able to notify every person within 10 miles of a nuclear power plant of accident within 15 minutes and evacuate population. Proposal to distribute potassium iodide pills mired in controversy. Nucleonics Week survey finds confused and uncertain response by states. No notable improvement in mass media capabilities despite an NRC pilot program. Assessment: Although utilities and NRC have improved their emergency response capabilities, the overall capacity of society to respond to a major accident remains in doubt.
Educate the Public (F4, G5)	NRC plans to investigate need for literature. No program instituted to date. Assessment: No substantive response despite widespread scientific belief as to need.

Institutional Changes

The Kemeny Commission reached a number of biting judgments concerning the primary institutions responsible for the assurance of nuclear safety, the most notable of which were:

The Nuclear Regulatory Commission: "With its present organization, staff, and attitudes, the NRC is unable to fulfill its responsibility for providing an acceptable level of safety for nuclear power plants" (p. 56) [1].

The Advisory Committee on Reactor Safeguards: The Committee is the only body independent of of the NRC staff which regularly reviews safety questions, but the Committee "has established no firm guidelines or procedures," its members are "part-time and have a very small staff," and it relies heavily upon the NRC staff for follow-up of concerns [1].

The Utility: The Utility (Metropolitan Edison) failed in a number of important cases "to acquire enough information about safety problems, failed to analyze adequately what information they did require, or failed to act on that information" (p. 43). "It did not have sufficient knowledge, expertise, and personnel to operate the plant or maintain it adequately" (p. 44) [1].

To deal with these institutional deficiencies, the Commission recommended a broad set of changes involving the Nuclear Regulatory Commission, the Advisory Committee on Reactor Safeguards, and industry.

The Nuclear Regulatory Commission. The Kemeny Commission found that the Nuclear Regulatory Commission lacked sufficient organizational and management capability to ensure safety, a judgment supported by the Rogovin Report. Unfortunately, the Commission recommended the rather shop-worn suggestion of agency reorganization, in this case a change from an independent regulatory commission to an executive branch agency with an Administrator, as the most prominent means of redress. The Kemeny Commission was the first accident postmortem to call for this change, though it subsequently also found favor in the Rogovin Report. The recommendation was unpopular from the start: the NRC staff opposed it, and all the current NRC Commissioners save one also opposed it. Congress was lukewarm to the idea, and the President, sniffing congressional opposition, never supported the recommendation. Instead, he called for, and Congress eventually approved, a strengthening of the chairman's role. Two years after the accident, top leadership in the Nuclear Regulatory Commission remains an outstanding problem recently described by one of the NRC Commissioners "as analogous to hitching fire horses at different points around a sled." The general weakening of regulatory agencies in the current Reagan Administration does not bode well for the prospect that the recent drift and indecision will halt and that coherent, effective leadership committed to safeguarding public health and the environment will emerge in the NRC. Other changes in the Commission have met with greater success and indicate some limited improvements in regulatory performance. Central to these responses has been a shift in Commission resources and emphasis on monitoring and assessing operating reactors. Within four months of the accident (and well in advance of the Kemeny Report), the NRC established a new Office for Analysis and Evaluation of Operational Data aimed at the serious deficiences apparent in the TMI accident in learning from past reactor incidents and malfunctions. The Commission also has established a program of resident inspectors who are stationed at individual power plants. The severity of licensing examinations for reactor operators has also been increased, producing a rise in failure rates from 5 to 30%. The Commission has improved its capability for crisis management by clarifying responsibilities and improving communication with and analytical strength for existing reactors.

Beyond these useful changes, however, is the more basic and difficult problem of attitudes and orientations throughout the professional staff of the Commission. The Kemeny Commission was quite specific about these problems:

> ...We have seen evidence that some of the old promotional philosophy still influences the regulatory practices of the NRC (p. 19).

> ...The evidence suggests that the NRC has sometimes erred on the side of the industry's convenience rather than carrying out its primary mission of assuring safety (p. 19).

> There seems to be a persistent assumption that plants can be made sufficiently safe to be "people-proof" (p. 20).

> We do not see evidence of effective managerial guidance from the top, and we do see evidence of some of the old AEC promotional philosophy in key officers below the top (p. 21) [1].

The Kemeny Commission was hopeful that the reorganization of the NRC would begin a change in attitudes from the top down. A coherent plan for dealing with these difficult behavioral problems has not been forthcoming, yet obviously substantial changes are critical to a strengthened regulatory performance. The behavior of the Commission since the accident suggests, unsurprisingly, that the pre-accident attitudes are proving difficult to extirpate. A scant five months after the accident and on the eve of the Kemeny Report, the NRC staff advised that technical fixes had so reduced the likelihood of a repeat of Three Mile Island that new operating licenses could be issued even though the design basis of new reactors might be inadequate to control the potential consequences of the estimated amount of hydrogen released into containment at Three Mile Island. This action led the new Nuclear Safety Oversight Committee to observe in a letter to the President that a "business as usual" mindset continued to exist at the NRC. Also of concern is an NRC licensing reform plan announced in March 1981 which would restrict the role of intervenors and limit their access to information.

The response of the NRC, in summary, while likely improving its regulatory performance in a number of limited areas, has failed to resolve the need for more effective top leadership or to ameliorate the ingrained attitudes inimical to safety in the professional staff. It continues to be preoccupied with formal, specific pro-nuclear regulators to individual problems, leading one Kemeny Commission member to conclude that the "the NRC shows little recognition of the fundamental flaws in its approach to reactor safety" [16].

The Advisory Committee on Reactor Safeguards. A second area of institutional change concerns the Advisory Committee on Reactor Safeguards. The Kemeny Commission called for a strengthening of the Committee's staff, the elimination of the requirement that the Committee review each license application, the provision of a statutory right of the Committee to intervene in hearings, and the right to initiate rulemaking on generic safety issues. By March 1981, significant actions had not been forthcoming to increase the capabilities of the staff, to enlarge the role of the Committe in generic safety issues, or to improve communications with the NRC itself. It is apparent that the Committee is experiencing continuing difficulties in making its views heard at the NRC. In February 1980, the Committee officially complained that the NRC had largely ignored its input on matters relating to NRC's post-TMI responses and the requirements for near-term operating licences [17].

The Kemeny Commission also recommended that a new independent committee by instituted whose purpose would be "to examine, on a continuing basis, the performance of the (NRC) and of the nuclear industry in addressing and resolving important public safety issues associated with the construction and operation of nuclear power plants, and in exploring the overall risks of nuclear power" (p. 62) [1]. President Carter established the Nuclear Safety Oversight Committee in March 1980 under the chairmanship of Governor Bruce Babbitt of Arizona. In the first year of its existence, the Committee concentrated on post-TMI responses, issuing letter reports on the NRC's Action Plan, iodine release in nuclear accidents, and emergency planning and response. The Committee has to date played a useful oversight function but its future in the Reagan Administration is in doubt.

The Utilities. Of equal or greater significance to the public institutions is the impact of the Kemeny Commission Report on in-

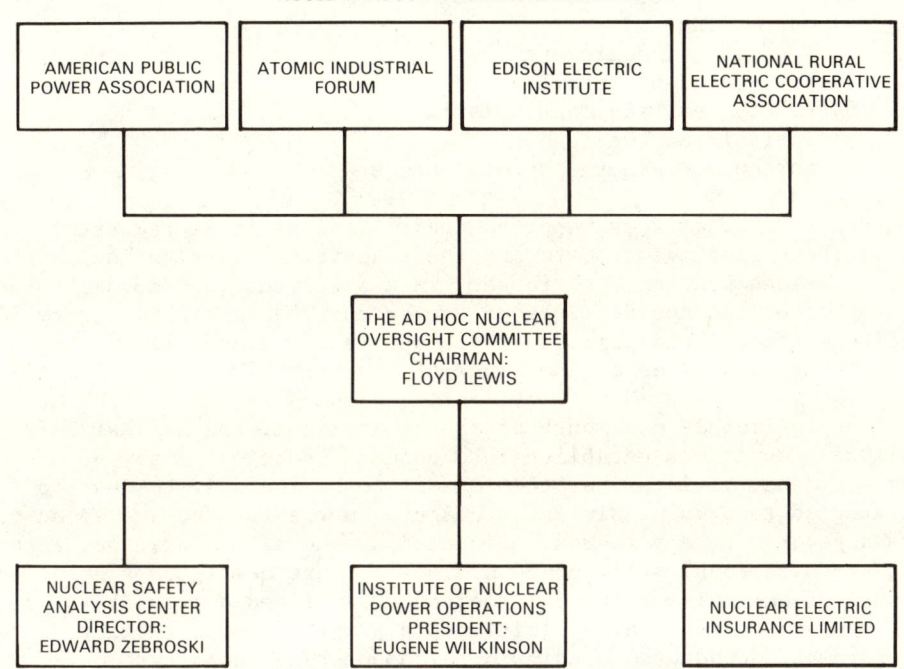

Fig. 1. Structure of nuclear industry response to the TMI accident.

dustry itself, especially the utilities that manage the operation of nuclear power plants. The Commission found far-reaching problems in the role of the utilities, warning that "the nuclear industry must dramatically change its attitudes toward safety and regulations" and that it must also "set and police its own standards of excellence to ensure the effective managment and safe operation of nuclear power plants" (p. 68) [1].

In fact, the major elements of industry response were set in motion well before the appointment of the Kemeny Commission. The structure of this response is shown in Fig. 1. Within two weeks of the accident, the four major industry groups - the American Public Power Association, the Atomic Industrial Forum, the Edison Electric Institute, and the National Rural Electric Cooperative Association - joined to create a policy task force (the TMI Ad Hoc Nuclear Oversight Committee) to address the safety issues presented by the accident. The seven subcommittees formed to develop policy recommendations indicated by the concerns immediately identified were:

1) emergency response planning
2) operations
3) systems and equipment
4) post-accident recovery
5) safety analysis considerations
6) control room design
7) unresolved generic safety issues

Most subcommittees reported their findings in September of 1979; these findings also formed the industry's contribution to the NRC's Lessons Learned Task Force. In a statement issued some three months after the accident, Floyd Lewis, chairman of the Industry Ad Hoc Committee, could point to three new institutions already begun or planned as well as a wide range of other utility responses.

Approximately one month after the accident, the Nuclear Safety Analysis Center was established to conduct technical analyses of the accident, to interpret the lessons to be learned, to develop strategies to prevent similar accidents in the future, and to address generic safety issues. Financed by the utilities, the Center has a professional staff of 50 and has to date completed a detailed technical analysis of the TMI accident, developed a priority system for needed safety changes, initiated a program on the testing of relief valves, conducted studies of a computerized data base of 22,000 operating failures (licensee event reports), instituted a computerized communication system linking 60 utilities for rapid dissemination of and requests for information, and conducted case studies of specific safety problems (e.g., the loss of electrical power to non-nuclear instrumentation at the Crystal River Nuclear Plant).

To deal with the "people problems" apparent in the TMI accident, the industry announced in June 1979 its intent to establish the Institute of Nuclear Power Operations. Chauncey Starr, with an advisory group drawn from the Navy's nuclear program, NASA, and airline safety, developed the mandate and structure of the Institute. With a 1980 budget of $11 million, a projected staff of 200, and participation by 55 utilities, INPO will develop benchmarks for excellence in nuclear operations, conduct evaluations of individual utilities (some six completed by January 1981) with the goal of an annual evaluation for each plant, formulate educational and training requirements for operating personnel, and accredit training organizations.

The utilities have also cooperated to create a mutual insurance plan to apply to the extraordinary costs accruing to a utility experiencing a major nuclear accident.

Taken together, these new institutions appear to represent a significant upgrading in industry's capacity and effort to manage nuclear power plant safety. It is noteworthy that the institutions emerged from industry's response to the accident and preceded the recommendations of the Kemeny Commission. As with the Nuclear Regulatory Commission, the extent to which attitudes have changed among the rank-and-file of industry professionals is unclear. Two years after the accident, a NRC commissioner, noting renewed industry complaints that the safety bureaucracy was "nitpicking" it to death, running up construction costs, and delaying licenses, warned that "memory fades and old habits reappear" [18]. Certainly, as the response to TMI becomes institutionalized, there is a temptation to believe that the outstanding problems are behind us and to declare nulcear power safe rather than to work to make it safe.

Risk Assessment and Management

An important impact of the Three Mile Island accident and the subsequent post-mortems was a reorientation in the overall risk assessment program of both government and industry. Three key changes involve new attention to a broader spectrum of reactor accidents, to man-machine interactions in risk, and to accident consequence mitigation.

Since the publication of WASH-1400 in 1975, it has been clear that most postulated accidents come from small reactor leaks, such as occurred at TMI, and not from large pipe breaks leading to catastrophies. Yet, WASH-1400 concludes that the greatest public risk is due to comparatively rare events where a postulated melted core releases a large fraction of its contents to the atmosphere. Only a minor portion of the public risk by contrast is assessed to result from the more frequent melts leading to small releases.

The Nuclear Regulatory Commission, prior to the Three Mile Island accident, oriented its programs to this structure of reactor risks. Thus, its risk assessment program and its criteria for design analysis prior to TMI focused heavily upon large pipe breaks leading to major loss-of-coolant accidents. Since the TMI accident, resources have been shifted substantially in the NRC's risk research program (a $231.9 million effort in FY-1981) away from big, pipe breaks and transients toward higher probability/lower consequence events. A similar change has also occurred in the research program of the Electric Power Research Institute's Nuclear Power Division (p. 338) [15].

A second major change is the allocation of significant new attention to human error as an ingredient in reactor accidents. WASH-1400 is quite inadequate in its attention to this issue, largely assuming human error rates taken from industries thought to be similar to nuclear power plants. In fact, a number of analyses, several of which were available before TMI, have demonstrated the importance of human error in reactor risks:

NRC official Merrill Taylor informed the Lewis Commission in 1978 that 50-85% of the hypothetical safety system failures he had examined in detail would be caused by humans (p. 62) [19].

The 1978 German reactor risk study found that human failure was responsible for two-thirds of all risks [19].

About 20-50% of all licensee event reports are due to human error (p. 63) [19].

The NRC reports that in about 1% of all licensee event reports (about 35 incidents per year), there are indications that a safety feature has been seriously compromised or made unavailable by human error (p. 63) [19].

Despite this need to conceptualize reactor operations as a man-machine system, the NRC's approach to reactor risks remained strongly equipment-centered. Also, the NRC preoccupation with large-break accidents ensured a neglect of human factors since such accidents require extremely fast reaction, therefore accentuating the role of automatic control through equipment.

It is not surprising, then, that a post-accident detailed review of the NRC's regulatory guides and standard review plan found "no examples of criteria written with a clear intent to include human engineering considerations in the licensing and regulatory system" (Vol. II, Part II, p. 345) [20].

The Kemeny Commission, along with the other accident post-mortems, was quite direct as to the significance of human behavior:

...As the evidence accumulated, it became clear that the fundamental problems are people-related problems and not equipment problems (p. 8).

...Wherever we looked, we found problems with the human beings who operate the plant, with the management that runs the key organization, and with the agency that is charged with assuring the safety of nuclear power plants (p. 8).

The most serious "mindset" is the preoccupation of everyone with the safety of equipment, resulting in the down-playing of the importance of the human element in the nuclear power generation. We are tempted to say that...what the NRC and industry have failed to recognize sufficiently is that the human beings who manage and operate the plants constitute an important safety system (p. 10) [1].

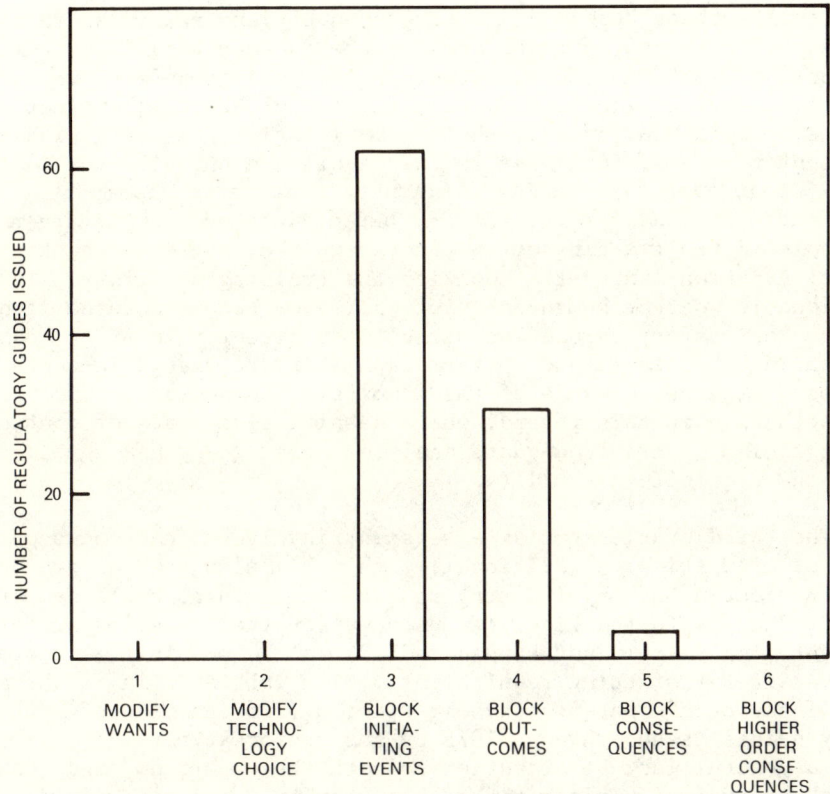

Fig. 2. Number of regulatory guides by hazard stage issued by the Nuclear Regulatory Commission (data through 1975). Note the paucity of guides under stages 5 and 6.

Since the accidents and the assessments, a number of efforts have been made to internalize human factors into nuclear risk assessment and management. The NRC has established a new Division of Human Factors and has restructured its risk assessment program to give greater emphasis to human error. The new Institute for Nuclear Power Operations has undertaken a major effort to improve the training of reactor operators, the single issue which most worried the Kemeny Commission. Since the accident, all 2500 licensed reactor operators in the United States have gone through the TMI accident sequence on training simulators. Industry's new Nuclear Safety Analysis Center is evaluating human response in making design and instrumentation changes in reactor control rooms. Again, most of these issues were apparent and the response begun prior to the Kemeny Commission Report.

Despite these encouraging changes, questions remain as to the adequacy of response, particularly in the Nuclear Regulatory Commission. In 1980, in testimony before the Nuclear Safety Oversight Committee, Saul Levine, the former director of NRC risk evaluation studies, complained that "...the Agency is still grappling with equipment problems, equipment is the be-all and end-all" and that the NRC continues to give insufficient attention to research on human factors" [21]. Also, the NRC has decided against taking a lead role in the training of operators, unlike, for example, the Federal Aviation Administration with the training of pilots. It is questionable whether human behavior will ever become internalized into nuclear safety regulation without the direct role of the Commission in such issues and a staff capability to enable in-depth analyses. The performance of the Commission in socioeconomic issues is instructive in this regard, where despite five years of continuing criticism of its inadequate analyses, the agency has yet to develop the requisite capability.

The third change in risk assessment involves a new focus on accident mitigation. Traditionally, this has also been an area badly neglected both by industry and the NRC. In no small measure, this is due (1) to the widespread assumption that a serious reactor accident simply would not happen and (2) to the possible expenses involved in retrofitting requirements. The imbalance in NRC hazard control has been apparent for some time; a 1976 analysis of regulatory guides issued through 1975 by the senior author (Fig. 2) reveals a general lack of attention to both the "upstream" and "downstream" control of reactor risks. The Kemeny Commission dealt primarily with emergency response, and the changes resulting from those recommendations will be considered below. Suffice it to note here that accident mitigation options are present and may have considerable potential for overall risk reduction. It is claimed, for example, that

> better containment building designs with filtered release systems to prevent containment failure due to internal overpressure could, according to Frank Von Hippel [22] be installed at both new and existing reactors at less than 1% of their replacement costs;
>
> means are available for interdicting the flow of contaminated water in the event of a meltdown from beneath reactor containment buildings to nearby water bodies, but no appropriate preparations have been made by the NRC;
>
> underground or more remote siting could lower substantially the consequences of a major accident; and

thyroid blocking by use of potassium iodide pills could reduce the number of thyroid tumor cases by as many as 100,000 over an area extending 200 miles downwind from the release source [22].

Belatedly, the NRC is giving consequences mitigation new emphasis in its risk assessment program, although most attention has focused to date on emergency planning. The recent findings, however, that substantially less volatile iodine may be released in a reactor accident than previously assumed may reduce attention to accident mitigation analyses.

Finally, an important issue not addressed by the Kemeny Report is the level of safety to which NRC and the industry should aspire. This issue, however, has received considerable attention from industry and was accorded high priority in the Rogovin Report and the NRC Lessons Learned Report. Recently, a draft 1981 Congressional Office of Technology Assessment Report argued that in the absence of a safety goal agreed upon by society, the adequacy of the NRC response to the TMI accident is impossible to assess and creates a large uncertainty in the licensing process [23]. And Congress is pressing for such a determination: The Bingham Amendment requires a proposed safety goal for nuclear reactor regulation by June 30, 1981, and the new Udall Omnibus nuclear legislation bill includes a similar mandate. The NRC has supported several studies of acceptable risk but has been generally reluctant to engage this issue, feeling (and perhaps rightly) that its results will only be seen as self-serving. Nevertheless, there currently is a draft plan [24] available for review and a process underway to comply with the congressional mandate. Yet, both the magnitude of the effort and the underdeveloped theory to support such a determination, especially in the nuclear area, do not promise likely achievement of such a goal.

Siting and Emergency Preparedness

The Three Mile Island accident elevated siting and emergency preparedness into primary areas of nuclear safety concern. In regard to siting, the Kemeny Commission concluded that the entire concept of a "low population zone" in NRC siting criteria was flawed and recommended:

> In order to provide an added contribution to safety, the agency (i.e., NRC) should be required, to the maximum extent possible, to locate new power plants in areas remote from concentrations of population. Siting determinations should be based on technical assessments of various classes of accidents that can take place, including those involving releases of low doses of radiation (p. 64) [1].

The siting problems of United States nuclear power plants are the result of the location of plants progressively closer to cities over time as a means of reducing electricity transmission costs. Engineered safety features were substituted for remote siting as the primary means of risk reduction. Catastrophic accidents (the so-called Class 9 accidents) were judged to be of such low probability as not to merit inclusion in the siting criteria. As a result, power plants came to be sited very close to major metropolitan areas.

Immediately following the TMI accident, the NRC appointed a special Siting Policy Task Force to reconsider siting regulations. The Task Force reported to the NRC commissioners in August 1979 (two months prior to the Kemeny Report), calling for an abandonment of the principle of basing siting on projections of potential dose commitments. Instead, there would be standardized fixed boundaries for exclusion areas and low population zones. The report also recommended divorcing reactor designs from siting, thereby reestablishing siting as a major "defense-in-depth" factor. Final action is yet to be taken on the proposal, but there is substantial industry resistance as well as opposition from Europe and Japan where such remoteness is unachievable, although West Germany is currently assessing underground siting.

Unresolved in the siting proposals are the power plants currently operating or on order. The NRC evaluation of 104 existing nuclear power plant sites found that about 30 failed to meet the new siting criteria, with Indian Point (near New York City) and Zion (near Chicago) particularly bad [25]. The new siting proposals recommend "grandfathering" existing sites, although compensating at these sites by emphasing emergency planning and additional engineered safeguards (e.g., core ladles). The changes, in short, will only marginally improve the overall siting of nuclear power reactors for the next several decades (or for a 100-150 GWe system).

Turning to emergency planning, the TMI accident demonstrated quite conclusively that none of the responsible parties was prepared for a major nuclear accident. The utility was unprepared to deal with the radiological aspects of the accident, the response by NRC was disorganized and confused, and the local governments in the power plant region had no emergency plans which provided adequately for evacuation. The neglect within the NRC is evident in the fact that only 3 full-time professionals and 1 secretary out of 2500 NRC employees worked on emergency preparedness issues prior to the TMI accident [26]. The Kemeny Commission found few grounds for optimism:

> The response to the emergency was dominated by an atmosphere of almost total confusion. There was lack of communication at all levels. Many key recommendations were made by individuals who

were not in possession of accurate information, and those who managed the accident were slow to realize the significance and implications of the events that had taken place (p. 17) [1].

Although the Commission noted a number of problem areas, it made no clear recommendations. Much of the response and outpouring of documents to the accident and the various post-mortems, including the Kemeny Report, have centered nonetheless upon improving emergency response. By the time the Kemeny Report appeared, the Atomic Industrial Forum had developed a model plan for emergency response by the nuclear industry. The plan calls for four well-coordinated but independent emergency centers near the site, interconnected with reliable communication. The Institute for Nuclear Power Operations is also focused upon training reactor operators in effective emergency response.

The Nuclear Regulatory Commission has instituted a series of changes to improve its capabilities in this area, the more important of which are:

Six teams have been established to assess the emergency planning and preparation of every operating nuclear power plant. In addition, the new resident inspector at each plant has well-defined duties during an accident.

The chairman of the NRC has been provided with clear lead responsibility during emergencies, and an emergency management coordinator has been established to work with him.

Guidelines have been provided to licensees and the states defining classes of emergencies and outlining appropriate actions.

The NRC's Operations Center in Bethesda has been upgraded, and dedicated reactor operations telephone lines exist to each facility, with extensions in each control room. The NRC has also prescribed a data link between the plant's control room and a technical support center located somewhere on site. Consideration is under way for the data link to extend also to the NRC Bethesda center.

The licensing of new plants has been made conditional on the development of acceptable emergency plans.

Substantial improvement has occurred, in short, in both industry's and the NRC's emergency response capability. It may be argued, however, that current NRC approaches unduly emphasize major accidents, thereby providing suboptimal protection for the public and cost-ineffective solutions [27].

More problematic is the role of the states and local governments. The Federal Emergency Management Agency has lead responsibility in this area and, in accordance with a presidential directive, completed a review in June 1980 of state emergency plans for operating nuclear reactors [28]. The Agency has mandated that a state be able to notify every person within a 10-mile radius of a nuclear power plant that a nuclear accident has occurred and indicate what actions the person must take for personal safety. State governors hold the authority, unless delegated to local governments, for ordering evaucations. By early 1981, it was apparent that, despite evident progress, many problems remain: The ability of utility officials to notify promptly relevant state and local authorities is in doubt. A 1981 survey by Nucleonics Week indicates that most states are not yet prepared for an emergency and have, in fact, taken "an uncertain series of steps in notification procedures," and confusion remains as to who has authority to do what during an emergency [29].

One final unresolved issue two years after the accident is the availability of potassium iodide pills in areas surrounding nuclear plants. The Kemeny Commission recommended that

> An adequate supply of the radiation protective (thyroid blocking) agent, potassium iodide for human use, should be available regionally for distribution to the general population and workers affected by a radiological emergency (p. 75) [1].

The NRC has, as yet, reached no decision on the issue. The Federal Emergency Management Agency's position is that some $100,000 worth of potassium iodide should be stockpiled in four locations but that distributional decisions be left to the states. In addition, the NRC staff paper on the issue advocates the availability of potassium iodide for hospitals and prisons near power plant sites. The lack of a clear policy some two years after the accident is regrettable, particularly given the routine distribution out to 10 miles about a reactor in England and the recent Swedish decision in favor of dissemination. Again, recent research on the release of volatile iodine may influence the response to this issue.

IMPACTS AND FUTURE TRENDS IN RISK ASSESSMENT

Although any definitive statements must await implementation of ongoing changes and responses to future crises, the authors reach seven major conclusions concerning the impacts of the Kemeny Report on nuclear safety in the United States:

1) The major elements of societal response were apparent several months after the accident and well in advance of the Kemeny Report. As a presidential commision, the Kemeny Commission had considerable symbolic value and doubtlessly contributed to the momentum for change, but it identified few issues not treated in other accident post-mortems.

2) The response of those managers responsible for nuclear power has addressed the range of concerns noted in the Kemeny Report and other assessments but has been uneven in resolving the relevant problems, especially those more fundamental to institutions and established processes.

3) The response by industry has been most timely and effective, regulatory responses have been more delayed and uneven, and the mass media have failed to respond to recommendations for change. The overall regulatory response has also been heavily dependent upon the role of industry. The long-term effect of the accident may well be to further self-regulation in nuclear power.

4) The changes instituted by industry and government have tended to address obvious gaps and specific problems apparent in the accident. The more fundamental and integrative problems of capability and attitudes which formed the primary concern of the Kemeny Commission and the need for new initiatives and ideas remain essentially unaddressed in the TMI response. One Kemeny Commission pronuclear member recently noted that "industry's concern with meeting the formalism of NRC regulations is still inhibiting and throttling new ideas and technical innovations more directly related to safety."

5) Although the record since the Kemeny Commission Report reveals an unprecedented imposition of new requirements on operating reactors, a comprehensive rationale for "backfitting" requirements is yet to be instituted. In fact, various new rules and regulations are being implemented in advance of the definition of the acceptable level of risk desired as a safety goal upon which such changes logically rest.

6) Many of the problems which emerged during the accident had previously been identified by managers of critics but lacked sufficient priority to command the attention of those responsible for nuclear safety. The TMI crisis sufficiently restructured the nuclear agenda and the balance of forces to overcome for a finite time the normal range of obstacles and political resistance to produce safety innovation.

7) The failure to date to resolve the basic institutional problems of the NRC in the context of an antiregulatory Reagan Administration points toward a continuing vacuum in societal leadership for nuclear safety in the United States.

Regarding the future of risk assessment in nuclear power, we have several observations:

1) All indications point to an expansion in the use of formal risk assessment in the United State both for safety policy-making and the management of individual nuclear plants. It is also likely that such assessments will be more broadly used for other energy sources as well.

2) Future risk assessments will likely become more comprehensive and balanced with increased treatment of human-machine interactions, slowly developing as well as rapidly developing accidents, consequence mitigation, and deliberate attempts to induce failure.

3) An increased politicization of such assessments appears likely because they provide visible opportunity targets in risk management, because they are occupying a more central place in risk issues, and because other institutional access is likely to be less available to opponents during the 1980s.

4) The political impacts of formal risk assessments appear to be more related to the conditions which give rise to them and to who prepares the assessment than the actual findings.

REFERENCES

1. United States President's Commission on the Accident at Three Mile Island, The Need for Change: The Legacy of TMI, USGPO (1979).
2. Robert A. Dahl, A Preface to Democratic Theory, University of Chicago Press (1956).
3. Michael W. Golay, "How Prometheus Came to be Bound: Nuclear Regulation in America," Technology Review, 83:29-41 (June/July 1980).
4. Leonard Bertin, "Harrisburg Revisited: TMI One Year Later," Ascent, 1:17-23 (Spring 1980).
5. Critical Mass Energy Project, TMI: One Year in Retrospect, special report, Washington (1980).
6. Edison Electric Institute, One Year After Three Mile Island: A Report to the President and the American People (1980).
7. U.S. General Accounting Office, Do Nuclear Regulatory Commission Plans Adequately Address Regulatory Deficiencies Highlighted by the Three Mile Island Accident? Washington (1980).
8. U.S. General Accounting Office, Three Mile Island: The Most Studied Nuclear Accident in History, Washington (1980).
9. U.S. Nuclear Regulatory Commission, "NRC Views and Analysis of the Recommendations of the President's Commission on the Accident at Three Mile Island," NUREG-0632, Washington (1979).
10. U.S. House Committee on Interior and Insular Affairs, Subcommittee on Energy and the Environment Oversight Hearings, Industry's Response to the Accident at Three Mile Island, 96th Congress, 1st Session, September 1979.

11. U.S. House Committee on Interstate and Foreign Commerce, Subcommittee on Energy and Power Hearings, NRC's Response to the Report of the President's Commission on Three Mile Island, 96th Congress, 1st Session (1980).
12. David E. Bacher and David D. Carlson, "Changes in the NRC's Emergency Response Capability Since Three Mile Island," Chapter 1 in U.S. Senate Committee on Environment and Public Works, Subcommittee on Nuclear Regulations, staff studies, Nuclear Accident and Recovery at Three Mile Island: A Special Investigation, 96th Congress, 2nd Session (1980).
13. U.S. House Committee on Science and Technology, Subcommittee on Energy Research and Production, Nuclear Powerplant Safety after Three Mile Island, 96th Congress, 2nd Session (1980).
14. U.S. House Committee on Science and Technology, Subcommittee on Energy Research and Production, Plans for Improved Safety of Nuclear Power Plants Following the Three Mile Island Accident, 96th Congress, 1st Session (September 19, 1979).
15. U.S. Senate Committee on Environment and Public Works, Subcommittee on Nuclear Regulation Hearings, President's Commission on the Three Mile Island Accident, 96th Congress, 1st Session (October 31, 1979).
16. Thomas H. Pigford, "The Management of Nuclear Safety: A Review of TMI After Two Years," Nuclear News, 24:41-48 (March (1981).
17. Nuclear News, 14 (mid-March 1980).
18. Victor Gilinsky, "Nuclear Power: Safety and Self-Interest," speech to the College of Natural Science Alumni Association, Michigan State University (April 9, 1981).
19. Robert Sugarman, "Nuclear Power and the Public Risk," IEEE Spectrum, 16:59-79 (November 1979).
20. U.S. Nuclear Regulatory Commission, Special Inquiry Group, "Three Mile Island: A Report to the Commission and to the Public,"NUREG/CR-1250, 2 vols., Washington (1980).
21. Saul Levine, Inside NRC, 13 (December 1, 1980).
22. Frank Von Hippel, "Physicist Suggests Way to Mitigate Effects of Nuclear Accident," IEEE Spectrum, 16:79 (November 1979).
23. Nucleonics Week, 4 (March 12, 1981).
24. U.S. Nuclear Regulatory Commission, "Plan for Developing a Safety Goal," NUREG-0735, Washington (1980).
25. Nucleonics Week, 3 (October 25, 1979).
26. Nucleonics Week, 10 (November 8, 1979).
27. Andrew P. Hull, "Emergency Preparedness for What? (Implications of the TMI Accident)," Nuclear News, 24:61-67 (April 1981).
28. U.S. Federal Emergency Management Authority, Report to the President: State Radiological Emergency Planning and Preparedness in Support of Commercial Nuclear Power Plants, Washington (1980).
29. Nucleonics Week, 2-3 (January 29, 1981).

Bibliography

Peter A. Bradford, "Reasonable Assurance, Regulation, and Reality," speech before the ALI-ABA Course of Study on Atomic Energy Licensing and Regulation (September 24, 1980).

Warren H. Donnelly and Donna S. Kramer, Nuclear Power: Three Mile Island Accident - Congressional Response and Investigations, U.S. Library of Congress, Congressional Research Service, Archived Issue Brief 1B79097, Washington (1980).

Donna S. Kramer, "Reactor Operator Training - The Utilities' Role," Chapter 4 in Part II, U.S. Senate Committee on Environment and Public Works, Subcommittee on Nuclear Regulations, staff studies, Nuclear Accident and Recovery at Three Mile Island" A Special Investigation, pp. 469-537, 96th Congress, 2nd Session (1980).

Lauriston S. Taylor, "Some Nonscientific Influences on Radiation Protection Standards and Practice: The 1980 Sievert Lecture," Health Physics, 39:851-873 (December 1980).

U.S. Nuclear Regulatory Commission, "Clarification of TMI Action Plan Requirements," NUREG-0737, Washington (1980).

U.S. Nuclear Regulatory Commission, Modification of the Policy and Regulatory Practice Governing the Siting of Nuclear Power Reactors, Washington (1980).

ROLE OF RISK IN TREATING ADVANCED LUNG CANCER

Kenneth Stanley

Harvard School of Public Health
677 Hungtington Avenue
Boston, Massachusets 02115

Cancer Unit, World Health Organization
Avenue Appia
Geneva, Switzerland

INTRODUCTION

Cancer of the lung continues to rank first in mortality and work years lost in individuals diagnosed with a malignancy [5]. In spite of knowledge about the primary causal agent, cigarette smoking, a few effective measures have been undertaken to erradicate the source of the problem. Approximately 20% of patients diagnosed with bronchogenic carcinoma will have disease localized to the thorax where surgical excision is possible. The 5-year survival rate for this group is approximately 25% [7]. In another 30% of the patients the disease will be regional and usually treated with radiation therapy, resulting in a 5-year survival rate of approximately 2-4% [2]. However, half of the diagnosed cases will initially have disease spread beyond the possibility for effective surgical excision or radiation therapy. Further, due to the metastatic course of the disease, approximately 90% of all cases with lung cancer will eventually fall in this category [6]. The 5-year survival for this class of patients with advanced bronchogenic carcinoma is less than 2%. Chemotherapy or supportive care is the current standard primary therapy. The median age at onset of disease is approximately 57 years.

This investigation was supported by Grant Number CA-23415-02 awarded by the National Cancer Institute, DHHS.

The purpose of this paper is to define and evaluate the various risks involved once an individual is known to have advanced lung cancer. We will consider the various elements of risk; the the population at risk, the types of risk, calssification of the risks by cause, perception of risk and factors that influence risk, the determinants. Next, we will look at the current clinical techniques for evaluating these risks and the methods for comparing risks. Lastly, we will discuss the possible means by which these risks can be modified.

Although the population chosen here is limited to individuals with advanced lung cancer, the principles are applicable to most adult malignancies subject to treatment with chemotherapy.

ELEMENTS OF RISK

Population at Risk

Risk may be defined as the exposure to the chance of injury or loss. In risk assessment one must carefully describe the population under evaluation. Risks may be evaluated separately for an individual, subgroup of individuals, or the entire population. Within a subgroup or population we look at the observed frequency to evaluate risk. For a specific individual we would make a more subjective assessment. In the remainder of this paper we will be discussing only those individuals with advanced lung cancer, taken collectively at their time of diagnosis.

Types of Risk

Major risks for individuals with advanced lung cancer are morbidity from disease, morbidity from treatment and psychological risks. Psychological risks are significant but the primary source of the risk is the patient's diagnosis with this fatal disease. Additional psychological risks are primarily a function of patient care, the extent of psychological assistance, the trauma over decisions concerning treatment and clinical trial options. Even though these may be perceived by the patient as major risks. Their effect is monor in the presence of the physical risks in this population.

Cause of Risk

Sometimes it is worthwhile to classify the risk by cause. In this instance we have two primary causes, disease and treatment. The effects of aging, accidents and other diseases are neglibible by comparison. The two primary causes can be classified as involuntary and voluntary, respectively, although the assignment of cause is difficult and the label as voluntary is often inappropriate.

Standard therapy is chemotherapy, most likely a 2 or 3-drug

combination, possibly combined with radiation therapy for additional control of the primary disease or palliation. If the disease is especially advanced, treatment may be of palliative intent with supportive care. Between these two poles of therapy aggressiveness we find a spectrum of strategies, however, most tend to collect at one end or the other. The basic strategy is one of attempted local control using radiotherapy and attempted systemic control using chemotherapy. In reality, neither modality is particularly effective. Chemotherapy has resulted in no more than a doubling of survival relative to supportive care for this class of patients [9] and chemotherapy has associated toxicity.

We shall turn to these issues later in this paper.

Perception of Risk

In cancer chemotherapy, one finds three separate perpsectives - the patient, the physician, and society. In general, both the individual and the physician take a short term view of risks. Physicians are trained to view the patients on a one-at-a-time basis, and naturally patients consider themselves as a population of one. Virtually all resources are brought to bear. In contrast, society takes a more global, or public health, view and is more concerned with a cost-effective approach based on limited resources.

There are also some differences between the perceptions of the physician and the patient. Physicians are mostly concerned with therapy side effects such as nausea, vomiting, hematologic problems, and hair loss; problems where treatment is identified as the cause. Patients are mostly concerned with difficulties surrounding "absence from home and work" and "parking the car."

Determinants of Risk

In general, risks are not uniform over time but rather are a function of aging and environmental exposure as well as hereditary characteristics. For advanced lung cancer the primary determinants are the disease characteristics and therapy aggressiveness.

Figure 1 gives the survival experience for more than 4000 individuals with advanced lung cancer [3, 8]. Survival is short with a median of approximately 14 weeks following diagnosis. Survival follows very closely to what is known as a negative exponential distribution which has a "memoryless" property. The chance that an individual will survive from diagnosis to week 1 is the same as the chance that they will survive from any week to the next. The risk of death in this case is said to be constant and the graph of the survival experience will form a straight line when the probability of survival is expressed on a logarithmic scale. This is seen in Fig. 2 for the same group of patients.

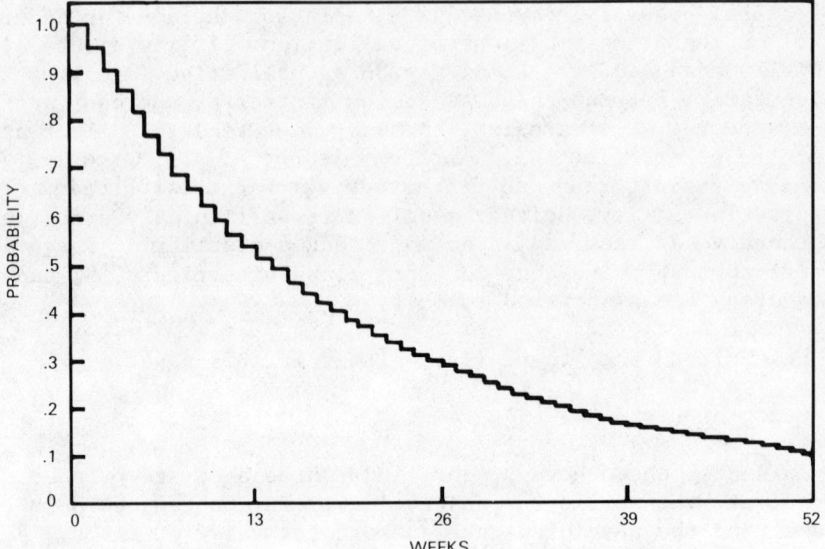

Fig. 1. Survival for individuals with advanced lung cancer. (47079 patients - VA Lung Group)

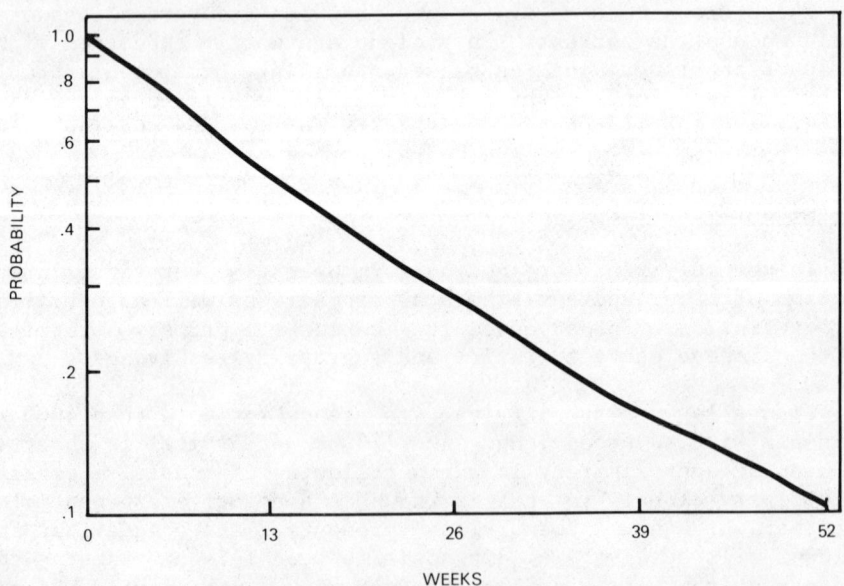

Fig. 2. Survival for individuals with advanced lung cancer expressed on a logarithmic probability axis. (47079 patients - VA Lung Group)

Table 1. Karnofsky Performance Status Criteria

%	
100	Normal: no complaints; no evidence of disease
90	Able to carry on normal activity; minor signs or symptoms of disease
80	Normal activity with effort; some signs or symptoms of disease
70	Cares for self. Unable to carry on normal activity or to do active work
60	Requires occasional assistance but is able to care for most of his needs
50	Requires considerable assistance and frequent medical care
40	Disabled; requires special care and assistance
30	Severely disabled; hospitalization is indicated although death not imminent
20	Very sick; hospitalization necessary; active supportive treatment necessary
10	Moribund; fatal processes progressing rapidly
0	Dead

There is a sizeable variation in survival among patient groups. The major prognostic factors are initial performance status, extent of disease involvement (confined to one hemithorax or beyond), recent weight change, and histologic characteristics of the tumor [8]. Performance status is a subjective assessment of ambulation and the ability to work expressed on a scale as the percentage of normal (see Table 1) [4]. The scale ranges from 100% - a normal active individual down to 0 - for death. This scale is clearly subjective, but its correlation with survival cannot be questioned. Survival by initial performance status is given in Fig. 3. Here we find the ordering of the performance status scale preserved, with 100% on top. Median survival varies by a factor of 10 depending upon the level of initial performance status.

Individuals with disease confined to one hemithorax have a median survival of seven months as compared to a median of three months for individuals with more advanced disease. A weight loss of more than 10 pounds in the six months prior to diagnosis and treatment will reduce the median survival by about one-half [8].

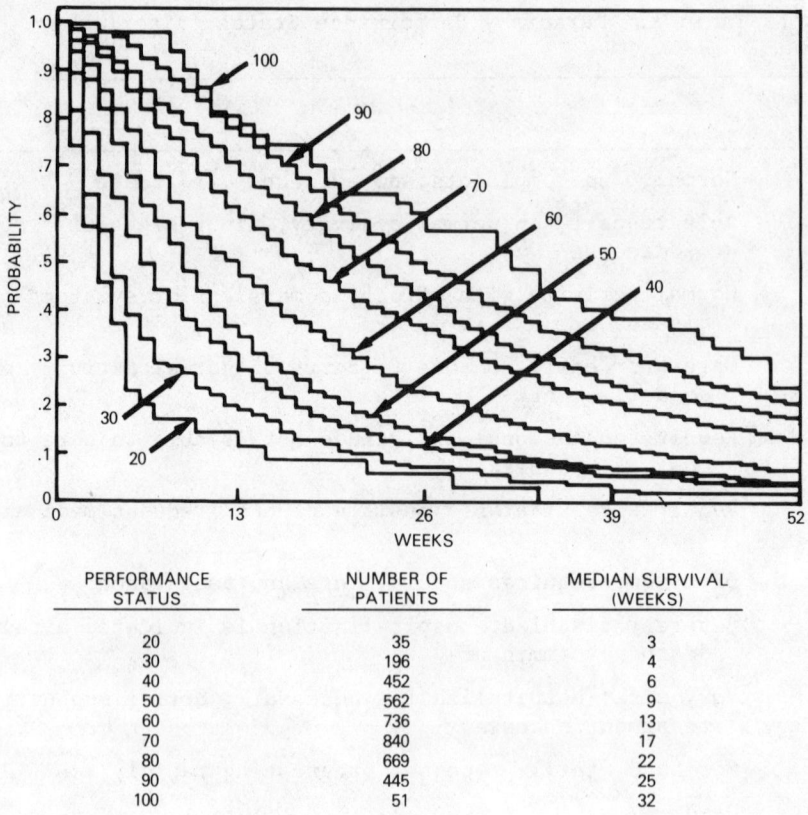

Fig. 3. Survival by initial performance status (3986 patients - VA Lung Group)

At this point we have identified the two primary risks, death from disease and death from treatment, and the factors which may influence these risks. Next we will consider the relationship between these risks and determining factors.

Relationship between the Risks and Factors

Because risks associated with advanced lung cancer are overshadowed by the risk of death, two opposing treatment strategies are found. In the first, the strategy is to balance the risks from treatment and the risks from the disease with aggressive therapy in an attempt to optimize the quality of life. However, the usual chosen point of balance is far from equating these risks. Death is perceived by many clinicians and patients as a natural risk, similar to acceptance of death due to old age. The risks and inconveniences associated with chemotherapy are frequently the focus of their attention. Even with "aggressive" therapies physicians are unwilling to induce, and patients unwilling to endure, more than minimal treatment morbidity because it is a voluntary risk. The risk of treatment morbidity is perceived as carrying greater weight. Whereas, in reality, the risk of disease morbidity is the overwhelming risk involved and the expected additional morbidity associated with aggressive treatment is minimal. Perhaps the optimal strategy is to intensify the therapy until risk of death from treatment is equal to risk of death from disease. However, current research is pursuing other approaches such as employing various combinations of agents with known activity and screening new agents for activity. Intensification to the point of equalizing the risks of treatment and disease morbidity will probably be attempted only if all else fails.

An opposing treatment strategy is also common. In this view, the patient is seen as being in a high risk population with little chance for effective treatment. Because chemotherapy will probably not provide a cure and there is a great likelihood that the inddividual will suffer ill effects from the attempt, the solution is to let the individual die without the physical insult associated with chemotherapy. Few, if any, treatment risks will be taken and palliative treatment and compassion will be the strategy. However attractive this approach might be, if adopted on a global basis this would mean that clinical cancer research would come to a standstill. Many of the successful chemotherapeutic treatments in childhood malignancies may not have been discovered under such a conservative approach. Clearly, the advance of medicine requires the acceptance of certain research risks.

The choice of treatment versus supportive care involves psychological in addition to physical risk considerations. Rather than say that nothing is effective and not treat the disease, many individuals and clinicians would prefer to attempt treatment. There is always the possibility of great benefit from a new chemotherapeutic program, although past experience would suggest that for this disease such a major advance is unlikely.

We can reformulate this dicussion of treatment strategies into an evaluation of the effect that variations in therapy aggressiveness will have on the two major types of risk, death from disease and death from treatment.

There is some evidence to suggest that intensive therapy is necessary to improve the survival and health of the individual. However, other reports indicate that the benefit is minor, if it exists at all. Obviously, an aggressive therapy will result in a greater risk of death due to the treatment, but we don't yet know if that will lessen the overall risk of death.

In a similar fashion, as the disease burden becomes smaller, some clinicians would argue that more intensive therapy is unwarranted because the patients are not as ill and overtreatment should be avoided. Others would take the opposite point of view and argue that chemotherapy has the greatest opportunity for cure when the disease burden is smallest. Hence, the therapy should be more intensive for patients with minimal disease. As the disease becomes more advanced, both the patient and clinician may be willing to take more of a chance for cure with an intensive or investigational chemotherapeutic program. It is unknown which of these relationships is correct, if either, using current chemotherapeutic agents; conflicting reports are found.

Clinical Evaluation of Risks

Present methods of evaluation of the effectiveness of chemotherapy include analysis of survival, objective tumor response (a measured decrease in tumor size based upon x-ray and clinical evaluations), and treatment toxicity (by type and degree).

The three endpoints, tumor response, survival, and treatment complications, convey the primary results of current clinical studies but are insufficient for proper risk assessment. Most clinicians and patients feel that quality of life is the primary issue. However, evaluation of this characteristic is difficult, mostly due to matters of perspective, lack of objective measurement, and lack of a standard. Even after a suitable objective measure is obtained there would still be a major question involving the evaluation of quality versus quantity of life. Clearly this is a personal decision.

Results of a typical report in the literature are summarized in Table 2 [1]. Overall median survival for the 283 patients was 19 weeks. Complete disappearance of the disease was observed in 3 patients while 3 other patients were determined to have died as a result of the chemotherapy. Two of the complete responses lasted only 13-17 weeks, the other is continuing at 61 weeks. The three-drug treatment, Hexamethylmelamine, Adriamycin, and Methotrexate, appeared to be the best of the available treatments. Measured against the past standard for this disease (Cyclophosphamide plus (CNU), it had a significantly superior response rate (p=.01), a marginally better survival, and about the same rate and type of complication. An optimist would conclude that progress had been made. In reality, the advantage is minimal, if existent. Patient selection factors, strong prognostic variables, and random variation limit the inferences that can be made.

Comparisons of Risks

Although the issue of whether the benefit of a complete responce in three patients somehow justified the three treatment re-

Table 2. Results of a Clinical Trial in Patients with Advanced Lung Cancer

Treatment	Complete response	Complete plus partial response rate	Median survival (weeks)	Rate of life threatening treatment complications	Drug deaths
Cyclophosphamide and Methotrexate	1/61	13%	23	13%	1
Baker's Antifol	1/62	15%	20	3%	0
Vincristine, Bleomycin and Methotrexate	0/20	10%	13	15%	0
Melphalan	0/18	0%	13	6%	0
Cyclophosphamide and CCNU	0/62	8%	16	2%	0
5-Fluorouracil and Procarbazine	0/19	11%	16	5%	2
Hexamethylmelamine Adriamycin and Methotrexate	1/41	27%	34	0%	0
Overall	3/283	13%	19	6%	3

lated deaths may be interesting, in fact death by disease would
have followed soon if death by treatment had not occurred. The
risk of death by treatment is not as important in this disease as
the risk of death by treatment in another disease, say early breast
cancer. The reason for this is the difference in rates of the competing risk of death by disease. Median survival is 5-6 years following diagnosis in early breast cancer. Death from treatment would
be a major loss of expected life. However, in advanced lung cancer,
death from the disease usually follows shortly after diagnosis. Premature death from treatment, although being a major concern, actually
amounts to minimal loss of expected life. The expected survival difference in advanced lung cancer between a death from treatment and a
death from disease would be on the order of 3 or 4 weeks. This difference might be as large as 5 or 6 years in early breast cancer.
In general, evaluation of risk should take into account the expected
loss due to the untoward event. The knowledge of the existance of a
risk of death from treatment of 1% is virtually useless without
knowledge of the competing risks involved. Such a rate might be considered unacceptably high in early breast cancer, and it might be
indicative of unaggressive therapy in advanced lung cancer.

In comparing risks we also need to be concerned with perspective. In this case, the physician/patient perspective is quite different from a public health perspective. An improvement in median
survival from 3 months to one year (a 4 fold increase) due to improvement in chemotherapy would be seen by the clinician and individual with this disease as a great scientific advancement. Their
attention is focused upon the time period following diagnosis. However, in terms of overall patient life an increase from $57^1/_4$ years
to 58 years is not nearly as striking (57 year median onset plus 3
months versus one year survival following treatment). Further, we
must take into account the chance of poor quality of life with hospitalization away from home, suffering of side effects such as nausea, vomiting and loss of hair, plus large medical expenses for this
additional period. The opportunity to decline the improved therapy
should exist.

Modification of Risks

What now can be done to modify the risks associated with advanced lung cancer? The two keys to substantial reduction of risks
would be the development of a therapeutic cure and the decrease of
the environmental and behavioral causes of this disease. Even
though cure will eventually come, none appears to be on the immediate horizon. The most effective current course of action would
be to reduce the risk of bronchogenic carcinoma through legal and
social action directed at cigarette smoking and other environmental
causes.

1. R. H. Creech, C. R. Mehta, M. Cohen, et al., Phase II master protocol for evaluation of new chemotherapeutic regimes in patients with inoperable non-small cell lung carcinoma. In press, Cancer, 1981.
2. S. J. Cutler, End results in cancer. Rep. No. 4, U.S. Dept. HEW, Washington, D.C., 1972.
3. E. L. Kaplan, P. Meier, Nonparametric estimation from incomplete observations, J. Am. Stat. Assoc., 1958:53, 457-481.
4. D. A. Karnofsky and J. H. Burchenal, The clinical evaluation of chemotherapeutic agents in cancer. in: "Evluation of chemotherapeutic agents," C. M. Macleod, ed., Symposium, Microbiology Section, New York Academy of Medicine, New York, N. Y., 1948. New York: Columbia University Press, 1949:191-205.
5. J. L. Murray and L. M. Axtell, Impact of cancer: years of life lost due to cancer mortality, J.N.C.I., 52:3-7, 1974.
6. O. S. Selawry, On chemotherapy of lung cancer. in: Lung cancer, natural history prognosis and therapy. L. Israel and A. P. Chahinian, 205-239, Academic Press, New York, 1976.
7. O. S. Selawry and H. H. Hansen, Lung cancer. in: "Cancer Medicine," J. R. Holland and E. Frei, eds., 1473-1518, Lea and Febiger, Philadelphia, 1973.
8. K. E. Stanley, Prognostic factors for survival in patients with inoperable lung cancer, J.N.C.I., 65:25-32, 1980.
9. J. Wolf, P. Spear, R. Yesner, and M. E. Patno, Nitrogen mustard and the steroid hormones in the treatment of inoperable bronchogenic carcinoma. American Journal of Medicine, 29:1008-1016, December 1960.

CIGARETTE SMOKE: CANCER RISK AT LOW DOSES

Charles E. Lawrence
Division of Epidemiology
New York State Department of Health
Albany, New York 12237
    and
Albert S. Paulson
OR & S Program
Rensselaer Polytechnic Institute
Troy, New York 12181

INTRODUCTION

The regulation of exposure to low-dose carcinogens is beset with difficulties and uncertainties. Ethical considerations rule out the use of human experiments to determine risk from low dose carcinogens. Even if animal experiments are used, a number of animals would be required to test for carcinogens at low doses. Nevertheless, safe levels of human exposure must be set. Thus, methods based on extrapolation from animal experiments with a relatively small number of animals exposed to high dosages of potential carcinogens are being adopted as the basis for regulating human exposure to low dosages of carcinogens [1]. Since it has been impossible to determine a dosage level at which there is no cancer hazard, an alternate "virtually safe" dose method has become common. The virtually safe dose is a dose at which the cancer risk is no greater than some very small risk, say, a one-in 10^5, 10^6, or 10^8 chance of developing cancer in a lifetime. For example, an action level of 100 µg/liter for total trihalomethanes in drinking water has been proposed by the EPA as prudent for these animal carcinogens [2]: the risk of cancer resulting from low-dose x-ray exposure has been determined in this way and used to establish acceptable x-ray dosage levels for mammography screening [3].

KEY WORDS: Virtually safe dose, environmental carcinogens, smoking and health, low dose carcinogens.

The regulation of carcinogens is usually intended to protect against unwitting, unavoidable or involuntary contact with carcinogens. In the case of cigarettes, smokers would be exempt from these regulations since smoking is done by a conscious, voluntary decision to do so. However, there is a sizeable fraction of the population exposed to cigarette smoke unwittingly and involuntarily: the nonsmokers. In comparison with the evidence for other reported carcinogens, the evidence that cigarette smoke is carcinogenic is overwhelming. Large-scale and long-term prospective epidemiologic studies such as the Doll report [4], the Hammond study [5], or the Dorn study [6, 7] clearly implicated cigarette smoking as a lung cancer agent. We report the results of extrapolating lung cancer risk from the high dose levels in one of these studies [6, 7] to the low dose levels experienced by nonsmokers.

Methods

Three main techniques have become established for low dose extrapolation from high dose experiments: the Mantel-Bryan probit method [8], the multistage methods of Guess and Crump [9, 10], and the linear extrapolation method [11]. Gori [12] has performed an extrapolation of risk for cigarette smoke. However, his focus on low tar cigarettes and extrapolation to the minimum epidemiologically detectable dose differs greatly from ours. Guess et al. [13] have shown that the upper confidence limit for additional risk from any continuous exposure direct-acting carcinogen is almost certainly linear. Thus the results from multistage models agree closely with those obtained from linear extrapolations. They further show that unless a linear component is excluded by assumption - as it is in the Mantel-Bryan method - it is most unlikely to be rejected on an empirical basis. Thus, the linear-based method has emerged as the primary means for these extrapolations. Two models were employed here for extrapolating the risk estimates to the low dose range. If P_d is the probability of contracting the disease (assumed rate), the linear extrapolation is obtained from

$$P_d = 1 - \exp(-rd)$$
$$\simeq rd$$

to first order for rd sufficient small, with d the dose and r a parameter to be estimated [7]. A non-linear extrapolation is obtained using the "improved Mantel-Bryan" method [4], which assumes that P_d is functionally related to the logarithm of the dose by

$$P_d = \phi(a + b \log d),$$

where ϕ represents the standard cumulative normal distribution evaluated at argument $a + b \log d$, and a and b are parameters to be estimated from the data.

We have used the Dorn study data to prepare extrapolations based on these two methods for deaths due to cancer of the lung and bronchus. In 1954, Dorn [6, 7] conducted a prospective epidemiologic study of the effects of smoking and mortality among 291,000 Veterans Administration (VA) life insurance policy holders. A high proportion of the cigarette smokers in the study begin smoking at an early age: 50% before age 21 and 80% before age 25.

To parallel animal experiments, we have restricted our analysis to data on those who never quit smoking and thus smoked for a majority of their adult life. Furthermore, we have only included deaths from the major smoking-related cancers (lung and bronchus). A report on Dorn's study by Kahn [7] provides mortality among these long-term smokers for an $8^1/_2$ year period following administration of the questionnaire and provides the basis for our extrapolation. In this group of smokers, mortality from cancer of the lung and bronchus was from 2.85 to 23.9 times higher than that of nonsmokers, depending on the number of cigarettes smoked per day and the age of the study subject.

Use of data from a study of humans requires modification of the usual extrapolation procedures. When laboratory animal data are employed, a correction for the ratio of surface area to body mass is made. In this example, the difficult problem of making inferences about human risk from animal data is avoided. In animal studies, incidence and mortality from disease is observed for the lifetime of the animals and a virtually safe dose is based on this lifetime risk. In our case, the period of observation for mortality is $8^1/_2$ years for individuals from a wide range of ages. Thus, the extrapolation, given here, of the risk for any particular age group will underestimate the lifetime risk.

Results

The safe doses of cigargettes/day according to each of these techniques are given in Table 1. Since mortality has been reported for four separate dose levels (1-9 cigarettes per day to 39+ cigarettes per day), there are several choices for extrapolation. It has been argued that one should extrapolate from combined levels (since this uses the most information), unless there is evidence of a flattening or decrease in risk at higher doses due to competing risks, as there appears to be in this case. Hence, extrapolation from the lowest dose level is indicated. Extrapolations from both levels are given in Table 1.

The large difference between mathematical models illustrates the sensitivity of these extrapolations to the inclusion of exclusion of a linear component. There is little reason in this circumstance for not assuming a linear component. Indeed, in plots of dose-response data in the observable range [14], the graphs appear

Table 1. Virtually Safe Doses of Cigarettes (cigarettes per day)

	Age	Linear Extrapolation	Mantel-Bryan		
			Risks		
			10^{-8}	10^{-6}	10^{-5}
Combined	55-64	$1.54 \times 10^3 \times r$*	.0083	.0598	.1845
Study	65-74	$1.03 \times 10^3 \times r$	.0053	.0384	.1183
Dose	75-84	$6.83 \times 10^2 \times r$	.003	.0214	.0661
Levels					
Lowest	55-64	$7.6 \times 10^2 \times r$	.0035	.0256	.079
Study	65-74	$4.9 \times 10^2 \times r$	.0025	.0178	.0550
Dose	75-84	$1.9 \times 10^2 \times r$	.0089	.0064	.0197

*r = acceptable risk level (10^{-5}, 10^{-6}, 10^{-8})

Table 2. Nonsmoker Cigarette Smoke Exposure in Various Environments [15, 16, 17]

Environment	Amount of tobacco burned	CO	Cigarette equivalents* of:	
			Nicotine	Total particulate matter
Airplane (U.S. DOT) 15-20 air changes/hr	—	.64	—	<.072
Bus (Seiff, HE)				
15 air changes/hr	23 cig.	6.048	—	—
15 air changes/hr	3 cig.	3.304	—	—
Theater (Godin, G., et al.)				
Foyer	—	.632	—	—
Auditorium	—	.248	—	—
"Public House" (Harmsen & Effenburger)	—	5.512-14.64	.0033-.152	—
Train (Harmsen & Effenburger)	—	7.336	.0248	—
Residence for 3 hours (17^3) (Hoegg)				
1 air change/hr	1 cig/hr	.9	—	.069
Meeting room 1 hour (8 m³/smoker) (Hoegg)				
10 air changes/hr	1 cig/hr	.1185	—	.013
Work space 8 hours (24 m³/smoker) (Hoegg)				
10 air changes/hr	1 cig/hr	.316	—	.034

*Cigarettes per 8 hour day for person breathing .6 m³/hour, excpet as noted.

to be nearly linear with the exception of higher than expected risk levels at the lowest dose. Furthermore, in the past it has been considered prudent to accept the more conservative linear extrapolation to low dose levels. As mentioned above, the lifelong risk will exceed the risk in any one of the three age groups given in Table 1. Thus, the virtually safe dose will be lower than that for the 75-84 year olds of .0068 cigarettes per day obtained from the combined dose estimate and .0019 cigarettes per day from the lowest dose estimate. Based on procedures used to regulate other substances, the virtually safe dose of cigarette smoke is less than about one two-hundredth (.005) of a cigarette per day.

DISCUSSION

Atmospheric levels of three constituents of cigarette smoke (total particular matter, carbon monoxide and nicotine) have been measured in a number of circumstances [15]. Table 2 gives some illustrative examples of the daily doses under the assumption that an adult breathes $.6m^3$ of air per hour. Studies in a number of controlled settings have also been conducted. Hoegg [15] has completed such a study and developed a formula for estimating cigarette equivalents. The last three entries in Table 2, derived from this formula, are given to show typical home, work and public assembly circumstances. As indicated by Table 2, the cigarette equivalents encountered in most common circumstances by nonsmokers exceed the virtually safe dose.

Granted that the nonsmoker's exposure to cigarette smoke is an unwitting, involuntary type of exposure, we have sought to estimate a safe dose for these nonsmokers along the same lines as those for worker exposure to carcinogens or public exposure to pollutants. A virtually safe dose of cigarette smoke is 1/200 of a cigarette per day. Reported exposures of nonsmokers exceed this dose in nearly all cases. Thus, there is a large discrepancy between the levels to which unwitting exposure to other carcinogens are increasingly being controlled and the levels of unwitting exposure to cigarette smoke.

ACKNOWLEDGEMENTS

We thank Mr. Jeffrey Arbuckel, Dr. Paul Daitch, Dr. Peter Greenwald, and Dr. Philip Taylor for their assistance and comments.

REFERENCES

1. National Research Council, Drinking Water and Health, Washington: National Academy of Sciences (1977).
2. The Federal Register, Interim primary drinking water regulations, Part II EPA: 5756-5780 (February 9, 1978).
3. Bailar, J. C., Mammography: A contrary view, Annals of Internal Medicine, 84:77-84 (1976).
4. Doll, R., and Hill, A. B., Lung cancer and other causes of death in relation to smoking. A second report on the mortality of British doctors, British Medical Journal, 2:1071-1081 (1965).
5. Hammond, E. C., and Horn, P., Smoking and death rates - report on 44 months of follow-up of 187,783 men - 1. Total Mortality, Journal of Amer. Med. Assoc., 166(10):1159-1172 (1978).

6. Dorn, H. D., Tobacco consumption and mortality from cancer and other diseases, Public Health Reports, 74:581-593 (1959).
7. Kahn, H. A., The Dorn study of smoking and mortality among U.S. veterans; report on $8^{1}/_{2}$ years of observation, in: Epidemiological approaches to the study of cancer and other chronic diseases, W. Haenszel, ed., National Cancer Institute Monograph 19, U.S. Department of Health, Education, and Welfare, Public Health Service, National Cancer Institute, 1 (1966).
8. Mantel, N., An improved Mantel-Bryan procedure for "safety testing" of carcinogens, Cancer Research, 35:865-872 (1975).
9. Guess, H. A., and Crump, K. S., Low-dose-rate extrapolation of data from animal carcinogenicity experiments - analysis of a new statistical technique, Math. Biosciences, 32 (1976).
10. Hartley, H. O., and Sielkin, R. L., Estimation of "safe dose" in carcinogenic experiments - Biometrics, 33:1-30 (1977).
11. Hoel, D. G., et al., Estimation of risks of irreversible, delayed toxicity, Journal of Toxic Environ. Health, 1:133-151 (1975).
12. Gori, G., Low risk cigarettes: a prescription, Science, 194:1243-46 (1976).
13. Guess, H., Crump, and Peto, R., Uncertainty estimates for low-dose-rate extrapolations of animal carcinogenicity data, Cancer Research, 37:3475-3483 (1977).
14. Hammond, E. C., Garfinkel, L., Seidman, et al., Some recent findings concerning cigarette smoking, in: Origins of Human Cancer, H. J. Hiat, J. D. Watson, and J. A. Winsten, eds., Cold Spring Harbor, New York, Cold Spring Laboratory (1977).
15. Hoegg, U. R., Cigarette smoke in closed spaces, Environmental Health Perspectives, 3:117-128 (1972).
16. The Health Consequences of Smoking, U.S. Department of Health, Education, and Welfare, Public Health Service, Center for Disease Control, Atlanta, Georgia (1975).
17. Schmeltz, D., Hoffman, D., and Wynder, W. L., The influence of tobacco smoke on indoor atmospheres - 1. An overview. Preventive Medicine, 4:66-82 (1975).

PERCEIVING THE RISKS OF LOW-YIELD VENTILATED-FILTER

CIGARETTES: THE PROBLEM OF HOLE-BLOCKING

Lynn T. Kozlowski

Clinicial Institute Addiction Research Foundation
33 Russell Street
Toronto, Ontario, Canada M5S 2S1
and
Associate Professor of Psychology
University of Toronto

Some smokers of low-yield cigarettes derive higher levels of 'tar', nicotine, and carbon monoxide (CO) than advertising or government smoking-machine assays have led them to believe [1-3]. My colleagues and I [4, 5] have been concerned about a specific technique of compensatory smoking - hole-blocking of ventilated filters - that appears to occur with alarming frequency. Ventilation holes in the filter cause each puff of smoke to be diluted with air. In a sample of 39 observed users of ventilated-filter cigarettes, 44% were found to block the ventilation-holes (with their lips or fingers) and hence defeat one of the major strategies that cigarette manufacturers have employed to produce low-yield cigarettes [5].

Ventilated-filter cigarettes are becoming increasingly popular. To my knowledge, every tobacco cigarette in the U.S. and Canada that delivers less than 7 mg tar is a ventilated-filter cigarette, and many brands below 13 mg tar benefit from the effects of at least slight air-dilution in the filter. The market share in the U.S. for ventilated-filter cigarettes was recently estimated at about 50% [6]. The reduction of 'tar', nicotine, and CO yields is a complex issue, and several manipulations are used in most modern cigarettes to reduce yields. The ventilated-filter has become especially important to manufacturers because (a) it can decrease CO yields (regular filter cigarettes can have higher CO levels than do unfiltered cigarettes) and (b) it is easily added to cigarettes in the final stages of production. This second aspect is important in part because tobacco is a variable organic material that can provide challenges - from batch to batch - for manufacturers who are charged with achieving particular tar and nicotine ratings [7].

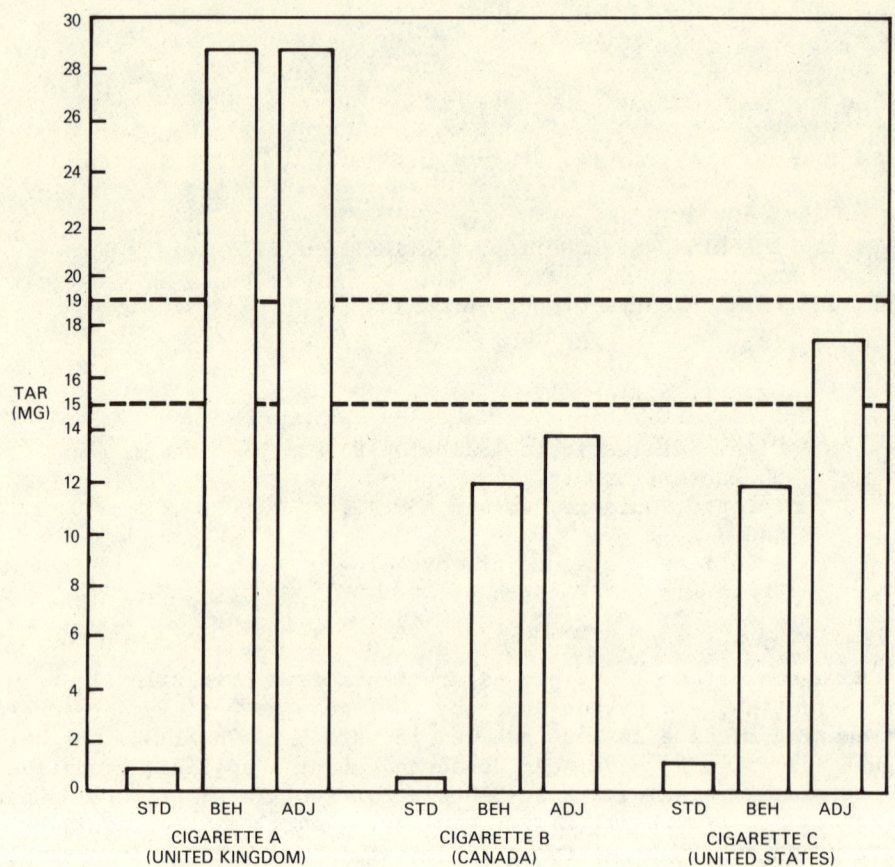

Fig. 1. A comparison of 'tar' yields under standard (STD), behavioral (BEH), and puff-adjusted (ADJ) smoking-machine conditions for three 'lowest-yield' brands. STD condition was: puff duration, 2 sec; puff interval, 58 sec; puff volume, 35 ml; and BEH was: 2.4 sec; 44 sec; 47 ml respectively, plus hole-blocking. The dotted lines indicate the 'tar' yields of a majority of cigarettes sold in the U.S. and Canada in 1979. See text for definition of ADJ and for brand information. (Adapted from [5].)

From the earliest stages of our research on the behavioral problem of hole-blocking, we have encountered two kinds of skepticism. To some individuals, it seemed 'far-fetched' that any smoker would ever block the ventilation holes. Though one might quibble about what percentage of smokers block the holes, the data that I have already cited answer these skeptics. Other individuals were readier to believe that hole-blocking occurred, but were unwilling to accept

that it made very much difference. After all, the cigarettes in
question were the mildest ones on the market; did't they contain
special, mild tobacco? We responded by showing that, if the standard smoking-machine assay were employed, hole-blocking could turn
a 4 mg tar cigarette into a 13 mg tar cigarette [4]. Furthermore,
when the smoking-machine assay was altered to better represent the
average behavior of smokers of low-yield cigarettes, hole-blocking
of the 'lowest-yield' cigarettes increased 'tar', nicotine and CO
yields by several times [5]. Figure 1 shows the results for 'tar'
yields for three of the 'lowest-yield' ventilated-filter brands, one
each from the U. K. (A = John Player King Size Ultra Mild), Canada
(B = Viscount No. 1 KSF), and U.S. (C = Cambridge KSF). (Nicotine
and CO yields show comparable patterns of results.) As will be described below, smoking-machines are set to take puffs at a specified
rate rather than to take a fixed number of puffs [1]. The ADJ
column in Fig. 1 shows an estimate of 'tar' yields, if the machines
had taken the same number of puffs in the behavioral (BEH) as they
had in the standard (STD) conditions (.e., Players = 10.6, Viscount =
10.1, Cambridge = 10.5 puffs). [ADJ = (BEH ÷ number of puffs BEH) ×
number of puffs STD.] Note that, although these brands start out by
giving almost identical 'tar' yields, these manipulations have substantial, but somewhat divergent, effects on 'tar' yields.

Why do smokers block the holes on ventilated-filter cigarettes?
If they have selected a low-yield brand, for example, out of concern
for their health, why would they intentially act to increase the
yields? First, those smokers who block the holes intentionally appear to do so out of an ignorance of the consequences of their actions. Just as some of our colleagues believed that 'the tobacco
was milder,' these smokers seem to believe that their ultra low-
yield cigarettes contain tobacco that is inherently so mild that,
even if they smoke more intensively, they are still better off with
these cigarettes than they were with their former, higher-yield cigarettes. We have encountered smokers who believe that the standard
ratings of 'tar' and nicotine yield indicate the total doses available per cigarette, in the way the dose on a pill bottle indicates
the total dose per pill. This problem of perception derives from
a poor understanding of how the 'tar' and nicotine ratings are measured by the smoking machine and misconceptions about how the reductions in yields have been achieved by the tobacco industry.

The smoking-machine is basically a syringe pump that takes a
35 cc puff on a cigarette once each minute until a fixed butt length
is reached [1, 2]. The butt-length to which a cigarette is smoked
is determined in part by the length of the 'filter plus overwrap'
and these butt-lengths do vary from brand to brand and have changed
through the years. Figure 2 illustrates some of the ways in which
the toxic yields of cigarettes are manipulated. Looking at the
smoking-machine as a behavioral system, there are only two ways to
reduce yields: a) decrease the number of puffs taken per cigarette

1. HIGH YIELD (MG) — 20T, 1.2N, 15CO

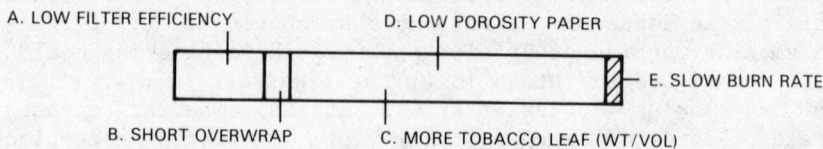

2. MEDIUM YIELD (MG) — 15T, 0.9N, 13CO

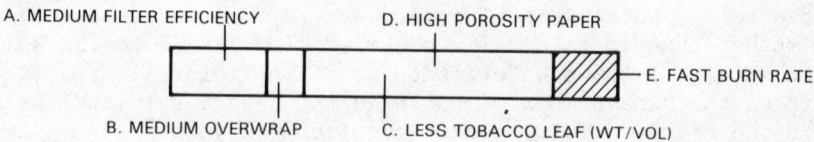

3. LOW YIELD (MG) — 3T, 0.3N, 3CO

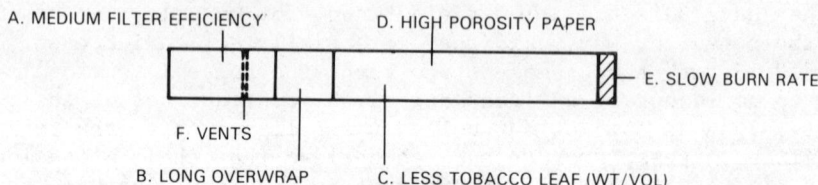

Fig. 2. Design factors influencing the 'tar' (T), nicotine (N), and carbon monoxide (CO) yields of cigarettes as determined by smoking machines. Factors B, C, and E tend to alter the number of puffs taken by smoking machines; factors A, C, D, F tend to influence the smoke concentration in each puff.

or b) decrease the dose per puff (i.e., reduce the concentration of smoke in each puff). We have argued that some of the highly-touted reductions in yields in the past 10 years have been due to reduction in the number of puffs taken per cigarette [1]. This reduction in number of puffs could be caused by increases in the burn-rate of the tobacco column and/or by increases in the butt-length due in part to increases in the length of the overwrap.* Comparison of cigarettes

---

*Grunberg et al., a caution for cigarette smokers and smoking researchers, in preparation.

1 and 2 of Fig. 2 illustrates some of the manipulations that have had a particular influence on the number of puffs taken per cigarette by the smoking machines. The behavior of smokers is not necessarily as responsive to these manipulations as is the behavior of the smoking machine [1].

Comparison of cigarettes 2 and 3 of Fig. 2 illustrates the role of air dilution techniques in the reduction of 'tar', nicotine, and CO yields. One manufacturer of ventilated filters has shown the extent to which filter ventilation can be used to reduce yields [8]; using the same tobacco column, non-perforated filters delivered smoke with about 14 mg 'tar'. All cigarette smoke is diluted with air. These researchers determined the effects on yields of added dilution: 15% air dilution in the filter was associated with yields of about 12 mg 'tar'; 1.0 mg nicotine and 11 mg CO, and 70% air-dilution was associated with yields of about 1.5, 0.2, and 1.8 mg, respectively. Based on a regression equation calculated from these published results, it can be estimated that a 0.01 mg 'tar' cigarette could be produced from this base of a 14 mg 'tar' cigarette by employing a filter that diluted smoke 79% with air [3]. It is easy to understand why smokers of the ultra-low tar cigarette complain that puffing on these cigarettes is like 'puffing on air'. (Incidentally, gram for gram of tobacco consumed, air-dilution filters have little effect on the sidestream smoke issuing to the atmosphere [3].)

The second major barrier to perceiving the consequences of hole-blocking concerns problems with immediate perception. Most hole-blockers do not appear to do it intentionally. About 80% of current hole-blockers reported that they did not currently block the holes [5]. We assume that these individuals were not trying to deceive us and they they had unwittingly developed a smoking style (e.g., holding the cigarette further back in the mouth so that their lips blocked the holes) that caused the holes to be occluded. (Since ventilated-filter cigarettes can be difficult to light, hole-blocking may sometimes arise more as a lighting aid than as a nicotine compensation technique.) This problem of awareness of hole-blocking is more difficult to deal with than is the ignorance about how low yield cigarettes are engineered.

One might think that, if smokers tried to 'catch themselves' while smoking and inspected their style of holding the cigarette, they should be able to eliminate the hole-blocking problem. One difficulty with implementing this procedure is that some smokers may not realize that their brand is a ventilated-filter cigarette. A look at recent advertising in a major tobacco trade journal (*Tobacco International*) shows that there is a demand for 'invisible' techniques of filter performation. (To mention two brands, <u>Merit</u> in the U.S. and <u>Accord</u> in Canada are not obviously ventilated-filter cigarettes.) Apparently, there is concern on the part of manufacturers about the image of ventilated-filter cigarettes as low-yield cigarettes. 'Invisible' perforating techniques permit the 'look' of a high-yield cigarette to be sported by a low-yield cigarette, and they can allow an established brand to change its yields without changing its appearance. (See the discussion in [7] for information on the techniques of perforation.)

'Invisible' hole-blocking should not be permitted. If ventilation-holes need to be avoided as one step (N.B., one step) in the less-hazardous use of cigarettes, then, it is necessary to inform smokers exactly where the perforations are to be found. Perhaps easily detectable texture changes or 'ridges' might be required on the outside of filters, so that a smoker could feel immediately when the holes were being blocked.

Smokers can perform a quick test of whether or not their brand has a ventilated filter. If they put the cigarette just far enough between their lips to take a puff and do not hold the fiter in their fingers as they puff, almost all ventilated-filter cigarettes (whether visibly or invisibly perforated) will show a characteristic stain pattern on the filter: there will be a 'bull's-eye' of tar stain on the end of the filter, surrounded by a ring of unstained filter. An unventilated-filter cigarette or a fully blocked ventilated-filter cigarette will produce a uniform tar stain on the end of the filter [4]. (This stain-inspection technique can also be used to give smokers 'feedback' as to whether they do block holes on ventilated-filter cigarettes [9].) Dissecting a spent filter with a razor blade and holding the filter plug wrap up to a strong light can also be used to disclose one or more discrete rings of vents or sometimes a diffuse array of small ventilation holes on the third of the filter wrap closest to the tobacco.

Some people have expressed concern that the placement of the ventilation holes on filters might ensure that they will be blocked by fingers or lips, that is, it has been suggested that one can not help but interfere with the perforations in the course of normal smoking. My first response was that the placement of holes left ample room for the proper placement of lips on the cigarette butt, however, some pilot research suggests that some individuals' lips may be large enough to make these smokers ill-suited to use ventilated-filter cigarettes under any circumstances. One style of smoking may create particular problems when combined with ventilated-filter cigarettes. Some smokers (probably not many) hold the cigarette between their teeth. (Actually, it was my witnessing this behavior in an airport waiting room that spurred me to include in our first series of interviews questions about the use of lips to block the holes [5].) A recent major advertising campaign for one of the best-selling (unventilated) cigarettes in Canada (Export A) features a truck driver who holds his cigarette with his teeth. In the U.S., a campaign of Winston (that includes all Winston brands, including the ventilated-filter Winston Ultra) features a burly young man also holding his cigarette with his teeth. If a smoker of low-yield cigarettes adopts this technique, does it increase the probability of hole-blocking?

Ten smokers (5 males and 5 females) were asked to put on a lip gloss or lipstick and then hold the cigarettes: a) in their mouths as they normally did, and b) comfortably between their teeth. They were asked to take a puff on the cigarette held in this manner, and the number of ventilation holes that were stained by the lip-covering indicated the number of holes blocked. (This brand of cigarette had one ring of 24 ventilation holes, 13 mm from the proximal end of the filter.) When the cigarettes were held as the subjects normally would hold them, 7 subjects did not block any holes and 3 subjects blocked some of the holes (13%, 38%, 46%). When held in the teeth, every subject blocked some of the holes ($\bar{X} = 60\%$, SD = 24). (No sex differences were found.) Based on the regression line calculated from the data published on the yields if 0%, 50%, or 100% of the holes of this same brand were covered [4], a standard smoking machine assay with 60% of the holes blocked would change this 4 mg tar cigarette into a 9 mg 'tar' cigarette.

Clearly, the smoking of ventilated-filter cigarettes offers particular challenges to those who want low yields from these cigarettes. Tobacco consumers need to be educated about the meaning of the standard 'tar' and nicotine ratings and about the role of the air-dilution holes in the creation of 'low-yield' cigarettes. Smokers of these cigarettes need to monitor the way in which they smoke, to be sure that they do not block the holes. It is especially important to realize that hole-blocking is only one form of compensatory smoking behavior. The fact that one does not block the holes does not also mean that one does not take more frequent, larger, deeper or longer inhalations from these cigarettes. In other words, ventilated-filter cigarettes are not alone in presenting a risk of compensatory smoking, rather they present an additional risk of compensating for reduced yields. Diagnostic techniques are needed to aid smokers in determining just what toxic substances they do derive from both ventilated and unventilated low-yield cigarettes [9].

REFERENCES

1. L. T. Kozlowski, W. S. Rickert, J. C. Robinson, and N. E. Grunberg, Have tar and nicotine yields of cigarettes changed? Science, 209:1550-1551, 1980.
2. L. T. Kozlowski, Tar and nicotine delivery of cigarettes: what a difference of puff makes? JAMA, 345:158-159, 1981.
3. L. T. Kozlowski, Smokers, non-smokers, and low tar smoke, Lancet, 1:508, 1981.
4. L. T. Kozlowski, R. C. Frecker, V. Khouw, and M. A. Pope, The misuse of 'less-hazardous' cigarettes and its detection: hole-blocking of ventilated filters, Am. J. Public Health, 70:1202-1203, 1980.

5. L. T. Kozlowski, W. S. Rickert, M. A. Pope, J. C. Robinson, and R. C. Frecker, Estimating the yields of smokers of tar, nicotine, and carbon-monoxide from the 'lowest yield' ventilated-filter cigarettes, Br. J. Addic., in press.
6. D. Hoffman and E. L. Wynder, The low yield cigarette, Am. J. Public Health, 70:1143-1144, 1980.
7. Evolving techniques of making cigarettes milder, World Tobacco, 64:93-101, 1979.
8. J. A. Parker and R. T. Montgomery, Design criteria for ventilated filters, Beitr. Tabakforsch, 10:1-6, 1979.
9. L. T. Kozlowski, Applications of some physical indicators of cigarette smoking, Addict. Beh., in press.

REACTIONS TO PERCEIVED RISK: CHANGES IN THE BEHAVIOR

OF CIGARETTE SMOKERS

>   Kenneth E. Warner
>
>   School of Public Health
>   University of Michigan

INTRODUCTION

Since publication of the 1964 Surgeon General's report on smoking and health [U.S. DHEW, 1964], Americans have been exposed to voluminous evidence linking cigarette smoking to a variety of serious illnesses. The Surgeon General's report collected evidence developed in the preceding decade and a half and, interpreting that evidence, it made the first widely publicized pronouncement that smoking was casually linked to lung cancer and represented a major risk factor in other debilitating diseases. Since then, scores of scientific studies have supported and extended understanding of the relationships between smoking and illness, and a variety of mechanisms have been used to convey the basic health risk message to the public. On the national scene, most notable among these were the Fairness Doctrine antismoking messages broadcast on television and radio from 1968 through 1970 [Warner, 1979] and the publicity surrounding then-HEW Secretary Califano's announced reinvigoration of the federal antismoking effort in 1978 [Califano, 1978]. For the last three years, new Surgeon General's reports have provided a comprehensive compendium of knowledge of smoking and health [U.S. DHEW, 1979] and focused on specific issues of current concern: the health consequences of smoking for women [U.S. DHHS, 1980] and the quest for a less hazardous cigarette and smoking behaviors [U.S. DHHS, 1981].

National "events" such as the publicity surrounding release of a Surgeon General's report are but the most conspicuous components

---

This project was supported by Grant Number HS 03634 from the National Center for Health Services Research, OASH.

of the nationwide antismoking campaign. The persistent efforts of
the major voluntary associations, organizations formed specifically
to address the smoking-and-health issue (e.g., ASH, GASP, Interagency Councils on Smoking and Health), boards of education, and
even commercial interests have vividly brought the smoking-and-health issue to the attention of millions of Americans.

Surveys administered over the past two decades demonstrate that
the public has heard the smoking-and-health message and believes it.
A 1975 survey found that 90% of the respondents agreed that "cigarette smoking is harmful to health," with over four-fifths of those
agreeing characterizing their agreement as "strong." Even among
current smokers, over 80% concurred that smoking was harmful [U.S.
DHEW, 1976]. Indeed, every major survey taken since the original
Surgeon General's report has found sizable majorities of the public
agreeing that cigarette smoking is a significant hazard to health
[e.g., U.S. DHEW, 1969, 1973].

The knowledge change has been converted into a national attitudinal change toward smoking. In 1964, a majority of the public
disagreed that "it is annoying to be near a person who is smoking
cigarettes" [U.S. DHEW, 1969]. By 1975, however, nearly three-quarters of survey respondents agreed with the statement, including
more than a third of current smokers (compared with only a fifth in
1964) [U.S. DHEW, 1976]. Another index of the change in attitudes
is the growth in the majority who believe that "the smoking of cigarettes should be allowed in fewer places than it is now": 52% of
survey respondents concurred in 1964, compared with 70% in 1975.
Even a majority of smokers supported this restriction on smoking in
1975 [U.S. DHEW, 1969, 1976].

Despite the clarity with which the smoking-and-health message
has reached the average citizen's ear, there are vast discrepancies
in the way different observers interpret the behavioral impact of
the antismoking campaign. "Optimists" point to the decreasing proportions of adults and teenagers who identify themselves as smokers,
while "pessimists" note that over 50 million Americans persist in
smoking and aggregate cigarette consumption is actually rising.

Assessment of the behavioral consequences of the perception of
risk associated with smoking is vastly more complex than simply measuring a change in the number of smokers or cigarettes. Smoking behavior is multidimensional, and the evidence shows that these dimensions have changed at significantly different rates and times. Furthermore, behaviors may have changed only in part due to a perception of risk; changing social mores could also play a role.

In the remainder of this paper, I will inventory many of the
behavioral changes which have occurred since the smoking-and-health
issue first received widespread public attention, relating these

Table 1. Percentages of Adult Males and Females Reporting Themselves to be Smokers, U.S., by Year

Year	% Males (1)	% Females (2)	% Total (3)	Ratio male to female rates (4) = (1) ÷ (2)
1955	52.6	24.5	37.6	2.15
1964	52.9	31.5	40.3	1.68
1966	51.9	33.7	42.2	1.54
1970	42.3	30.5	36.2	1.39
1975	39.3	28.9	33.8	1.36

Source: U.S. DHEW, 1979, Appendix Table 1.

changes to the various stages of the antismoking campaign and examining how they have followed changes in knowledge about and attitudes toward the risks of smoking. The intent is to develop a comprehensive picture of the behavioral response to the campaign for reducing the toll of smoking.

COMPOSITION OF THE SMOKING POPULATION

Adults

Table 1 indicates the percentages of adult males and females who have reported themselves to be smokers in various years. Two important phenomena are discernible in these data: 1) the rate of self-reported smoking has been declining significantly and 2) the predominance of males in the smoking population has been receding. The latter reflects two trends: a) the rate of smoking among adult males has decreased persistently and significantly since the 1964 Surgeon General's report, b) while the rate of smoking among women actually rose through much of the 1960s, falling slowly in the 1970s. The combined influence of these trends is reflected in column (4) of Table 1, which shows that the male smoking rate was more than double the female rate prior to the original Surgeon General's report, whereas by 1975 the differential was little more than a third.

A number of factors help to explain this fundamental shift. An important one is the simple fact that the smoking habit began to diffuse within the female population years later than it did among males. In large part, this represented the perceived "impropriety" of women smoking prior to recent decades. As the slogan of a cigarette oriented toward the female market states it, women have "come a long way" toward achieving "equality" with men in the right to smoke. Thus women were "catching up" with men's smoking in the 1950s and 1960s when the antismoking intervened in this "liberation" movement.

Changes in the cigarette product may have contributed to the relative increase in the female proportion of the smoking population. For example, introduction of filter-tipped cigarettes after World War II made smoking less "dirty" and may have encouraged women to take up the habit.

It also appears that the public perceived that smoking was a greater threat to the health of men and that the antismoking campaign was directed primarily at men. Data presented in the original Surgeon General's report and in many subsequent documents provided a logical basis for this. For example, the early data showed a much stronger connection between smoking and lung cancer in men than in women, with the mortality ratio of the male smoker cited as from 10-1 to 30-to-1, compared with a maximum 5-to-1 mortality ratio for female smokers.* It appears that this difference primarily reflected differences in the predominant smoking behaviors in earlier decades (e.g., number of cigarettes per day, depth and frequency of inhalation), rather than inherent differences in the sexes' susceptibility to disease. Only in very recent years have the health hazards of smoking women received specific emphasis [U.S. DHHS, 1980].

From data such as those presented in Table 1, it appears that the rate of smoking among men may have leveled off in the 1950s at slightly over 50% even in the absence of adverse smoking-and-health publicity. If this is true, an impact of the antismoking campaign would have been visible directly in decreases in the proportion of males smoking. By contrast, the fact that the rate of smoking among women was increasing in the 1950s and 1960s suggests that an impact of the campaign would have been more difficult to detect; most probably, it would have been reflected in a decrease in the rate of increase of smoking. Given the pressures pushing the rate of female smoking upward, the decrease in the rate since the late 1960s suggests a distinct response to the health-risk message.†

At the close of this glimpse at adults' smoking rates, it is important to emphasize that the self-reporting of smoking behavior introduces biases toward underestimation of actual cigarette consumption [Warner, 1978]. The nature and magnitude of these biases are discussed near the end of this paper, but they are mentioned here simply to indicate that self reports on smoking status may misrepresent the size of the true smoking population. This becomes

---

*The mortality ratio compares the probability of smokers' dying from lung cancer to that of nonsmokers. For men, the 10-to-1 figure was cited as an upper limit for smokers of less than one pack per day and 30-to-1 for smokers of more than one pack per day.
†Of course, it is possible that decreases in consumption rates reflect a response to the decreasing social desirability of smoking as much as or more than a perceived threat to health. I have not managed to isolate explanations, nor am I aware of any research which has.

Table 2. Percentages of Teenagers Reporting Themselves to be Regular Smokers, U.S., by Age, Sex, and Year

Year:	% Males					% Females			
	1968	1970	1972	1974	1979	1968	1970	1972	1974	1979
Age										
12-14	2.9	5.7	4.6	4.2	3.2	0.6	3.0	2.8	4.9	4.3
15-16	17.0	19.5	17.8	18.1	13.5	9.6	14.4	16.3	20.2	11.8
17-18	30.2	37.3	30.2	31.0	19.3	18.6	22.8	25.3	25.9	26.2

Source: National Institute of Education, 1979, Exhibit 1.

particularly important in examining data such as those in Table 1 when one recognizes that the 1964 estimates come from a survey taken within months of issuance of the original Surgeon General's report and 1970 estimates are derived from a survey administered in the third (and last) year of the Fairness Doctrine antismoking messages on television and radio [Warner, 1979]. To the extent that survey respondents were sensitive to the antismoking publicity, the possibility of underreporting the true smoking status becomes a concern.

Teenagers

Table 2 presents data on smoking by teenagers, from surveys administered from 1968 through 1979. The caveat in the preceding paragraph is particularly applicable for scrutinizing survey data on teenage smoking, since teens are even more prone than adults to misrepresent their smoking status [Luepker et al., forthcoming]. Subject to this caveat, however, the data suggest two important phenomena: 1) following years of stable or upward movement in the percentage of adolescents who report themselves to be smokers, the most recent survey suggests a substantial downturn in the rate of smoking in almost all age-sex classes;* and 2) teenage females have reached a parity with teenage males in the rate of smoking. The latter suggests that the gap between adult male and female smoking rates will continue to diminish, while the former, if true, implies future decreases in the overall percentage of confirmed adult smokers.

The reasons underlying changes in teenage smoking patterns are more obscure than those affecting adult patterns, and the possibility of gross misrepresentation of smoking status further complicates the issue. Nevertheless, the most recent survey gives reason for cautious optimism that the epidemic of adolescent smoking is subsiding, sobered by realization that "women's liberation" appears to have achieved a disquieting victory in the realm of teenage smoking.

AGGREGATE CIGARETTE CONSUMPTION

Column (1) of Table 3 shows that total U.S. cigarette concumption has increased fairly steadily over the past two decades, though there were notable downturns in 1964, the year of the first Surgeon General's report, and 1968 and 1969, the first two years of the Fairness Doctrine antismoking messages on television and radio. The continuing growth in cigarette consumption is cited by the tobacco industry as evidence that smoking remains popular, and serves as a source of concern among antismokers that the industry is right.

---

*The slight increase in the rate for the oldest group of girls (17-18 years) may be attributable to a cohort effect, since in the preceding survey year (1974), the rate for 12-14 year-olds jumped considerably.

Table 3. Total and Adult Per Capita Cigarette Consumption by Year

Year	Total consumption (billions of cigarettes) (1)	Percentage increase (decrease) from preceding year (2)	Consumption per adult (3)	Percentage increase (decrease) from preceding year (4)
1960	484.4	—	4171	—
1961	502.7	3.8	4266	2.3
1962	508.4	1.1	4265	—
1963	523.9	3.0	4345	1.9
1964	511.2	(2.4)	4194	(3.5)
1965	528.7	3.4	4263	1.6
1966	541.2	2.4	4287	0.6
1967	549.2	1.5	4280	(0.2)
1968	545.7	(0.6)	4186	(2.2)
1969	528.9	(3.1)	3993	(4.6)
1970	536.4	1.4	3985	(0.2)
1971	555.1	3.5	4037	1.3
1972	566.8	2.1	4043	0.1
1973	589.7	4.0	4148	2.6
1974	599.0	1.6	4141	(0.2)
1975	607.2	1.4	4123	(0.4)
1976	613.5	1.0	4092	(0.8)
1977	617.0	0.6	4051	(1.0)
1978	616.0	(0.2)	3967	(2.1)
1979	620.0	0.6	3924	(1.1)
1980	630.0	1.6	3880	(1.1)

The growth in total consumption, however, has not kept pace with the growth in the smoking-age population. This is reflected in column (3) of Table 3 in which the aggregate data are converted into cigarettes per adult (defined as individuals over 17 years of age). As columns (3) and (4) show, by this measure cigarette consumption has fallen steadily, if gradually, since 1973. It is important to recognize that cigarettes per capita - a common index of "average" cigarette consumption - is an imperfect index for it fails to distinguish changes in per smoker consumption from changes in the percentage of smokers. The evidence presented in the preceding section, as well as that discussed later, suggests that the gradual drop in per capita consumption reflects a more rapid decrease in the proportion of smokers and an increase in cigarettes consumed by the average continuing smoker.

Comparison of per capita consumption figures for recent and more distant years is a common basis for assessments of the effec-

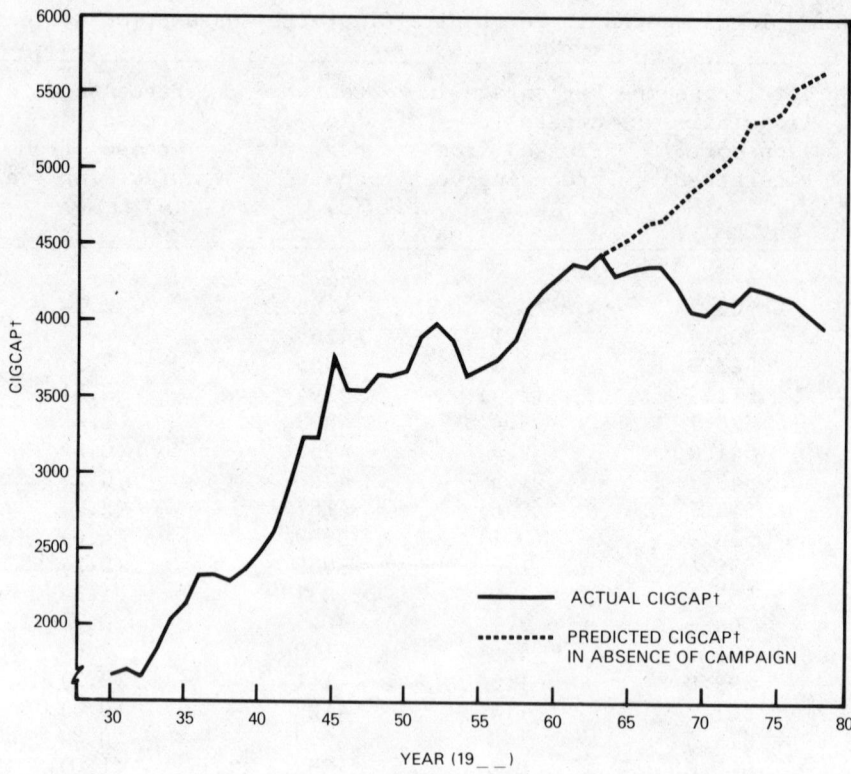

Fig. 1. Percapita cigarette consumption, actual and predicted in absence of the antismoking campaign. (Source: Warner, 1981.)

tiveness of the antismoking campaign. Typical use of these numbers is illustrated by a journalist's conclusion in the mid-1970s that, since per capita consumption was not then substantially below its 1963 level, "the Surgeon General, the American Cancer Society and militant anti-smokers have been wasting their breath" [San Francisco Chronicle, 1976]. But to measure the impact of the antismoking campaign, the appropriate comparison is not current consumption with past consumption, but rather current consumption with an estimate of what current consumption would have been in the absence of the campaign. As earlier discussion suggested, the rate of cigarette smoking was rising in the early 1960s, with years of continued increases likely, given the diffusion of the smoking habit among women. Furthermore, from the 1950s onward, per cigarette tar and nicotine (t/n) were falling steadily. (See the next section.) The nicotine compensation hypothesis (discussed later) argues that smokers will consume more cigarettes if they switch to lower t/n cigarettes in order to maintain, or reduce less rapidly, their dosage of nicotine. If the hypothesis holds, the continual decreases in per-cigarette t/n should have pushed per-smoker consumption higher.

The combined effect of the increasing popularity of smoking among women and decreasing nicotine per cigarette most probably would have been an increase in per capita consumption through the 1960s and 1970s. This would have continued the general upward trend which has prevailed throughout the century. (See Fig. 1.) Indeed, the only two-year decrease in per capita consumption prior to 1964 occurred in the early 1950s during the first widely-publicized smoking-and-health scare. Thus, it seems logical that the stable and then falling rates of per capita consumption represent a major shift in smoking behavior from that which would have been anticipated in the absence of adverse smoking-and-health publicity.

Elsewhere [Warner 1977, 1981] I have estimated the pattern per capita consumption would have followed had the forces influencing cigarette demand not been affected by the anti-smoking campaign. This pattern is represented by the dotted line in Fig. 1. Thus an estimate of the consumption impact of the campaign is the difference between the estimated consumption (the dotted line) and actual consumption (the solid line). In 1978, the former was over 40% greater than the latter. While the precise magnitude of this difference is subject to estimation error, the qualitative finding is not: in the absence of the anti-smoking campaign, per capita consumption of cigarettes ($CIGCAP_t$ in the graph) would have been much higher than it is at present.

CHANGES IN THE CIGARETTE PRODUCT

Measures such as the percentage of smokers or the number of cigarettes sold are the most visible indicators of changing smoking behavior, but probably the most profound changes relate to the nature of the cigarette itself. For example, while everyone knows that tar and nicotine (t/n) per cigarette have been falling, few people are aware of the extent of the decrease: average nicotine per cigarette has dropped by 30% since the original Surgeon General's report and by 60% since the mid-1950s. Tar per cigarette has fallen some 40% since the report and, like nicotine, 60% since the mid-1950s. (See Fig. 2.) Similarly, the market share of filter-tipped cigarettes has risen from under 60% prior to the Surgeon General's report to well over 90% today. Filter cigarettes first claimed 1% of the market in 1952, so they grew even more rapidly through their first decade. (See Table 4.)

Decreases in per-cigarette t/n automatically accompanied the introduction and spread of filters in the 1950s and 1960s as smokers from very high t/n unfiltered cigarettes. The shift seems to have been predominantly a response to the health concern, but filters

Table 4. Filter-Tip Share of Cigarette Market

Year	% Filters	Year	% Filters
1949	0.3	1964	60.9
1950	0.6	1965	64.4
1951	0.7	1966	68.2
1952	1.3	1967	72.4
1953	2.9	1968	74.9
1954	9.2	1969	77.5
1955	18.7	1970	80.1
1956	27.6	1971	82.4
1957	38.0	1972	83.7
1958	45.3	1973	85.4
1959	48.7	1974	86.7
1960	50.9	1975	87.7
1961	52.5	1976	88.5
1962	54.6	1977	89.4
1963	58.0	1978	90.9

Source: U.S.D.A.

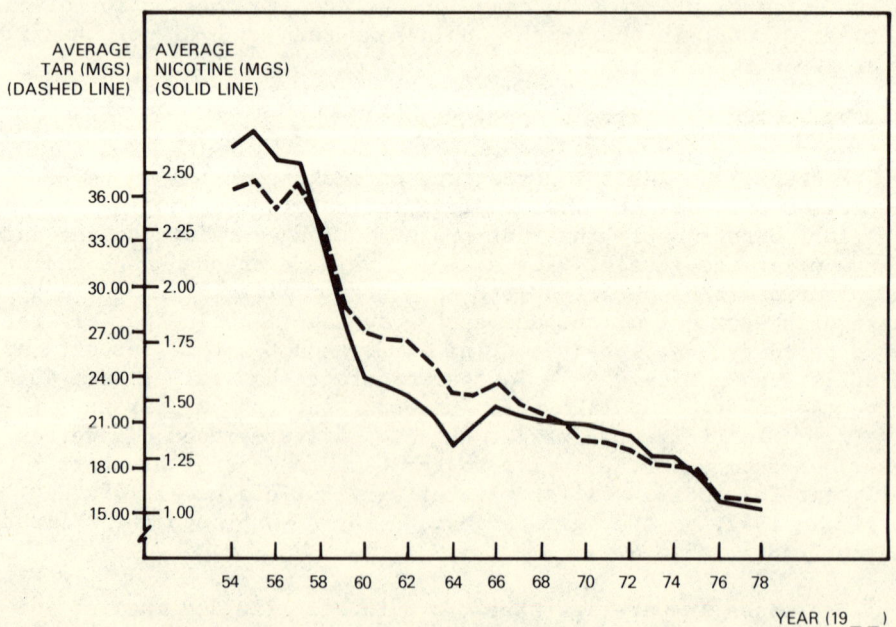

Fig. 2. Sales-weighted average tar and nicotine per cigarette. (Source: Data from Philip Morris, Inc.)

also offered a "neater" way to smoke. Indeed, some observers attribute the rapid growth in smoking among women in part to the availability of a "less dirty" means of indulging in the habit. In the most recent years, by contrast, the still-growing low t/n share of the market appears to reflect a conscious switching by smokers from filtered high t/n cigarettes. Publicity on t/n, including the most recent Surgeon General's report [DHHS, 1981], has converted some of the general perception of smoking being hazardous to health into a more specific attribution of risk to tar and nicotine.

The decrease in t/n has an intriguing implication. If all other things were equal, Americans would have to more than double their cigarette consumption in order to ingest the same amount of t/n as they did 30 years ago, when per capita cigarette consumption was comparable to today's level. Of course, the "if" in this observation represents an important and probably unsupportable assumption. As is discussed in the next section, there is considerable reason to believe that smoking behaviors differ for low and high t/n smokers; for example, smokers of low t/n cigarettes may pull harder on their cigarettes and inhale more deeply, negating some of the potential comparative advantage of the low t/n cigarette. Conversely, the very high t/n cigarettes of the 1950s were so strong that many smokers may not have inhaled with each puff on a cigarette. Consequently, it seems most unlikely that the decrease in tar and nicotine ingestion is nearly as large as suggested by simply multiplying per cigarette t/n by per capita consumption. Nevertheless, it does seem likely that the "national intake" of tar and nicotine is lower today than it was three decades ago. And to the extent that tar and nicotine are the constituents of tobacco smoke responsible for illness, or that they occur in proportion to those constituents, the decrease would appear to be a salutary one.*

Other changes in the cigarette product are not so encouraging. In order to enhance flavor, improve burning and so on, cigarettes now include literally thousands of additives, several of which are known or suspected carcinogens. The additives are generally present in trace amounts, but given the quantities of cigarettes which confirmed smokers consume, the potential for new health hazards would

---

*Mitigating this conclusion is the possibility, frequently suggested, that low t/n cigarettes have made it easier for nonsmokers, especially teenagers and women, to begin smoking. Data from a recent survey indicate that low-tar cigarettes (15 mg or less) are smoked by a large fraction of teenage smokers while only a very small fraction smoke high-tar cigarettes (21 mg or more) [National Institute of Education, 1979]. Reviewing these and other data, Harris concluded "I know of no evidence contradicting the hypothesis that the availability of lower-tar and nicotine cigarettes enhanced the rate of initiation of smoking among younger females" [Harris, 1980].

appear to be real. This is a risk which is perceived by only a very
small minority of smokers, since manufacturers have never been obli-
gated to list ingredients on cigarette packages and governmental concern with additives is of very recent vintage [DHHS, 1981].

There are numerous other changes in the product which have occurred throughout the period of the antismoking campaign. Cigarettes
sizes and styles have varied, changes designed to differentiate the
product and attract a share of the market. Several less obvious al-
teractions have aided manufacturers in lowering t/n in order to attract customers in this fast-growing segment of the market. For example, different mixes of tobaccos produce lower t/n yields, as do
variations in filtration, including placing pin-holes in the cigarette before the filter. And, due to a process called "puffing" which
expands the volume of a given weight of tobacco, cigarettes contain
less tobacco than they used to. Efforts to lower t/n represent a
clear market response to the consumers' perception that they can reduce risk by switching to a lower t/n brand of cigarettes.

BEHAVIORS OF INDIVIDUAL SMOKERS

The preceding sections have catalogued major changes in Americans' smokers behaviors, including the shift by continuing smokers
toward lower t/n cigarettes and decreases in the percentages of
adults and, more recently, teenagers who smoke, accounted for by
many individuals quitting smoking and many others failing to initiate
the habit. Other, more subtle, changes have also been noted. In
this section, I will examine a few of these as they relate to continuing smokers, i.e., those who have chosen not to cease smoking.

From responses to surveys, one can conclude that the majority
of continuing smokers have conflict about their smoking. They
acknowledge the health threat but are unwilling or unable to give
up smoking. Consequently many such individuals turn to lower t/n
cigarettes as a "compromise." Were their other smoking behaviors
unchanged, this would represent a reduction in health risk, possibly
proportional to the change in per-cigarette t/n for many specific
risks. Unfortunately, however, many such smokers appear to alter
their habits inadvertently, in deleterious ways, to "compensate"
for the reduced nicotine per cigarette. This is consistent with the
nicotine regulation (or compensation) hypothesis which suggests that
smokers develop a need for a given dosage of nicotine which they
can ingest through a variety of means [Schater, 1978; Russel, 1980].
Among these means are the following: smoking more cigarettes; smoking more of each cigarette, by puffing more frequently or smoking
further down the cigarette; inhaling more deeply; and "subverting"

t/n reduction technologies, for example by covering up filtration holes with the fingers holding a cigarette.*

The evidence on the extent to which these behaviors are adopted is equivocal. Of necessity, much of it derives from survey data which, as is discussed in the next section, are subject to inaccuracy. A series of four Public Health Service surveys, covering the years 1964, 1966, 1970, and 1975, offered little evidence of improvements in the more subtle aspects of smoking behavior (e.g., amount and depth of inhalation) and a mixed picture regarding the most basic aspects: both men and women indicated significant shifts away from unfiltered cigarettes, but large numbers of both sexes adopted the longer (100 mm) cigarettes in the 1970s; and while men reported to change over the period in the number of cigarettes they smoked on a daily basis, women reported an increase in daily consumption from 17 to 19 cigarettes [U.S. DHEW 1969, 1973, 1976]. Given the bias toward increased underreporting of deleterious smoking behaviors (discussed in the next section), these findings cannot be construed as indicative of substantial adoption of less hazardous smoking behaviors.

There is objective evidence that nicotine compensation does occur. Recently, Russell [1980] reported that smokers' blood nicotine levels did not vary nearly as much as the per-cigarette nicotine of the smokers' brands. While the average nicotine of the brands varied by over 80%, the average blood nicotine levels varied by only 20%, and the latter was not statistically significant. Cigarettes' nicotine yield, accounted for only 4.4% of the variation in smokers' blood-nicotine levels. The small differences in blood nicotine levels might be attributable primarily to self-selection; i.e., smokers of low-nicotine cigarettes may have started out with lower blood nicotine before switching brands.

There is substantial evidence that smokers' daily cigarette consumption has risen significantly. As noted above, surveys show this phenomenon [U.S. DHEW 1969, 1973, 1976], though they understate both the level of daily consumption and the rate of its increase [Warner,

---

*The nicotine regulation hypothesis is not universally accepted. Garfinkel [1979] presents some evidence suggesting that smokers who switch to lower t/n cigarettes do not significantly increase the number of cigarettes they smoke. Garfinkel's data, however, are based on questionnaire responses and thus are subject to the hazard of increased underreporting [Warner, 1978]. In addition, Garfinkel did not examine other techniques of nicotine regulation.

1978]. Harris has estimated that average daily consumption in the U.S. rose from 22 cigarettes in 1954 to 30 in 1978 [Harris, 1980]. It seems probable that a significant portion of the increase in daily consumption has resulted from nicotine compensation by smokers who have switched to lower t/n cigarettes, though it should be noted that average per-smoker consumption would rise simply if those who quit smoking disproportionately represented light smokers.

In summary, the combination of objective and subjective (survey) data paint a picture of continuing smokers trying to reduce risk by adopting the most publicized strategy - switching to low t/n cigarettes - but then negating much of the potential benefit by engaging, perhaps unconsciously, in compensating risk-increasing behaviors, such as smoking more cigarettes and inhaling more deeply. It must be emphasized that this portrait of the low t/n smoker is like an impressionist painting: the general image is discernible, but the details are blurred. Our knowledge of the nature and extent of nicotine compensation remains imprecise.

REPORTING OF SMOKING BEHAVIOR

Above I have examined a variety of changes in cigarette smoking behavior apparently deriving from the widespread perception of health risk attributable to smoking. As dramatic as some of these changes have been, perhaps the most intriguing evidence on the degree to which the perception of risk has been internalized has nothing to do with smoking behavior per se. Comparing data from a series of four surveys with objective data on aggregate cigarette consumption,* I have found that 1) survey respondents underreport their smoking by a significant amount and 2) the amount of underreporting appears to have increased over the years since the 1964 Surgeon General's report [Warner, 1978].

There is a substantial body of evidence demonstrating that the "reporting of an event is likely to be distorted in a socially desirable direction" [Cannell, Oksenberg, and Converse, 1977]. Events or behaviors perceived to be desirable are often overreported [Ferber, et al., 1969; Lansing and Blood 1964; Parry and Crossley 1950], while those perceived to be socially undesirable or personally sensitive or threatening are often underreported [Cannell, Fisher, and Bakker 1965]. In the present instance, both the social undesirability and the perceived personal threat associated with smoking have increased steadily since 1964, the year of the first survey [U.S. DHEW 1969, 1973, 1976]. The existence of the personal

---

*The objective data come from the U.S. Department of Agriculture. They are based on production and sales data for both manufactured and "roll-your-own" cigarettes, corrected for inventories and accounting for imports and exports.

Table 5. Actual and Reported Cigarette Consumption

Year	Actual consumption (billions)[a] (1)	Percent change in actual cons. (2)	Reported consumption (billions) (3)	Percent change in rep. cons. (4)	Rep. const. as fraction of actual cons. (=(3)/(1)) (5)
1964	509.95		370.68		.7269
1966	538.50	+ 5.60	386.97	+ 4.39	.7186
1970	533.35	- 0.96	350.27	- 9.48	.6567
1975	604.10	+13.27	388.25	+10.84	.6427
1964 to 1975		+18.46		+ 4.74	

[a] Average of U.S. Department of Agriculture series for calendar and fiscal years.
Source: Warner, 1978.

threat and social undesirability would be expected to be reflected in some underreporting, and the increase in these negative features of smoking logically might produce an increase in the amount of underreporting.

Table 5 summarizes the findings of my analysis of underreporting. Column (1) shows actual aggregate U.S. consumption in each of the survey years and column (2) shows the percentage change from the preceding year. Column (3) gives aggregate consumption estimated from consumption reported by smokers in the surveys and column (4) indicates the percentage change in these numbers from one survey year to the next. Comparison of columns (1) and (3) demonstrates that self-reported consumption consistently fell short of actual consumption, while comparison of columns (2) and (4) indicates that reported consumption consistently rose by smaller amounts (or, from 1966 to 1970, fell by a larger amount) than did true consumption. These phenomena are also reflected in column (5) which gives the percentage of actual consumption reported in each survey year. The most striking evidence of the degree of increase in underreporting is found in the bottom line of the table: from 1964 to 1975, total cigarette consumption actually increased by 18.5%, but survey respondents' self-reports translate into an increase of only 4.7%.*

It is interesting to note that the largest survey-to-survey discrepancy between changes in actual and reported consumption occurred between 1966 and 1970, years spanning the Fairness Doctrine media messages, when reported consumption dropped by 9.5% while actual consumption fell by only 1%. There is a substantial body of evidence suggesting that the Fairness Doctrine antismoking messages served as effective deterrents to smoking [e.g., Hamilton, 1972; Warner, 1977], but these figures seem to indicate that another strong effect of the messages was to convey a perception of risk and undesirability sufficient to produce a substantial increase in smokers' misrepresentation of their habits.

As noted above, teenagers' reporting of their smoking habits deviates from truth even more than that of their elders [Luepker et al., forthcoming]. There is reason to believe, however, that their underreporting represents more a fear of parental (or other adult) reaction than a direct perception of health risk.

Conclusion

Thorough awareness of the hazards of smoking is shared by smokers and non-smokers alike. While smoking remains a persistent

---

*See Warner [1978] for a discussion of the procedure for estimating reported consumption and description of sensitivity analyses used to test the strength of the findings.

and troublesome national health problem, there are numerous indicators that this awareness has been translated into efforts by smokers to eliminate or reduce the extent of the hazard. Two decades ago, a majority of adult males engaged in the habit and the rate of increase in smoking by women suggested that a majority of females would be smokers within a few years. Today smoking is a minority habit, claiming only a third of the adult population, and among that third are many who would like to join the non-smoking majority. If recent survey data reflect reality, the proportion of teenagers who smoke has finally begun to drop and to do so rapidly. This argues further decreases in the adult rate of smoking as the current generation of adolescents grows into adulthood.

For years the failure of per capita consumption to fall significantly served as a remainder of the stubborness of the smoking habit. Annual decreases in this measure since 1973, however, suggest that knowledge and attitudinal changes dating from the original Surgeon General's report are now being converted into significant behavioral change. These annual decreases are particuarly impressive in light of the fact that the post-war baby boom now comprises a significant proportion of the adult population, and baby boomers are of prime (i.e., heaviest) smoking age.

Perhaps the most dramatic response to the smoking-health scare has been the shift toward filtered cigarettes and the decrease in the tar and nicotine content of cigarettes which modern smokers consume. For many such individuals, the attractions of smoking, or the difficulties of quitting, account for persistent of the habit, though awareness of the hazards associated with smoking has motivated behavioral responses intended to reduce the risk. The shift toward low t/n cigarettes is the clearest manifestation of this objective. Yet ironically and inadvertently, many of these same smokers appear to have adapted to the lower yield of these cigarettes by adopting compensating behaviors: they may smoke more of these "less hazardous cigarettes," or more of each of them; they may inhale more deeply; they may discover ways of "defeating" their cigarettes' technological means of delivering low t/n (e.g., covering up filtration holes).

If this assessment of nicotine compensation is accurate, a challenge for the 1980s will be conveying to continuing smokers the nature of the risk associated with the more subtle smoking behaviors (e.g., length of cigarette consumed and depth of inhalation). Objective evidence on the degree of nicotine compensation suggests that many low t/n smokers may have an exaggerated perception of the degree of risk reduction they have achieved. It also suggests that the quest for a "less hazardous cigarette" may have to focus on a strategy discussed by proponents of the nicotine regulation hypothesis: seeking means of delivering nicotine with little or no tar and other potentially hazardous components of cigarette smoke (e.g., a high-nicotine, low-tar cigarette) [Russel, 1980].

On balance, the positive changes in smoking behavior since 1964 seem to outweight the negative ones, perhaps considerably. The most significant change is the reversal of the once-seemingly inexorable increases in the percentage of the population smoking, and with them the almost monotonic increases in per capita consumption which characterized the entirety of the century prior to the 1960s. Thus the antismoking campaign has surely reduced the amount of smoking and smoking-associated illness from that which would have been anticipated in the absence of the campaign. The reduced proportions of smokers and the shift toward low t/n cigarettes by continuing smokers combine to suggest that the rising epidemic of smoking-related illness and death may reverse direction. Nevertheless, the magnitude of the future smoking-induced disease problem will remain substantial, reflecting the "bad news" in this "good news/bad news" story: the basic health message has been delivered effectively - only a small fraction of the population questions it - but a sizable minority has not managed to eliminate the risk it perceives and generally fears. Furthermore, two factors suggest that the gap between perceived and actual risk for continuing smokers may have been growing: 1) nicotine regulation may mean that many smokers have not achieved the risk reduction they sought, and believe they have achieved, by switching from high to low t/n cigarettes; 2) the additives used by manufacturers in pursuit of a good-taste, low t/n cigarette have added a new variable to the smoking-and-health equation: do smokers of low t/n cigarettes confront new health hazards? The answer is currently unknown.

REFERENCES

Califano, J., Address delivered to the National Interagency Council on Smoking and Health, January 11 (1978).

Cannell, C., Fisher, G., and Bakker, T., "Reporting of Hospitalization in the Health Interview Survey," Vital and Health Statistics, PHS Pub. No. 1000, Ser. 2, No. 6, Washington, U.S. Government Printing Office (1965).

Cannell, C., Oksenberg, L., and Converse, J., "Striving for Response Accuracy: Experiments in New Interviewing Techniques," J. Marketing Res., 14:306-315 (1977).

Ferber, R., et al., "Validation of a National Survey of Consumer Financial Characteristics: Savings Accounts," Rev. Econ. Stat., 51:436-444 (1969).

Garfinkel, L., "Changes in the Cigarette Consumption of Smokers in Relation to Changes in Tar/Nicotine Content of Cigarettes Smoked," Am. J. Pub. Health, 69:1274-1276 (1979).

Hamilton, J., "The Demand for Cigarettes: Advertising, the Health Scare, and the Cigarette Advertising Ban," Rev. Econ. Stat., 54:401-411 (1972).

Harris, J., "Public Policy Issues in the Promotion of Less Hazardous Cigarettes," in: "Banbury Report 3: A Safe Cigarette?," G. Gori and F. Bock, eds., Cold Spring Harbor, N. Y., Cold-Spring Harbor Laboratory, pp. 333-340 (1980).

Lansing, J., and Blood, D., The Changing Travel Market, Ann Arbor, Michigan, Survey Research Center, Monograph No. 38.

Leupker, R., et al. "Saliva Thiocyanate: A Clinical Indicator of Cigarette Smoking in Adolescents," Am. J. Pub. Health (forthcoming).

National Institute of Education, Teenage Smoking: Immediate and Long Term Patterns, Washington, U.S. Government Printing Office (1979).

Parry, N., and Crossley, H., "Validity of Responses to Survey Questions," Public Opinion Quaterly, 14:61-80 (1950).

Russell, M., "The Case for Medium-Nicotine, Low-Tar, Low-Carbon Monoxide Cigarettes," in: "Banbury Report 3: A Safe Cigarette?," G. Gori and F. Bock, eds., Cold Spring Harbor, N. Y., Cold Spring Harbor Laboratory, pp. 297-308 (1980).

San Francisco Chronicle, "Smokers Keep Up a Steady Puff," February 13, p. 6 (1976).

Schacter, S., "Pharmacological and Psychological Determinants of Smoking," Annals of Internal Med., 88:104-114 (1978).

U.S. Department of Health, Education, and Welfare, Smoking and Health: Report of the Advisory Committe to the Surgeon General of the Public Health Service, Washington, U.S. Government Printing Office (1964).

——Use of Tobacco: Practices, Attitudes, Knowledge, and Beliefs, United States - Fall 1964 and Spring 1966, Washington, U.S. Government Printing Office (1969).

——Adult Use of Tobacco - 1970, Washington, U.S. Government Printing Office (1973).

——Adult Use of Tobacco - 1975, Washington, U.S. Government Printing Office (1976).

——Smoking and Health - A Report of the Surgeon General, Washington, U.S. Government Printing Office (1979).

U.S. Department of Health and Human Services, The Health Consequences of Smoking for Women, Washington, U.S. Government Printing Office (1980).

——The Health Consequences of Smoking - The Changing Cigarette, Washington, U.S. Government Printing Office (1981).

Warner, K., "The Effects of the Anti-Smoking Campaign on Cigarette Consumption, Am. J. Pub. Health, 67:645-650 (1977).

——"Possible Increases in the Underreporting of Cigarette Consumption," J. Am. Stat. Assoc., 78:314-318 (1978).

——"Clearing the Airwaves: The Cigarette Ad Ban Revisited," Pol. Analysis, 5:435-450 (1979).

——"Cigarette Smoking in the 1970's: The Impact of the Antismoking Campaign on Consumption," Science, 211:729-731 (1981).

PERCEIVED VS. ACTUAL RISKS:

THE PROBLEM OF MULTIPLE CONFOUNDING*

>Theodor D. Sterling
>
>University Research Professor
>Faculty of Interdisciplinary Studies
>Simon Fraser University
>Burnaby, B.C. V5A 1S6

A critical evaluation of a number of discussions, each of which attempts to apportion a certain amount of risk to smoking, is neither easy nor rewarding. Claims about the consequences of smoking are issues that have the potential for flaring tempers, evoking emotions, and calling forth moral convictions that convert scientific exchanges into adversarial encounters. Thus, before engaging in my analysis of the preceding papers, I note three important guideposts I will follow:

1. All criticisms which I shall voice are scientific and not personal.

2. I do not intend to speculate but lean heavily on data (and in doing that I shall bring in data, some of them new but others that ought to have been but were not referred to by the preceding speakers).

3. I insist on evaluating observations of effects to exposure to cigarette smoke within a spectrum of all the different types of microchemical environments to which we are exposed.

Perhaps the most striking observation about the papers in this session is that the word occupation has been singularly absent.

---

*Based on the discussion of presentations in the session on Smoking Cigarettes of the International Workshop on the Analysis of Actual vs. Perceived Risks, Washington, D.C., June, 1981.

Nowadays the importance of such an omission hardly needs discussion, especially when attempts are made to assess risk due to smoking. Because it is well known that smokers are subject much more than nonsmokers to occupational hazards [1, 2, 3, 4, 5, 6]. The statistical confounding between occupation, on one hand, and smoking, on the other, has now been established with sufficient research that variables related to occupation somehow must be adjusted for, or standardized, before computation of risks due to cigarette smoking may be attempted. Only now only do we know that among smokers there is a much higher proportion of blue-collar workers than among nonsmokers, a much higher proportion of workers who in their occupation are exposed to toxic fumes, dusts, and other hazards at the workplace (see especially Ref. 6). We also know that this is true for wives of smokers (whether the wives smoke or not) and for children of smokers. Thus even nonsmoking wives and children whose husbands and fathers are smokers also may be exposed to toxic dusts brought home on the hair, skin, and clothing of that member of the family. Also, occupation is a major determinant of social class with attendant differences in lifestyle, nutrition and exposure to other hazards. Thus, the occupation (and socioeconomic class) of the smoker and nonsmoker and their family are major factors for any calculus of risk.

We are able to demonstrate the effect of the statistical confounding on the evaluation of smoking risks due to occupation by a convincing example. This example was made possible by the recent work of Bonham and Wilson [7], who evaluated age adjusted rates of restricted activity days of children in homes in which there was smoking. We shall use the identical data file (the Public Use Tape of the Health Interview Survey of the National Center for Health Statistics for 1970, which includes information on respondents' occupation and smoking).

The following tables come from our study on familial effect of occupational exposure of head of household [8]. They are based on 25,647 children from 10,980 households in which there were 0, 1, or 2 smokers, and on 3,698 households with nonsmoking and nonworking women whose husbands' smoking habits were known. All households came from two occupational groups:

Group BC: These were blue-collar occupations in which the employee was likely to bring home varieties of dusts on clothing, hair, or skin. (For example: laborers, construction.)

Group PM: These were professional and managerial occupations in which the employee was not likely to be exposed to any dusts (For example: architectural, draftsman.)

The numbers of restricted activity days for acute or chronic respiratory or nonrespiratory conditions were used as indices of disease. (These indices are identical to those used by Bonham and Wilson, to which the reader is referred for further explanations.)

Table 1. Apparent Effects of Parental Smoking on Children's Health (age adjusted rates for days of restricted activity)

Smokers in household	Due to respiratory symptoms		Due to other than respiratory symptoms	
	Acute	Chronic	Acute	Chronic
No	4.16	1.32	3.97	.74
Yes	4.77	1.31	3.71	.93

Table 2. Apparent Effect of Husband's Smoking on Nonsmoking Wife's Health (age adjusted rates for days of restricted activity)

Husband smokes	Due to respiratory symptoms		Due to other than respiratory symptoms	
	Acute	Chronic	Acute	Chronic
Never	2.44	2.16	3.07	12.16
Current	4.91	.78	4.06	15.91

Table 3. Index of Confounding I. Percent of Children from PM and BC Households by Smoking in Household

Smokers in household	Percent PM	Percent BC	Percent ALL
No	47	53	100.0
Yes	31	69	100.0

Table 4. Index of Confounding II. Percent of Nonsmoking and Non-working Wvies with Husbands in PM or BC occupations by Husband's Smoking

Husband smokes	Percent PM	Percent BC	Percent ALL
Never	54	46	100.0
Current	32	68	100.0

We start with Tables 1 and 2. For some disease categories we seem to observe a sort of general relationship in which children and wives who live with heavy smokers seem to have more disease than children and wives not associated with such smokers. This association is not constant. It does not hold for restricted activity days among children for chronic respiratory and acute - other than respiratory causes, and among wives for chronic respiratory symptoms.

Tables 3 and 4 summarize some of the aspects of possible confounding providing information about index cases by smoking habit for the head of household, stratified by occupation. As expected, the proportion of cases in BC and PM households was not the same for series defined by the smoking behavior of the head of household. We note, for instance, that the percentage of children in BC households increase from 53% to 69% when we compare households with no smokers with that of smokers, while the proportion children in PM households decreases simultaneously from 47% to 31%. Table 4 shows the different distributions of nonsmoking women with husbands in PM and BC occupations, stratified by husband's smoking habits. Again we see that marked shift in blue-collar households between those where the husband smokes and where he does not.

But such shifts are the essence of misleading observations and inferences based on them by confounding a number of variables with each other. To illustrate, if we compare disease rates among nonsmoking wives of smokers to that of wives of nonsmokers, we simultaneously compare two groups in which among wives of current smokers, 32% have husbands with professional/management and 68% with blue-collar occupations, while among wives of never smokers these proportions are 54% and 46% respectively. If wives of smokers now report a higher disease rate than those of nonsmokers, that increase could well be due to a variety of differences related to socioeconomic status, but most likely to exposure to industrial dusts brought home on clothing, skin, and hair of the employed member of the family. The same conditions and arguments apply to the differences in the proportions with which children in smoking and nonsmoking households have a parent in a blue collar or a professional/management occupation. In fact, Tables 5 and 6 show in outcome of fitting a log linear model to variables likely associated to occupation [9]. The analysis is the same as reported by Bonham and Wilson. The log linear model obtains a satisfactory fit for both children and wives data. In case of the children, primary sources of effects are education of head of household, number of children in household and age of child. (Blue-collar households contain significantly more children than PM households.) Effect of occupation falls short of statistical significance but is still substantially larger than that of smoking. In case of wives, the most significant variables are husband's occupation and wife's education and age. No effect due to smoking can be observed at all.

Table 5. χ^2 Values for Independent Additive Effects by Source on Restricted Activity of the Child, by Source of Effect

Source	Degrees of freedom	χ^2
Smoking in household	1	.5
Occupation of head of household	1	2.15
Education of head of household	2	30.99
Number of childrin in household	3	29.49
Age of child	2	11.74
Residual	132	142.83

Table 6. χ^2 Values for Indepencent Additive Effects by Source on Restricted Activity of Nonsmoking Wives, by Source of Effect

Source	Degrees of freedom	χ^2
Amount smoked by husband	2	.03
Occupation of husband	1	3.18
Husband's education	2	1.20
Wife's education	2	15.35
Wife's age	4	10.28
residual	222	230.80

Thus the assessment of risk of smoking changes radically once that risk is placed in the context of variables confounded with smoking - occupation, education, age, socioeconomic status.

With that background, let me now turn to the discussions. Dr. Rush does consider the social class as a possible confounding factor (although he does nothing in his analysis to take this into account) but ignores the wealth of information linking occupation to smoking. Social class, by itself, is of course an important confounding variable. But what in social class relates to disease? Social class membership is, after all, determined by sociologists on the basis of various social indices such as source and amount of income, education, living area, etc. [10]. Disease is most often due to an effective contact with a disease causing agent, especially diseases which Dr. Rush uses for his data base. I find it more reasonable and in line with current knowledge to look to exposure of occupational dusts in the home as a hazard to children than membership in an imprecise entity such as social class. Our analysis of the HIS data support that argument. On the other hand, there are only limited data linking maternal nutritional factors to the results of the numerous perinatal mortality surveys in affluent Western societies to which Dr. Rush refers. (Which is not the same as the severe malnutrition of mother and child due to extremes of poverty in many Third World countries.) Not only are there only limited data to link assumed poor nutritional habits to "lowered social status," but I find the use Dr. Rush makes of social class to be an evaluation of a way of life rather than an assignment to a category. Nutritional deficiencies may be among the causes of perinatal mortality but to assume that "lower class" individuals suffer from such deficiencies because of different diet which is judged as unhealthy by standards of more upper social classes may be indicative of the kind of class bias which seems to be prevalent among many health scientists [11, 12]. Perhaps it is an indication of such bias that, in selecting a confounding variable, Dr. Rush chose another one that is self inflicted, to use a phrase coined by Labor leaders [13]. Food served at the table is very much an individual choice in Western countries, except for instances of bottomline poverty.

The paper by Drs. Lawrence and Paulson again draws its conclusions on disease rates observed for different levels of smoking, without paying any attention to the occupation of the smoker. I repeat that when smokers are ordered by the amount they smoke, that order is also positively related to the proportion of blue-collar workers within each one of these ordered groups. The greater the amount of smoking in a series of respondents, the greater the proportion of blue-collar workers among them [5]. To what extent the carcinogenic potential in the Lawrence and Paulson paper is due to hazardous work exposures rather than to smoking among indexed smoking group is a question open to speculation.

Also, Lawrence and Paulson refer to recent Japanese and Greek studies that reported higher risks of lung cancer for wives of smokers as compared to wives of nonsmokers. But these studies fail to consider that in developing countries, including Japan, women have been for many years exposed to emissions from kerosene stoves used for heating and cooking, that these exposures have been linked to lung cancer [14], that there appears to be a class and income difference in smoking rates, at least in Japan [15], and that the recently reported study by the American Cancer Society fails to find a significantly increased risk for lung cancer among nonsmoking wives of smoking husbands [16].

The paper by Dr. Harris again discusses mortality risks due to smoking. But the odds of dying before a given age, be it 50 or 65 or any other, are not independent of occupation, and of course of some other variables, all of which are statistically associated with smoking. Unfortunately, mortality statistics on smoking, especially those originating with life insurance companies, have consistently failed to relate mortality tables to occupation [17]. Yet evidence does exist that mortality experience differs for different occupational groups. For instance, this author reported as early as 1964 that mortality rate of coal tar workers over the age of 65 was twice that expected on the basis of standard mortality tables or of the mortality experience of males in the same communities where the steel industry was located [18].

When the odds of dying are based on "differences between the smoker's and nonsmoker's lifespans," another source of confusion is introduced.

Nonsmokers tend to be older than smokers. Thus the expected average age at death of nonsmokers is bound to be higher than that of smokers. That error in statistical reasoning was exemplified by Warren's paper about the differences in mortality among radiologists and other physicians [19] and was discussed at length that type of comparison would by now be passe. One can no more compare average age at death of smokers to nonsmokers than one can that of radiologists to other physicians and obtain anything meaningful.

One final comment on both the Harris and the Lawrence and Paulson papers. Both of them make heavy use of the U.S. Veterans study. Unfortunately, that study has been beset by a large number of errors in classification. Quite accidentally, in trying to use these data, we discovered that approximately 50,000 out of 200,000 smokers had been misclassified. This finding led to a lengthy exchange with the National Cancer Institute, an exchange which was publicly discussed by Drs. Jablon, Shapiro and this author at the Annual Meeting of the American Statistical Association in 1979 [21]. Subsequently these errors were verified by an NCI reviewer and computer programer on one category of 8632 smokers in which some 2028 were found to be in error*

---

*For reasons never made clear, the NCI review of our finding that approximately 25% of smokers had been misclassified was limited to one smoking category only - that of "Current Smokers of both Cigars and Pipes." At least we never were told of any other check or its results by NCI. I quote from the reviewer/programmer's summary [22]: "Kahn classified 8632 vertans into a category of current smokers of both cigars and pipes."

"I find that 2028 of these people are actually former, not current smokers of pipes. The error occurs for 1954 and 1957 data."

How cursory this analysis was can be seen from the fact that no effort was made to verify that 1875 current smokers of both cigars and pipes were not included in the category in which they belonged. For further details, the reader is referred to Ref. 21.

[22]. It is unfortunate that two discussions in this conference were based on data that possibly are very incorrect. Our own failure to persuade NCI to institute procedures that would have corrected the numerous errors in the U.S. Veterans data may have had the unfortunate consequences that some very good work by some very able investigators will have been done for naught.

Dr. Kozlowski's observations are indeed interesting. I might point out that in our review of occupational distribution among people with different smoking habits, we found that among smokers of unfiltered cigarettes there was a higher proportion of blue-collar workers than among smokers of filtered cigarettes [3, 4]. In general, such observations as were made by Dr. Kozlowski and by our own work would indicate the need for a better understanding of why people smoke, a topic which best may be called the sociology of smoking.

My colleague, Professor Weinkam, and I had finished a report on underreporting in cigarette smoking surveys when Dr. Warner's fine paper was published showing the same results of approximately 40% underreporting of smoking rates [23]. Thus we are familiar with Dr. Warner's data base. These data show a certain trend in per capita smoking. But these observable trends do not carry with them explanations of why they take on the shape they do. Perhaps anti smoking campaigns mounted through public and private agencies had an effect? Nevertheless, the fact that people continue to smoke despite a concerted effort to have them quite by public and private anti smoking campaigns needs to receive more mature thinking than it has in the past. Especially noteworthy is the relationship between the persistence of smoking and social class and, within social class, with certain occupations.

The paper by Pinney and Luoto is based not so much on data as on opinion. Most social scientists who are interested in the relationship between information and public policy will deny that "it is perhaps axiomatic that public health policy is heavily dependent upon a base of scientific evidence." The history of statues and legislation on environmental issues strikingly demonstrates the priority of political and economic factors over arguments adduced by health scientists.

And so I must summarize my disappointment with the papers in this session. By exhibiting conceptual blindness toward the confounding effects of occupation, the contributors have shown the force of the prevailing middle-class bias among health workers that leads toward victim blaming rather than lending itself to ferreting out social, industrial and personal factors contributing to risk. There is a need among scientists to be aware of our own biases. I suspect that the contributors to this session have failed to seek such an awareness. And such an awareness is called for not only to make assessment of risk a more accurate business. The respect for the human quality of life ought to compel life scientists to keep uppermost in mind the many sources of injury to which individuals in our society are subject, especially if these individuals are blue-collar workers who have little say about work related hazards. It appears to be almost a lack of decency if we discuss a problem which is so deeply affected by hazards in the workplace without paying due heed to these hazards.

REFERENCES

1. J. E. Dunn, G. Linden, and L. Breslow, Lung cancer mortality experience in certain occupations in California, Amer. J. Pub. Hlth., 50:1475-1487 (1960).
2. M. W. Higgins, M. Kjelsberg, and H. Metzner, Characteristics of smokers and nonsmokers in Tecumseh, Michigan, Amer. J. Epidemiol., 86:45-49 (1967).
3. T. D. Sterling and J. J. Weinkam, Smoking Characteristics by type of employment, J. Occup. Med., 18:743-753 (1976).
4. T. Sterling and J. J. Weinkam, Smoking patterns by occupation, industry, sex, and race, Arch. Environ. Hlth., 33:313-317 (1978).
5. T. D. Sterling, Does smoking kill workers or working kill smokers? or The mutual relationship between smoking, occupation, and respiratory disease, Int. J. of Hlth. Sve., 8(3):437-452 (1978).
6. G. B. Friedman, A. B. Siegelaub, and C. C. Seltzer, Cigarette smoking and exposure to occupational hazards, Amer. J. Epidemiol., 98:175-183 (1973).
7. S. Bonham and R. W. Wilson, Children's health in families with cigarette smokers, Amer. J. Pub. Hlth., 71:290-293 (1981).
8. T. Sterling and J. Weinkam, Children's and spouses' health in families of blue-collar as compared to families of professionals and managers, in process of publication. (Copy of report available by request from authors.)
9. S. J. Haberman, The Analysis of Frequency Data, Chicago, Ill., University of Chicago Press (1974).
10. T. Parson, The Social System, New York, The Free Press (1951).
11. D. E. Beauchamp, Public health as social justice, Inquiry, 13:3-14 (1976).
12. R. M. Veatch, What is just health care delivery? in: Ethics and Health Policy, R. Brauson and R. M. Veatch, eds., Cambridge, Massachusets, Ballinger Publishing Co. (1976).
13. W. Ryan, Blaming the Victim, New York, Vintage Books (1971).
14. J. S. M. Leung, Cigarette smoking, the kerosene stove, and lung cancer in Hong Kong, Brit. J. Dis. Chest, 71:273-276 (1977).
15. Statistics Bureau, Japan Statistical Yearbook, Tokyo, Prime Minister's Office (1980).
16. L. Garfinkle, Time trends in lung cancer mortality among non-smokers and a note on passive smoking, J. Nat. Cancer Inst., 66:1061-1066 (1981).

17. M. J. Cowell and B. L. Hirst, Mortality differences between smokers and nonsmokers, Worcester, Massachusets, State Mutual Life Assurance Co. of America, October 22, 1979.
18. T. D. Sterling, J. J. Phair, and J. Rustagi, New developments in chronic disease epidemiology, Amer. Indust. Hyg. Assoc. J., 23:433-446 (1962).
19. S. Warren, Longevity and causes of death from irradiation in physicians, J.A.M.A., 162:464-468 (1956).
20. R. Seltser and Ph. E. Sartwell, Ionizing radiation and longevity of physicians, J.A.M.A., 166:585-587 (1958).
21. T. D. Sterling and J. J. Weinkam, What happers when major errors are discovered long after an important report has been published? American Statistics Association Annual Meeting, Washington, D.C., August, 1979. (Copy by request from author.)
22. J. J. Bailar, Personal communication (1980).
23. K. E. Warner, Possible increases in underreporting of cigarette consumption, J.A.S.A., 73:314-318 (1978).

THE PUBLIC PERCEPTION OF RISK

D. Litai,* D. D. Lanning, and N. C. Rasmussen

Department of Nuclear Engineering
Massachusetts Institute of Technology
Cambridge, Massachusetts

I. Introduction

There has been a very rapid increase in the use of probabilistic risk analysis (PRA) in recent years. Typically these studies calculate risk as the product of probability (or more correctly frequency) and consequences. This format of course does not account for any differences in the characteristics of the risk so that direct comparison of the results of such studies for different risks can be misleading. An example would be comparison of auto fatalities with commercial airline fatalities. In the United States there are annually about 50,000 auto fatalities and about 200 commercial airline fatalities. Yet it is clear that in our society there are many more people apprehensive about commercial air travel than driving in automobiles. There may be many reasons for this, among them being a risk aversion for large-consequence events. There are of course other characteristics that lead people to feel differently about risks for which the products of frequency and consequence are the same.

Several authors [1-4] have discussed these issues and suggested risk conversion factors (RCF) for some of the perceived differences in risks. One of the most complete discussions of these factors can be found in Rowe's book "Anatomy of Risk" [1]. This paper reports on further work in this area. A wide variety of risk data for the United States has been analyzed and used as the basis for inferring a set of risk conversion factors for 8 different risk characteristics. The RCFs discussed here are based on actuarial data about risks and

---
*Currently at Israel Atomic Energy Commission, Tel Aviv, Israel.

Table 1. Risk Characteristics Included in Study

Volition	Voluntary	—	Involuntary
Severity	Ordinary	—	Catastrophic
Origin	Natural	—	Man-made
Effect manifestation	Immediate	—	Delayed
Exposure pattern	Continuous	—	Occasional
Controllability	Controllable	—	Uncontrollable
Familiarity	Old	—	New
Benefit	Clear	—	Unclear
Necessity	Necessary	—	Luxury

therefore based upon the way people have behaved rather than on any psychological analysis of how people express their beliefs. A more detailed account of the work presented here can be found in Ref. 5.

II. The Assumptions

It is evident from reviewing the literature that the inference of RCFs is a very complicated process and to make headway in this problem we have found it necessary to make some simplifying assumptions. Although these assumptions seem reasonable to us the main justification for using them is to make the analysis tractable. The reader should keep in mind that the final results are based upon these assumptions and thus one must be cautious in applying them too broadly. Let us now review these assumptions.

1. The factors that affect people's willingness to accept risk can be summarized using 9 characteristics. A review of the literature identified more than 25 characteristics proposed by different authors. Many of these were identical or very similar. The 9 characteristics used here include in our judgement most of the ideas proposed in the literature. Those left out we felt were of minor importance. The specific characteristics used are given in Table 1.

2. A dichotomous scale is used in describing the risk characteristic. Thus a risk is either immediate or delayed, catastrophic, or ordinary, etc.

3. The analysis considers only the consequence of death. This is based upon our belief that concern over being killed dominates concern over being injured. Further, most activities that have a risk of being killed also have a risk of being injured, and the ratio of deaths to injuries is of the same order for a large majority of activities.

4. Risks that involve the same factors in the same way may be directly compared.

5. Observation of human behavior may be used to identify the above factors and to quantify their relative importance.

6. By trial and error society has arrived at a balance between benefit and risk associated with an activity. Although this balancing process changes with time, by definition the current levels of risk are acceptable or the activity would be stopped.

Although these assumptions are somewhat restrictive, we have found them necessary to keep the problem manageable.

III. The Approach for Quantifying the Risk Conversion Factors

The first step in quantifying the RCF was to categorize all the identified risks by assigning to them 9 characteristics from the dichotomous scale of Table 1. Thus automobile fatalities would have the following 9 characterists: voluntary, ordinary, man-made, immediate, continuous, controllable, old, clear, and necessary. It is clear that not everyone would assign exactly these characteristics. Some may feel for example, that in our society today the car is no longer voluntary. Others may use an automobile only occasionally. However, in most cases there is a rather clear choice as to what we believe the vast majority would assign as the proper characteristic and we have used our judgement. A few cases such as homicide were not clear and were not used in developing the RCFs. An incomplete classification for some common risks is given in Table 2 to illustrate the process. Since some risks have more than one form of consequence they can appear in more than one box. For example, nulcear energy has the potential for both immediate and delayed fatalities so it appears in two boxes. It is interesting to note that since there are 9 characteristics and each has 2 possibilities there are $2^9 = 512$ possible boxes in the complete table of which Table 2 is a part. However, when all the known risks are entered many of the boxes have no entry and in fact only 40 boxes had an entry after assessing the various risks.

The next step was to take risks that differed in only one characteristic and use the ratio of these two risks to determine the RCF. To illustrate this process we use "immediate occupational" and "delayed occupational" risks of fatality. Clearly not all workers are exposed to the same risk. The data for various occupations is summarized in Tables 3 and 4 for "immediate" and "delayed" respectively. Each set of data can be approximated by a log normal distribution quite well, and since the quotient of log normal is also log normal, the division is simplified. The result is a risk aversion of a factor of 30 for immediate death with an error factor of about 10. The error factor might reasonably be interpreted as the

Table 2. (Incomplete) Classification of Some Common Risks

		Voluntary		Involuntary	
		Immediate	Delayed	Immediate	Delayed
Man-made	Catastrophic	• Aviation		• Dam failures • Chlorine release • Sabotage • Nuclear energy	• Some industrial pollution
	Ordinary	• Occupational risks • Sporting activities • Surgery • Driving cars	• Smoking • Saccharin • Occupational risks	• Aircraft crashes – people on ground	• Food additives • Pesticides • Nuclear energy (cancer) • Coal energy • Industrial pollution
Natural	Catastrophic			• Earthquakes • Hurricanes • Epidemics	
	Ordinary			• Lightning • Animal bites • Acute diseases	• Various diseases

Table 3. Occupational Risks in the U.S.A. for Immediate Fatalities (1973-76)

Risk [$\frac{\text{fatalities}}{10^5 \text{ people} \times \text{year}}$]	Exposed population (millions)	% of population at risk
<1	5	8.0
1-3.1	18	28.4
3.1-9.9	26	41.0
10-31	10.5	16.6
31.1-99	4	5.7
100-310	0.3	0.3
311-	~0.01	0.01
	65.5	100.0

Unimodal distribution skewed to the right.

Mode:	$\sim 5 \times 10^{-5}$	5% Point:	8×10^{-6}
Median:	6×10^{-5}	95% Point:	4×10^{-7}
Mean:	1.4×10^{-4}	Error Factor:	7

Table 4. Occupational Risks in the U.S.A. for Delayed Fatalities

Risk [$\frac{\text{fatalities}}{10^5 \text{ people} \times \text{year}}$]	Exposed population (millions)	% of population at risk
10-30	0.3	5
31-99	1.1	18
100-309	3.2	54
310-999	1.2	20
1000-3099	0.2	3
	6.0	100

Unimodal distribution skewed to the right.

Mode:	1.7×10^{-3}	5% Point:	3.1×10^{-4}
Median:	1.9×10^{-3}	95% Point:	9×10^{-3}
Mean:	3.0×10^{-3}	Error Factor:	6.6

Table 5. Comparisons Used in Estimating Risk Conversion Factors

Risk factor	Data used
Delayed – Immediate	Occupational Immediate – Occupational Delayed Recreation Immediate – Smoking
Necessary – Luxury	Occupational Delayed – Smoking Occupational Immediate – Recreation
Ordinary – Catastrophic	Pedestrian – Industrial Catastrophy Mining Ordinary – Mining Catastrophy Water Transport Ordinary – Water Transport Catastrophy
Uncontrollable – Controllable	People killed on ground in – Industrial Catastrophy Aircraft Catastrophy
Voluntary – Involuntary	Occupational Immediate – Pedestrian Occupational Delayed – Industrial Pollution
Natural – Man-Made	Weather Catastrophy – Industrial Catastrophy Earthquake – People killed on ground in Aircraft Catastrophy
Occasional – Continuous	Elective Surgery – Occupational Immediate
Old – New	Recreation – Oral Contraceptive
Clear – Unclear	No data applicable

Table 6. Comparison of RCF Values

RCF	Value (E. F.)				
	This study	Rowe (1)	Starr (2)	Kinchin (3)	Otway and Cohen (4)
Natural/Man-Made	20	10(2)			
Ordinary/Catastrophic	30	50			
Voluntary/Involuntary	100	100(10)	~1000		1-1000
Delayed/Immediate	30(11)	20%/yr†		30	
Controllable/Uncontrollable	5-10	100(10)			
Old/New	10				
Necessary/Luxury	1(7)				
Regular/Occasional	1				

*Where no E.F. is given, a value of ~10 may be assumed.
†Must be compounded by number of years of delay.

Table 7. Mean Values for Risk Accepted by U.S. Society for Major Risk Categories

			Controllable risk				Uncontrollable risk			
			Ordinary		Catastrophic		Ordinary		Catastrophic	
			immediate risk	delayed risk	immediate risk	delayed risk	immediate risk	delayed risk	immediate risk	delayed risk
Man-made hazard	Involuntary	Old risk	1.3×10^{-6}	4×10^{-5}	5×10^{-8}	1.5×10^{-6}	3×10^{-7}	10^{-5}	10^{-8}	3×10^{-7}
Man-made hazard	Involuntary	New risk	1.3×10^{-7}	4×10^{-6}	5×10^{-9}	1.5×10^{-7}	3×10^{-8}	10^{-6}	10^{-9}	3×10^{-8}
Man-made hazard	Voluntary	Old risk	1.3×10^{-7}	4×10^{-3}	5×10^{-6}	1.5×10^{-4}	3×10^{-5}	10^{-3}	10^{-6}	3×10^{-5}
Man-made hazard	Voluntary	New risk	1.3×10^{-5}	4×10^{-4}	5×10^{-7}	1.5×10^{-5}	3×10^{-6}	10^{-4}	10^{-7}	3×10^{-6}
Natural hazard	Involuntary	Old risk	3×10^{-5} (?)	10^{-3} (?)	10^{-6}	—	6×10^{-6} (?)	2×10^{-4} (?)	2×10^{-7} (?)	—

An error factor of ∼10 is associated with each of the quoted values.
Mortality rates are per person per year in the exposed population

range of values for various groups in society. A similar result was obtained from the compison of smoking (delayed) and immediate death from recreational activities. In many cases the data was not sufficient to obtain distributions for each risk, so the RCF was determined from the quotient of the mean values. In the limited cases where sufficient data existed it appeared that an error factor of 10 was typical so this value was assigned to all cases. Table 5 summarizes the risks used to develop the values for the RCFs.

IV. Results

In Table 6 the RCFs obtained in this study are presented and compared to those of other authors. It is interesting to note how well they agree, especially with those of Rowe [1] which were obtained by a different method. The value for RCF for benefit - clear or unclear - was not assigned. However, we found no evidence that anyone takes risks for no benefit so an RCF could not be determined.

The next step was to create Table 7. This table represents our best estimate of the average risk accepted by society in the United States today in each of the specific categories. Note that the values could be generated in one of two ways. One way would be to use the actuarial data to fill in one box and then fill in all the other boxes by using the risk conversion factor values. The other way would be to use the actuarial data to fill in all the boxes. Since the RCFs were determined from some of the data, these are not totally independent procedures. Actually both methods were used and then several iterations were made to adjust the values slightly so that the values in Table 7 are in the exact ratios implied by the RCFs but agree quite well with the actuarial data in most cases. These then are our estimates of the average risk levels accepted by society today.

It is very tempting to suggest these as the acceptable values. In fact they are a logical starting point for discussions of acceptable risk levels. However, several factors should be kept in mind. The first is that these are average values and many groups accept substantially lower risks. As noted, the error factor of about 10 on these numbers is really not the uncertainty of the mean value but rather an expression of the range of risk values accepted by different groups in society. In setting any broad safety goals it would seem that regulatory bodies may well wish to consider these preferences of risk averse groups in society. A second issue is that the accepted levels are not static in time. As societies become more affluent they can and do expend more resources on risk aversion. In addition, as risks are better understood, better methods to reduce them become available. Thus the numbers in Table 7, derived mainly from the decade of the 1970s for the United States, should not be considered as fixed values for all time.

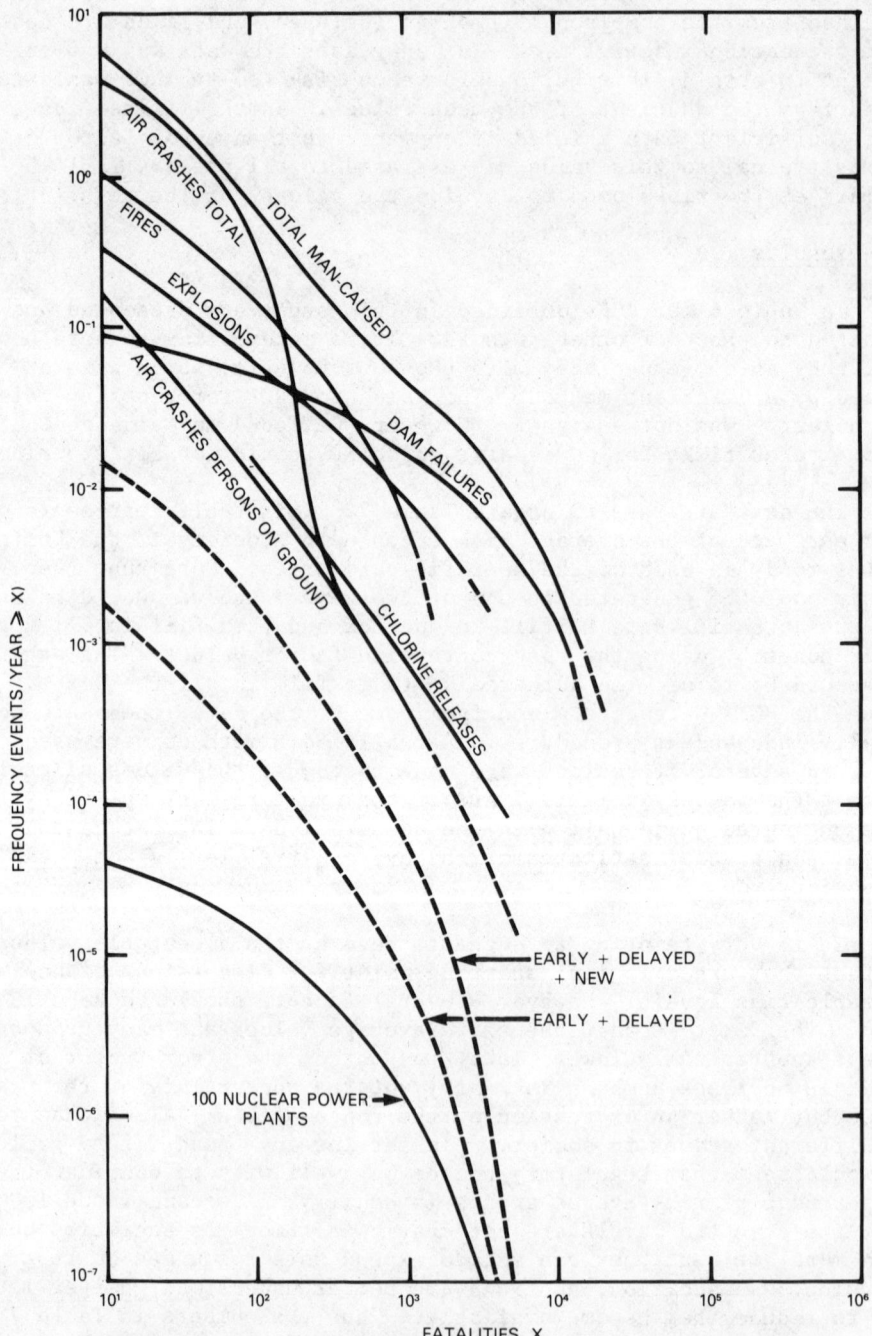

Fig. 1. Comparison of early fatalities of nuclear versus other man-caused risks (solid lines from WASH-1200 (6); dashed curves added (see text).

PUBLIC PERCEPTION OF RISK   223

There is a second way the RCFs might be used. To illustrate this we use the results from the Reactor Safety Study [6], also known as WASH-1400. One result of that study was the comparison of early fatalities from potential nuclear plant accidents to early fatalities from other sources of potentially large accidents. These results are represented by the solid lines on Fig. 1. One of the criticisms of this figure was that large nuclear accidents are expected to have a significant number of delayed fatalities as a result of increases in the cancer rate which will occur 10 to 40 years after the accident. Although these cancers were calculated in the WASH-1400 study, they are of course not included in the early fatalities curve. Although the other curves in the comparison do not include any delayed fatalities either, with the possible exception of chlorine releases, they would be expected to have very few. Thus the curves for the other accidents are a good representation of the total risk, whereas the nuclear curve represents only a part of the nuclear risk.

At the time of the WASH-1400 study this problem was recognized but there was no generally accepted way for combining early and delayed fatalities. If we use the RCF of 30 for "delayed" vs "early," then one should add the "delayed" fatalities divided by 30 to the "early" as was done to produce the dashed curve marked "early + delayed." Further, if you believe the upper curves are old risks and "nuclear" a new one, the public response would penalize "nuclear" by another factor of 10 as shown on the curve "early + delayed + new." If the risk conversion factors correctly quantify the perceived characteristics of nuclear power, then the upper dashed curve would represent the way the average citizen would view nuclear risks relative to the other risks. Since nuclear risks are not part of the data base there is no way to be sure from our analysis that these assumptions are actually correct. However, if true they would imply that nuclear is close enough so that the error factor would make it overlap with the lower of the other risks. Thus, one should expect to find that some members of the public would not feel it is safe enough, while the majority should believe nuclear is safe enough. This seems to be the case and although it by no means proves that the RCFs are correct, they are at least consistent with the apparent public reaction to reactor safety.

V. Conclusion

Although the above results are based on some restrictive assumptions and a limited amount of data, nevertheless we feel that they form an interesting starting point for some new insights into the difficult problem of how to compare risks that have different characteristics. Today almost all regulatory agencies are facing the difficult question of "how safe is safe enough?". Although risk conversion factors may be able to play a useful role in setting safety criteria, it must also be recognized that RCFs as developed here are only a part of what needs to be considered. The RCFs involve only risk comparisons, whereas a safety criterion must surely include not only risk but cost and benefit as well.

REFERENCES

1. Rowe, W. D., An Anatomy of Risk, New York, Wiley (1977).
2. Starr, C., Social Benefit vs. Technological Risk, Science, 165: 1232-38 (1969).
3. Kinchin, G. H., Assessments of Hazards in Engineering Work, Proc. Instn. Civ. Engrs., 64:431-38 (1978).
4. Otway, H. J., and Cohen, J. J., Revealed Preferences: Comments on the Starr Benefit-Risk Relationships, Vienna, International Institute for Applied System Analysis (1975).
5. Litai, D., A Risk Comparison Methodology for the Assessment of Acceptable Risk, Ph.D. Thesis, Massachusetts Institute of Technology, Cambridge, Massachusetts (1980).
6. Nuclear Regulatory Commission, Reactor Safety Study, WASH-1400, Washington, D.C., Government Printing Office (1975).

IMPACT OF THE THREE MILE ISLAND ACCIDENT AS PERCEIVED

BY THOSE LIVING IN THE SURROUNDING COMMUNITY

Anne D. Trunk* and Edward V. Trunk†

Pennsylvania State University
The Capitol Campus
Middletown, Pennsylvania 17057

The accident at Three Mile Island Unit 2 on March 28, 1979, presented the world with a unique historical example for study. It illustrated how people react to what the press labeled a "nuclear disease." This was a "disaster" without noise or evidence of damage. The TMI community could not see or feel or sense anything other than what was transmitted in the form of news reports. Under such circumstances people seem to react to the flavor (upbeat, downbeat) or pattern (relaxing, worsening) of the news reports and particularly on the basis of preconceptions of the risks involved. On March 30th, when approximately one-third of Middletown left the area, many were unable to utilize a logical thought process in arriving at a course of action. Still today, two years after the accident, many residents have mixed feelings about nuclear energy and TMI. It is difficult to accept the fact that off-the-island, for all practical purposes, the accident had no effect. Even more difficult to accept is the conclusion that had the accident progressed further than it did (melt down) the containment would still not have been breeched. There would have been no "China Syndrome."

The authors live in Middletown within the critical 5-mile region of TMI, with six children (one preschool), and remained in the area until the Accident was officially over. Through personal sources of information and a close watch on news reports it was possible to evaluate the seriousness of the accident. Those that remained could witness events and receive opinions as neighbors left and returned. Many opinion surveys and impact studies have been conducted since

---

*President's Commission on the Accident at Three Mile Island.
†Division of Science, Engineering & Technology, Pennsylvania State University.

the accident. Fortunately, the large majority of the community have moved in the forward direction. There are many improvements and signs of progress, but some people are still mentally upset and feed on any news that may be alarming. At the moment, these are actively working against all proposals to cleanup Unit 2 and restart Unit 1.

ATTITUDES PRIOR TO THE ACCIDENT

Three Mile Island is located near two small cities, Middletown (12,000 people) and Goldsboro (475 people). The intermediate townships are low density farmland. Three medium size cities, Harrisburg, York, and Lancaster, are located outside a 10 mile radius.

Electric service to this area is divided between Metropolitan Edison Company and Pennsylvania Power and Light Company. Met Ed services the Middletown area. Like most utilities, the early generating stations it built used coal (plus one small hydroelectric plant). The 116 MW Crawford Station in Middletown was dirty. In response to complaints and regulatory pressures, Met Ed converted the low pressure boilers to oil (1970). About that time oil prices began rising and utilities were looking to nuclear energy for cleaner and lower cost power generation. Met Ed planned for two nuclear units on Three Mile Island with a combined capacity of 15 Crawford Station (1700 MW). Unit 1 went on line on September 2, 1974 and Unit 2 followed suit on December 30, 1978. Meanwhile, Crawford Station was retired in 1977 after 55 years of service. It had become too expensive to operate on oil and the coal portion of the plant was still not clean enough for EPA Standards.

No one in the community missed these historical events, but they caused no great concern. The community's unhappiness about coal dust layers on anything left outdoors had been eased by the conversion to oil. Construction of the nuclear station made news whenever a milestone was reached, such as the shipment overland of the reactor shell. The many license hearings were also publicized and the public was aware that there was an organized movement that contested each stage of the process. Where these people came from, no one seemed to know. Most of the arguments pertained to safety and the fact that TMI was located along the flight path for Harrisburg International Airport. Safety systems were designed with back up controls and the public was assured that the containment building was designed to withstand direct impact of a jet liner. The positive "nothing can go wrong" attitude displayed later by the utility was a direct result of these hearings. That seems to be one of the side effects of the system.

The utility made an attempt to sell the nuclear energy concept to the public by holding an open house for the community during construction. They also constructed an observation tower with visitor's center near the island where one could see films and models of the plant, and obtain educational literature. Guided tours were available during the construction stage. The public obviously received a false understanding that an accident as occurred with TMI 2 would not happen. What the utility was trying to communicate was that the plant was designed to withstand any conceivable eventuality (accident). But the utility obviously did not dwell on accidents. So it is easy to see that this misunderstanding could arise.

We enter the accident period with a public that has a false sense of security and a utility that is labeled "cocky." Residents adjacent to the island are unhappy with the noise of construction and indiscriminate use of the plant's bull horn at all hours. In general, the community welcomes this "clean" generating station and the wealth it brings.

THE ACCIDENT

Looking back, there were obviously some fears about atomic energy that did not surface until after the accident. The accident focused attention on the many risks that residents had been living with since 1974 when Unit 1 started up. It was the manner in which the crisis developed and the manner in which the news was transmitted to the public that had most to do with shaping present attitudes towards the plant. The details of the accident and the role of the media have been thoroughly covered elsewhere [1, 2].

Wednesday, March 28, passed as a minor event. Yet this was the true day of crisis. There was a slight radioactive gas release. The televised CBS Evening News labeled it deadly radiation.

By Thursday the news media started assembling in Middletown. The plant had a second radioactive gas release and began discharging waste water into the river. Both releases were announced to the public beforehand. The plant was stable.

On Friday, the crisis off the island started. A third radioactive gas release, misinterpretation of radiation levels, rumors spread by the media and by word of mouth involving possible evacuation, all gave a false impression that the plant situation had worsened. The Governor's advisory at 12:30 p.m. for pregnant women and preschool-age children to leave the 5-mile radius area capped the rumors. The schools in this area closed. Residents were frightened by the appearance of a new crisis. Most residents examined the facts and weighed the Gornernor's words in the spirt of "excess of caution" and stayed. About one-third of the residents of Middletown left.

Those who remained behind endured a sustained environment of crisis and witnessed how the media was shaping the minds of people. Eight examples follow.

1. The hydrogen bubble crisis was a predictable phenomenon. It could not explode for lack of oxygen, yet the public was held in fear for five days. The main concern most had was that this gas bubble might be slowing down the cooling rate. Yet one could not miss the media's preoccupation with an explosion.

2. The press conferences were an education in how not to communicate. The prepared statements were generally directed at progress in bringing the reactor to cold shut down. The questions were generally directed to determine what the worst possible outcome might be. The conferences were shouting matches and much of the data and units (person-rem, etc.) were not understood. Neither was the nomenclature of the plant understood and that the dissipation of decay heat after shutdown is a normal procedure. All this served to heighten tension. Had the news media sent their science reporters rather than their feature story reporters the situation might have been different.

3. Meltdown or "China Syndrome" became keywords of fear. Any melting had already occurred on Wednesday, no one knows how much. There was a preoccupation with guessing how much and an impression that this was still a danger. The movie, The China Syndrome, could not have been worse timed for the TMI accident. The Presidential Commission Study showed that the containment could not have been breeched even with a complete melting of the core.*

4. A nuclear reactor cannot explode like an atomic bomb. Yet this likelihood continued to enter discussions and reports. Why? We know today that people still equate nuclear reactors with atomic bombs.

5. Opinions were also shaped by wilful distortion of the news. Phoney "For Sale" signs were placed in front of homes and photographed. A staged empty street gave the impression that Middletown was a ghost town. Selective interviews with frightened people were highlighted on TV. Confident people interviews were edited out.

6. Two local medical doctors were advising their patients that if they were nauseous or had stomach aches or sore throates, there was a good chance they had radiation poisoning.

7. The accident presented the opportunity for so called "experts" to vent their views. Typically, a news report would present opposing views, both given by "authorities" on the subject.

---

*This was not included in the final report.

The way the reports were presented, it was difficult for the public to differentiate between fact and opinion. Antinuclear sentiment began to dominate the news.

8. The local environment gave one the impression that a crisis still existed even during the following week: schools remained closed, a curfew existed, the news was saturated with TMI, and the area was overflowing with reporters from all over the world.

Those who left and returned after the crisis ended endured a greater psychological stress than those who remained. It was noted particularly in the reaction of their children. The trauma of leaving one's possessions with the possibility of never returning left a deep impression. For example, there was short run on banks to empty savings accounts. Upon returning, most displayed a hostility toward the utility, the NRC, and even toward those who did not leave.

The accident ended with the utility struggling to clean up the damage and the community with divided feelings. The anti-nukes were saying, "I told you so." The pro-nukes were saying, "The system worked."

IMPACT OF THE ACCIDENT - AS PERCEIVED

The news remained saturated with TMI material for a period of time following the accident. Several anti-nuclear groups surfaced in the Harrisburg-Lancaster area and took advantage of the media interest. The prophets of doom gave residents something to be afraid of. The tourist industry (Lancaster Amish Country and Hershey) was expected to collapse. The radiated area would keep tourists away. The farming industry was also expected to fold up. Radiated crops would not be salable and farm animals would have to be slaughtered. The bottom was predicted to drop out of the real estate market. A gradual exodus of the population was to occur and "For Sale" signs would be everywhere. Recreational fishing in the Susquehanna river would be dead for many years into the future. Capital for expansion and new business would be rechanneled into other areas. The subjects of personal health and infant mortality were much more sensitive. Anti-nuke "experts" are still predicting an outbreak of cancer cases within the next 20 years. It was also predicted that in 1979 the infant mortality rate would radically increase. You know that when a person is told to be afraid often enough, he finally becomes afraid. These predictions represented the impact of the accident as perceived by many in the TMI area. Even when presented with facts to the contrary, they still cling to a negative outlook.

A continuing lack of education about radiation produces fear of the unknown. When a tiny amount of radiation in our local milk supply was discovered, 12 picocuries per liter, this was immediately reported in the press as a cause for concern. Much or our milk supply was dumped in the interest of caution. But no one seemed to realize that a picocurie is an insignificant radiation level, several thousand times less than an amount to justify precautionary measures. As a matter of interest, the natural amounts of radioactive potassium and carbon present in a normal human being would require that person, if he were a dead laboratory animal, to be disposed of as nuclear waste! Even though we are now using nuclear energy for the benefit of society, we remain uneducated and fearful of it. No wonder! In examining television coverage of nuclear energy for the 10-year period prior to the accident, the Media Institute detected fear as a recurring theme. The Institute's report also reviewed TV coverage during the accident and details a case for sensationalism and anti-nuclear bias [3].

The first opinion survey immediately after the accident was conducted by the Social Research Center of Elizabethtown College [4]. It found that 62% of the respondents supported "the use of nuclear power as a source of energy for our nation." Males were more supportive than females. Those that left were three times as likely to consider moving as those who remained. After half felt that they had "not been told the truth about the situation."

A more extensive survey conducted for the NRC showed some shifting attitudes with time [5]. The drawn out clean up process had a negative effect. There was concern about emissions from the plant during cleanup. Four percent reported they would move. Over one-third changed their opinions about nuclear power. Close to the plant, the opinion was evenly split. Outside of a 40-mile radius the opinion was about two to one that the advantages of nuclear power outweight the disadvantages. Sixty percent of the respondents felt the accident would hurt the economy of the area and about 50% felt that the TMI plant's disadvantages outweighed its advantages.

During 1979, the TMI issue was still too emotional for public evaluation and discussion. Twon meetings and forums were disrupted by well organized anti-nuclear demonstrations. One particular fire house meeting in Middletown was so boisterous it made National headlines. When the citizens of Middletown realized that outsiders were giving them a bad name, a counter movement was organized, "Friends and Family of TMI." This group has sponsored monthly meetings to educate the public and for the first time residents are learning to separate knowledge and emotion.

IMPACT OF THE ACCIDENT - AS MEASURED

The perceived impact of the accident was quite the opposite of the actual impact as reported in various studies released in 1980. The Greater Harrisburg Board of Realtor's report covered a 20-mile radius of TMI [6]. Despite the serious mortgage money crunch, high interest rates, and gasoline shortage, both the number of units sold (up 6.2%) and the total dollar value (up to 16.1%) increased in 1979 over 1978. This was better than the national average. An NRC study concluded the following statements [7].

"The presumption was made frequently by those at a distance from the plant site that real estate values would plummet, that tourism and agriculture would be adversely affected, and that the entire economic future of the area would be in question. Yet in the vicinity of the plant, real estate transactions continued to take place, dairy products were produced and sold, visitors came to have their pictures taken against the background of the Three Mile Island cooling towers, and industrial developments continued to move forward. A conspicuous characteristic of the post-accident environment was the discrepancy between the presumed severity of impact suggested by persons with little direct familiarity with conditions in the area, and the absence of continuing effects alleged by many living in the area."

Two prominent motels have been constructed in this area since the accident (Marriott and Sheraton) and Middletown gained a new supermarket (Giant Foods). The Governor's final impact report concluded that the area's business firms "have no reservations on expanding their business in the aftermath of the accident and do not see the reactivation of both TMI reactors as having an impact on their operations" [8]. The major effect found was the higher electric rates for Met Ed's customers. While tourism was off in June, 1979, the data for the entire year showed growth over similar 1978 figures. The concensus of the industry attributed the early summer depression more to factors such as the gasoline shortage, the polio scare and national economic conditions, rather than TMI. As a point of interest, the TMI visitors' center has received over 122,000 people in the past two years since the accident. This rivals the nine year total prior to 1979. Also, over 10,000 have taken tours of the island facilities. TMI has become a sightseer's stopping point.

An NRC recreational fishing impact study detected no measurable radioactivity in Susquehanna river fishes [9]. "The post accident depression in monthly fishing harvest indices and a nearly full recovery by July followed the same general trend as the perception of threat and concern with emission... ." Most of the anglers who fish near TMI are local residents and were sensitive to environmental factors. The release of 4000 gallons of water during late July was widely publicized by the media. This produced numerous inquiries from anglers concerned about the safe consumption of river fish. As a result, the harvest indices dropped in August and rose again in September. The report concluded that the brief depression in recreational fishing was a result of the anglers' perception of threat rather than any measurable threat from TMI. Investigations of reported plant and animal health effects were unable to establish the nuclear plant as the cause [10]. Most of the animal problems analyzed were traced to nongenetic factors, such as malnutrition and infection. Bird counts in the area have shown the bird population to be stable. All vegetation stress found was attributable to natural causes. The question of health effects on people in the TMI area cannot be resolved with satisfaction to all because of potential long terms effects of low level radiation. The Presidential Commission report concluded the radiation releases were so small that health effects would not be measurable [1]. Potential cancer cases had to be reported in fractions of a person. Infant mortality is a current topic of debate. Infant deaths rose after the accident but do did births. As a matter of fact, 1979 Pennsylvania births reached a seven-year high. When deaths are reported statistically as deaths per thousand births, the TMI area rate for the six month period following the accident was 15.7, a decrease from 17.2 for the previous six months. Mortality figures fluctuate widely. Despite Dr. Sternglass' contention to the contrary, the figures do not indicate that the accident increased infant mortality. If anything, we had an increase in births.

SUMMARY

The foregoing has pointed out that the public had a false sense of security about TMI. The plant was designed to withstand an accident. What the public heard was that an accident could not happen. A lack of knowledge and misconceptions about nuclear reactors contributed to an interpretation that the accident would have a devastating effecting on such things as real-estate, tourism, farming, infant mortality, health, business and jobs. This perception changed as data was reported. The community is alive and prospering. Some still have strong apprehensions. The drawn out cleanup process does not help. The large majority have learned to accept the presence of TMI and are assured that the impact has been more psychological than real.

REFERENCES

1. J. G. Kemeny, B. Babbitt, P. E. Haggerty, C. Lewis, P. A. Marks, C. B. Marrett, L. McBride, H. C. McPherson, R. W. Peterson, T. H. Pigford, T. B. Taylor, and A. D. Trunk, Report of the President's Commission on the Accident at Three Mile Island, U.S. Government Printing Office, Washington, D.C. (1979).
2. A. D. Trunk and E. V. Trunk, Three Mile Island: A Resident's Perspective, Annals of the New York Academy of Sciences, Vol. 365, New York, N.Y. (1981).
3. The Media Institute, Television Evening News Covers Nuclear Energy, Washington, D.C. (1979).
4. D. B. Kraybill, Three Mile Island: Local Residents Speak Out: A Public Opinion Poll, Elizabethtown College, Elizabethtown, Pennsylvania (1979).

5. C. B. Flynn, Three Mile Island Telephone Survey: Preliminary Report on Procedures and Findings, U.S. Nuclear Regulatory Commission, Washington, D.C. (1979).
6. D. P. Shearer, Three Mile Island Nuclear Accident Community Impact Study on Real Estate, Greater Harrisburg Board of Realtors, Harrisburg, Pennsyalvania (1980).
7. C. B. Flynn and J. A. Chalmers, The Social and Economic Effects of the Accident at Three Mile Island: Findings to Date, U.S. Nuclear Regulatory Commission, Washington, D.C. (1979).
8. W. H. Plosila, The Socio-Economic Impacts of the Three Mile Island Accident: Final Report, Governor's Office of Policy and Planning, Commonwealth of Pennsylvania, Harrisburg, Pennsylvania (1980).
9. C. R. Hickey, Jr., Impact of the 1979 Accident at Three Mile Island Nuclear Station on Recreational Fishing in the Susquehanna River, U.S. Nuclear Regulatory Commission, Washington, D.C. (1979).
10. G. E. Gears, G. LaRoche, J. Cable, B. Jaroslow, and D. Smith, Investigations of Reported Plant and Animal Health Effects in the Three Mile Island Area, U.S. Nuclear Regulatory Commission, Washington, D.C. (1980).

"THE PUBLIC" VS. "THE EXPERTS":

PERCEIVED VS. ACTUAL DISAGREEMENTS ABOUT RISKS OF NUCLEAR POWER*

        Baruch Fischhoff,† Paul Slovic,†
        and Sarah Lichtenstein

        Decision Research
        A Branch of Perceptronics
        1201 Oak Street
        Eugene, Oregon 97401

A recent public opinion survey (Harris, 1980) reported the following three results:

a) Among four "leadership groups" (top corporate executives investors/lenders, Congressional representatives and federal regulators), 94-98% of all respondents agreed with the statement "even in areas in which the actual level of risk may have decreased in the past 20 years, our society is significantly more aware of risk."

b) Between 87% and 91% of those four leadership groups felt that "the mood of the country regarding risk" will have a substantial or moderate impact "on investment decisions - that is, the allocation of capital in our society in the decade ahead." (The remainder believed that it would have a minimal impact, no impact at all, or were not sure.)

c) No such consensus was found, however, when these groups were asked about the appropriateness of this concern about risk. A majority of the top corporate executives and a plurality of lenders believed that "American society is overly sensitive to risk," whereas

---

*This research was supported by the National Science Foundation under award PRA-8116925 to Perceptronics, Inc.
†Correspondence may be addressed to Baruch Fischhoff at Decision Research, 1201 Oak St., Eugene, Oregon 97401

a large majority of Congressional representatives and federal regulators believed that "we are becoming more aware of risk and taking realistic precautions." A sample of the public endorsed the latter statement over the former by 78% - 15%.

In summary, there is great agreement that risk decisions will have a major role in shaping our society's future and that those decisions will, in turn, be shaped by public perceptions of risk. There is, however, much disagreement about the appropriateness of those perceptions. Some believe the public to be wise; others do not. These contrary beliefs imply rather different roles for public involvement in risk management. As a result, the way in which this disagreement is resolved will affect not only the fate of particular technologies, but also the fate of our society and its social organization.

The views about risk perceptions given by the respondents to this poll, like those offered by other commentators on the contemporary scene, are, at best, based on intense, but unsystematic observation. At worst, they represent attempts to bias the political process by promulgating self-serving beliefs. Such happens, for example, when one claims that people are so poorly informed (and uneducable) that they require paternalistic institutions to defend them or that they would be better off surrendering some of their political rights to technical experts. It also happens, at the other extreme, when one claims that people are so well informed (and offered such freedom of choice) that they can fend for themselves in the marketplace and need no governmental protection.

Like speculations about chemical reactions, speculations about human nature need to be disciplined by fact. To that end, various investigators have been studying how and how well people think about risks. Although the results of that research are not definitive as yet, they do clearly indicate that a careful diagnosis is needed whenever "the public" and the "the experts" appear to disagree. It is seldom adequate to attribute all such discrepancies as reflecting public misperceptions. From a factual perspective, that assumption is often wrong; from a societal perspective, it is generally corrosive by encouraging disrespect between the parties involved. When the available research data do not allow one to make a confident diagnosis, a sounder assumption is that there is some method in anyone's apparent madness. The present essay suggests some ways to find that method. Specifically, it offers six reasons why disagreements between the public and the experts need not be interpreted as clashes between actual and perceived risks.*

---

*Fuller expositions of the research upon which this summary is based may be found in sources such as Fischhoff, Slovic, and Lichtenstein (1980; 1982), Green (1982), Slovic, Fischhoff, and Lichtenstein (1980), Vlek and Stallen (1980), and Warner and Slater (1981).

Reason 1: The Distinction between "Actual" and "Perceived" Risks is Misconceived

Although there are actual risks, nobody knows what they are. All that anyone does know about risks can be classified as perceptions. Those assertions that are typically called "actual risks" (or "facts" or "objective information") inevitably contain some element of judgment on the part of the scientists who produce them.* The element is most minimal when judgment is needed only to assess the competence of a particular study conducted within an established paradigm. It grows as one needs to integrate results from diverse studies or to extrapolate results from a domain in which they are readily obtainable to another domain in which they are really needed (e.g., from animal studies to human effects). Judgment becomes all when there are no (credible) available data, yet a policy decision requires that some assessment of a particular fact be made.

The expert opinions that comprise the scientific literature are typically considered to be "objective" in two senses, neither of which can ever be achieved absolutely and neither of which is the exclusive province of technical experts. One meaning of objectivity is reproducibility: one expert should be able to repeat another's study, review another's protocol, reanalyze another's data, or recap another's literature summary and reach the same conclusions about the size of an effect. Clearly, as the role of judgment increases in any of these operations, the results become increasingly subjective. Typically, one would expect reproducibility to decrease (and subjectivity to increase) to the extent that a problem attracts scientists with diverse training or to the extent that the field entrusted with a problem has yet to reach a consensus on basic issues of methodology.

The second sense of "objectivity" means immunity to any influence by value considerations. One's interpretations of data should not be biased by one's political views or pecuniary interests. Applied sciences naturally have developed great sensitivity to such problems and are able to invoke some penalties for detected violations. There is, however, little possibility of "regulating" the ways in which values influence other acts, such as one's choice of topics to study or ignore. Some of these choices might be socially sanctioned, in the sense that one's values are widely shared (e.g., deciding to study cancer because it is an important problem); other choices might be more personal (e.g., not studying an issue because

---

*From this perspective, the title of this conference, "The Analysis of Actual vs. Perceived Risk," is a misnomer. A more accurate, and more clumsy, title would be "The Analysis of Risks as Perceived by Ranking Scientists within Their Field of Expertise vs. as Perceived by Anybody Else."

one's employer does not wish to have a troublesome data base created on that topic). Although a commitment to separating issues of fact from issues of value is a fundamental aspect of intellectual hygiene, a complete separation is never possible (Bazelon, 1979; Fischhoff et al., 1981; Sjöberg, 1979).

At times, this separation is not even desired - that happens when experts are asked for (or volunteer) their views on how risks should be managed. Because they mix questions of fact and value, such views might be better thought of as the opinions of experts rather than as expert opinions, a term that should be reserved for expressions of substantive expertise. Often the reasons for eliciting such opinions are obscure. It would seem as though members of the public are the experts when it comes to striking the appropriate tradeoffs between costs, risks, and benefits. That expertise is best tapped by surveys, hearings, and political campaigns (Hammond and Adelman, 1976; Mazur, 1981).

Of course, there is no all-purpose public any more than there are all-purpose experts. The ideal expert on a matter of fact has studied that particular issue and is capable of rendering a properly qualified opinion in a form useful to decision makers. Using the same criteria for selecting value experts might lead one to philosophers, politicians, psychologists, sociologists, clergy, intervenors, pundits, shareholders, or bystanders, depending upon how these criteria were interpreted. Thus, one must ask, "in what sense," whenever someone says, "expert" or "public" (Schnaiburg, 1980; Thompson, 1980). We will use "expert" in the restrictive sense and "public" or "laypeople" to refer to every one else, including scientists in their private lives.

Reason 2: Laypeople and Experts Are Talking Different Languages

Explicit risk analyses are a fairly new addition to the repertoire of intellectual enterprises. As a result, the risk experts are only beginning to reach consensus on terminology and methodology. Their communications to the public are only beginning to express some coherent perspective and to help the public sort out the variety of meanings that "risk" could have (Crouch and Wilson, 1981). Experimental studies (Slovic, Fischhoff, and Lichtenstein, 1979; 1980) have indicated that when expert risk assessors are asked to assess the "risk" of a technology on an undefined scale, they tend to respond with numbers that approximate the number of recorded or estimated fatalities in a typical year. When asked to estimate "average year fatalities," laypeople produce fairly similar numbers. When asked to assess "risk," however, laypeople produce quite different responses. These estimates seem to be an amalgam of their average year fatality judgments, along with their appraisal of other features, such as a technology's catastrophic potential or the equity with which its risks are distributed. These catastrophic potential judgments match those of the experts in some cases, but differ in others (e.g., nuclear power).

On semantic grounds, words can mean whatever a population group wants them to mean, as long as that usage is consistent and does not obscure important substantive differences. On policy grounds, the choice of a definition is a political question regarding what a society should be concerned about when dealing with "risk." Whether we attach special importance to potential catastrophic losses of life or convert such losses to expected annual fatalities (i.e., by multiplying the potential loss by its annual probability of occurrence) and add them to the routine toll is a value question - as would be a decision to weight those routine losses equally rather than giving added weight to losses among the young (or among the non-beneficiaries from a technology).

For other concepts that recur in risk discussions, the question of what they do or should mean is considerably murkier. It is often argued, for example, that different standards of stringency should apply to voluntarily and involuntarily incurred risks (e.g., Starr, 1969). Hence, for example, skiing could (or should) legitimately be a more hazardous enterprise than living below a major dam. Although there is general agreement among experts and laypeople about the voluntariness of food preservatives and skiing, other technologies are more problematic (Fischhoff et al., 1978; Slovic et al., 1980). We have found considerable disagreement within expert and lay groups in their ratings of the voluntariness of technologies such as prescription antibiotics, commercial aviation, hand guns, and home applicances. These disagreements may reflect differences in the reference groups considered; for example, use of commerical aviation may be voluntary for vacationers, but involuntary for certain business people and scientists. Or they may reflect disagreements about the nature of society or the meaning of the term. For example, each decision to ride in a car may be voluntarily undertaking and may, in principle, be foregone (i.e., by not traveling or by using an alternative mode of transportation); but in a modern industrial society, these alternatives may be somewhat fictitious. Indeed, in some social and professional sets, the decision to ski may have an involuntary aspect. Even if one makes a clearly volitional decision, some of the risks that one assumes voluntarily may be indirectly and involuntarily imposed on one's family or the society that must pick up the pieces (e.g., pay for hospitalization due to skiing accidents).

Such definitional problems are not restricted to subjective terms such as voluntary. Even a technical term such as "exposure" may be consensually defined for some hazards (e.g., medical x-rays) but not for others (e.g., handguns). In such cases, the disagreements within expert and lay groups may be as large as those between

them. For debate to proceed, one needs some generally accepted
definition for each important term - or at least a good translating
dictionary. For debate to be useful, one needs an explicit analysis
of whether each concept, so defined, makes a sensible basis for
policy. Once they have been repeated often enough, ideas such as
the importance of voluntariness or catastrophic potential tend to
assume a life of their own. It does not go without saying that
society should set a double standard on the basis of voluntariness
or catastrophic potential, however they are defined.

Reason 3: Laypeople and Experts Are Solving
Different Problems

Many debates turn on whether the risk associated with a particular configuration of a technology is acceptable. Research (Slovic, Fischhoff, and Lichtenstein, 1981) has found substantial disagreements not only between people belonging to different population groups, but also within groups when the question is posed in different ways. Although these disagreements may be interpreted as reflecting conflicting social values or confused individual values, closer examination suggests that the acceptable-risk question itself may be poorly formulated.

To be precise, one does not accept risks. One accepts options that entail some level of risk among their consequences. Whenever the decision-making process has considered benefits or other (non-risk) costs, the most acceptable option need not be the one with the least risk. Indeed, one might choose (or accept) the option with the highest risk if it had enough compensating benefits. The attractiveness of an option depends upon its full set of relevant positive and negative consequences (Fischhoff et al., 1981).

In this light, the term "acceptable risk" is ill-defined, without specifying the options and consequences to be considered. Once options and consequences are specified, "acceptable risk" might be used to denote the risk associated with the most acceptable alternative. When using that designation, it may be quite difficult to remember how context dependent it is. That is, people may disagree about the "acceptability of risks" not only because they disagree on how to evaluate the consequences (i.e., they have different values), but also because they disagree about what consequences and options are to be considered.

A number of well-known policy debates might be speculatively attributed, at least in part, to differing conceptions of what the set of possible options is. For example, the risks (or possible risks) of saccharin may look unacceptable when compared with the risks of (the option of) life without sweeteners. They may however, seem more palatable when the only alternative option considered is another sweetener that appears to be more costly and more risky. Or, nuclear power may seem acceptable when compared with alternative sources of generating electricity (with their risks and costs), but not so acceptable when aggressive conservation is added to the option set. Technical people from the nuclear industry seem to prefer the narrower definitions of the problem, perhaps because they like the light it casts on their energy source, perhaps because they prefer to concentrate on the kinds of solutions most within their domain of expertise. Citizens involved in energy debates may feel themselves less narrowly bound; they may also be more comfortable with solutions such as conservation that require their kind of expertise (Bickerstaffe and Pearce, 1980).

People who agree about the facts and share common values may still disagree about the acceptability of risks because they have different notions about which of those values are relevant to a particular decision problem. All parties may think that equity is a good thing in general without also agreeing that energy policy is the proper arena for resolving inequities. For example, some may feel that both those new inequities caused by a technology and those old ones endemic to a society are best handled separately (e.g., through the courts or with incomes policies).

Thus, when laypeople and experts disagree about the acceptability of a risk, one must always consider the possibility that they are addressing different problems, with different sets of alternatives or a different set of relevant consequences. Assuming that each group has a full understanding of the implications of its favored problem definition, the choice between definitions is a political question. When the public's definition is adopted in whole or in part, then this aspect of public perceptions has been accommodated in the decision-making process without any specific component of that process being labeled as such (Stallen, 1980).

Reason 4: Debates over Substance May Disguise Battles over Form - and Vice Versa

In most political arenas, the conclusion of one battle often sets some of the initial conditions of its successor. Insofar as risk management decisions are shaping the economic and political future of a country, they are too important to be left to risk managers (Wynne, 1980). When people from outside the risk community enter into risk battles, they may try to master the technical details, or they may concentrate on monitoring and shaping the risk management process itself. The latter strategy may exploit their political expertise and keep them from being outclassed (or mislead) on technical issues. As a result, their concern about the magnitude of a risk may emerge in the form of carping about the way it is studied. They may be quick to criticize any risk assessment that does not have such features as eager peer review, ready acknowledge-

ment of uncertainty, or easily accessible documentation. Even if
those features are consonant with good research, scientists may resent being told by laypeople how to conduct their business even more
than they resent being told by novices what various risks really are.

Lay activists' critiques of the risk assessment process may be
no less irritating, but somewhat less readily ignored, when they
focus on the way in which scientists' agendas are set. As veteran
protagonists in hazard management struggles know, without scientific
information, it may be hard to arouse and sustain concern about an
issue, to allay inappropriate fears, or to achieve enough certainty
to justify any action. However, information is, by and large, created only if someone has a (professional, political, or economic)
use for it. Thus, we may know something only if someone in a position to decide feels that it is worth knowing. Doern (1978) proposed that lack of interest in the fate of workers is responsible
for the lack of research on the risks of uranium mining; Neyman
(1979) wondered whether the special concern over radiation hazards
has restricted the study of chemical carcinogens; Commoner (1979)
accused oil interests of preventing the research that could establish solar power as a viable energy option. In some situations,
knowledge is so specialized that all relevant experts may be in the
employ of a technology's promoters, leaving no one competent to discover troublesome facts (Gamble, 1978). Whether the cause is fads
or finances, failure to study particular topics can thwart particular parties and may lead them to impugn the scientific process.

At the other extreme, debates about political processes may
underlie disputes that are ostensibly about scientific facts. As
mentioned earlier, the definition of an acceptable-risk problem
circumscribes the set of relevant facts, consequences and options.
This agenda setting is often so powerful that a decision has effectively been made once the definition is set. Indeed, the official
definition of a problem may preclude one from advancing one's point
of view in a balanced fashion. Consider, for example, an individual
who is opposed to increased energy consumption but is only asked
about which energy source to adopt. The answers to these narrow
questions provide a de facto answer to the broader question of
growth. Such an individual may have little choice but to fight
dirty, engaging in unconstructive criticism, poking holes in analyses supporting other positions, or ridiculing opponents who adhere
to the more narrow definition. This apparently irrational behavior
can be attributed to the rational pursuit of officially unreasonable
objectives.

Another source of deliberately unreasonable behavior arises
when participants in technology debates are in it for the fight.
Many approaches to determining acceptable-risk levels (e.g., cost-benefit analyses) make the political-ideological assumption that
our society is sufficiently cohesive and common-goaled that its

problems can be resolved by reason and without struggle. Although such a "get on with business" orientation will be pleasing to many, it will not satisfy all. For those who do not believe that society is in a fine-tuning stage, a technique that fails to mobilize public consciousness and involvement has little to recommend it. Their strategy may involve a calculated attack on what they interpret as narrowly defined rationality.

A variant on this theme occurs when participants will accept any process as long as it does not lead to a decision. Delay, per se, may be the goal of those who wish to preserve some status quo. These may include environmentalists who do not want a project to be begun or industrialists who do not want to be regulated. An effective way of thwarting practical decisions is to insist on the highest standards of scientific rigor.

Reason 5: Laypeople and Experts Disagree about What Is Feasible

Laypeople are often berated for misdirecting their efforts when they choose risk issues on which to focus their energies. However, a more careful diagnosis can often suggest a number of defensible strategies for setting priorities. For example, Zentner (1979) criticizes the public becuase its rate of concern about cancer (as measured by newspaper coverage) is increasing faster than the cancer rate. One reasonable explanation for this pattern is that people may believe that too little concern has been given to cancer in the past (e.g., our concern for acute hazards like traffic safety and infectious disease allowed cancer to creep up on us). A second is that people may realize that some forms of cancer are the only major causes of death whose rates are increasing.

Systematic observation and questioning are, of course, needed to tell whether these speculations are accurate (and whether the assumption of rationality holds in this particular case). False positives in divining people's underlying rationality can be as deleterious as false negatives. Erroneously assuming that they understand an issue may deny them a needed education; erroneously assuming that they do not understand may deny them a needed hearing. Pending systematic studies, these error rates are likely to be determined largely by the rationalist or emotionalist cast of one's view of human nature.

In lieu of data about specific cases, perhaps the most reasonable general assumption is that people's investment in problems is determined by their feelings of personal efficacy. That is, they do not get involved unless they feel that they can make a difference, personally or collectively. In this light, their decision-making process is dominated by a concern that is known to dominate other psychological processes: perceived feelings of control (Seligman,

1975). As a result, people will deliberately ignore major problems
if they see no possibility of effective action; some reasons why they
might reject a change of "misplaced priorities" when they neglect a
hazard that poses a large risk:

   a) the hazard is needed and has no substitutes;

   b) the hazard is needed and has only riskier substitutes;

   c) no feasible scientific study can yield a sufficiently clear
      and incontrovertible signal to legitimate action;

   d) the hazard is distributed naturally, hence cannot be con-
      trolled;

   e) no one else is worried about the risk in question, hence,
      no one will heed messages of danger or be relieved by evi-
      dence of safety;

   f) no one is empowered to or able to act on the basis of evi-
      dence about risk.

   Thus, the problems that actively concern people need not be
those whose resolution they feel should rank highest on society's
priorities. For example, one may acknowledge that the expected
deaths from automobile accidents over the next century are far
greater than those expected from nuclear power, yet still be active
only in fighting nuclear power out of the conviction that "Here, I
can make a difference. This industry is on the ropes now. It's
important to move in for the kill before it becomes as indispensible
to American society as automobile transportation."

   Where the priorities of experts and laypeople differ, it may
not reflect disagreements about the size of risks, but differing
opinions on what can be done about them. At times, the technical
knowledge or can-do perspective of the experts may lead them to see
a broader range of feasible actions. At other times, laypeople may
feel that they can exercise the political clout needed to make some
options happen, whereas the experts feel constrained to doing what
they are paid for. In still other cases, both groups may be silent
about very large problems because they see no options. That might
be the most charitable explanation of the relative silence (in 1981)
of scienctists and citizens regarding the threat of nuclear war.

Reason 6: Laypeople and Experts See the Facts Differently

   There are, of course, situations in which disputes between laypeople and experts cannot be traced to disagreements about objectivity, terminology, problem definitions, process or feasibility. Having eliminated those possibilities, one may assume the two groups

really do see the facts of the matter differently. Given that laypeople and experts are talking about the same thing, it may be useful to distinguish between two situations: those in which laypeople have no source of information other than the experts, and those in which they do have such sources. The reasonableness of disagreements and the attendant policy implications look quite different in each case.

How might laypeople have no source of information other than the experts, yet come to see the facts differently? One way is for the experts' message not to get through intact, perhaps because: a) the experts are unconcerned about disseminating their knowledge or hesitant to do so because of its tentative nature; b) only a biased portion of the experts' information gets out, particularly when the selection has been influenced by those interested in creating a particular impression; c) the message gets garbled in transmission, perhaps due to ill-informed or sensationalist journalists; d) the message gets garbled upon reception, either because it was poorly explicated or because the recipients lacked the technical basis for understanding it (Friedman, 1981; Hanley, 1980; Nelkin, 1977).*

A second way of going astray is to misinterpret not the substance, but the process of science. For example, unless an observer has reason to believe otherwise, it might seem sensible to assume that the amount of scientific attention paid to a risk is a good measure of its importance. Science can, however, be more complicated than that, with researchers going where the contracts, limelight, blue ribbon panels, or juicy controversies are. In that light (and in hindsight), science may have done a disservice to public understanding by the excessive attention it paid to saccharin. A second aspect of the scientific process that may cause confusion is its frequently disputatious nature. It may be all too easy for observers to feel that "if the experts can't agree, my guess may be as good as theirs" (Handler, 1980). Or, they may feel justified in picking the expert of their choice, perhaps on spurious grounds, such as assertiveness, eloquence, or political views. Indeed, we suspect that it is seldom the case that the distribution of lay opinions on as issue does not overlap at least a portion of the distribution of expert opinions. At the other extreme, laypeople may be baffled by the veil of qualifications that scientists often cast over their work. All too often, audiences may be swayed more by two-fisted debators (eager to make definitive statements) than by two-handed scientists (saying "on the one hand X, but on the other hand Y," in an effort to achieve balance).

---

*For example, Lord Rothschild (1978) has noted that the BBC does not like to trouble its listeners with the confidence intervals surrounding technical estimates.

In each of these cases, the misunderstanding is excusable, in the sense that it need not reflect poorly on the intelligence of the public or on its ability to govern itself. It, however, would seem hard to justify using the public's view of the facts instead of or in addition to the experts' view. A more reasonable strategy would seem to be attempts at education. These attempts would be distinguished from attempts at propaganda by allowing for two-way communication, that is, by being open to the possibility that even when laypeople appear misinformed, they may still have some defensible reason for seeing things differently than do the experts.

For laypeople to disagree reasonably, they would have to have some independent source of knowledge. What might that be? One possibility is that they have a better overview on scientific debates than do the active participants. Laypeople may see the full range of expert opinions and hesitations, immune to the temptations or pressures that actual debators might feel to fall into one camp and to discredit skeptics' opinions. In addition, laypeople may not feel bound by the generally accepted assumptions about the nature of the world and the validity of methodologies that every discipline adopts in order to go about its business. They may have been around long enough to note that many of the confident scientific beliefs of yesterday are confidently rejected today (Frankel, 1974). Such lay skepticism would suggest expanding the confidence intervals around the experts' best guess at the size of the risks.

Finally, there are situations in which the public, as a result of its life experiences, is privy to information that has escaped the experts (Brokensha, Warren, and Werner, 1980). To take three examples: 1) The MacKenzie Valley Pipeline (or Berger) Inquiry discovered that natives of the far north knew things about the risks created by ice-pack movement and sea-bed scouring that were unknown to the pipeline's planners (Gamble, 1978). 2) Post-accident analyses often reveal that the operators of machines were aware of problems that the designers of those machines had missed (Sheridan, 1980). 3) Scientists may shy away from studying behavioral or psychological effects (e.g., dizziness, tension) that are hard to measure, yet still are quite apparent to the individuals who suffer from them. In such cases, lay perceptions of risk should influence the experts' risk estimates.

CONCLUSIONS

There are many reasons for laypeople and experts to disagree. These include misunderstanding, miscommunication, and misinformation. Discerning the causes underlying a particular disagreement requires a combination of a) careful thought, to clarify just what is being talked about and whether agreement is possible given the disputants' differing frames of reference, and b) careful research, to clarify just what it is that the various parties know and believe. Once the situation has been clarified, the underlying problem can be diagnosed as calling for a scientific, educational, semantic, or political solution.

The most difficult situations will be those in which the participants cannot agree on what the problem is (and have no recourse to an institution that will resolve the question by arbitration or by fiat), and those in which education is called for, yet fails (after some reasonable, diligent effort). Policy makers then face the hard choice either of going against their own better judgment by using the public's assessment of risk (in which they do not believe) or of going against the public's feelings by imposing policies that will be disliked. Such policies may seem overly cautious (e.g., motorcycle helmet laws - to some people) or insufficiently cautious (e.g., nuclear power - to some people). When fears are ignored, the result can be stress or psychosomatic effects, which can be as real in their impact as they are illusory in their source. When strong public opinions are ignored, the result can be hostility, mistrust, and alienation. Since a society does more than manage risks, the policy maker must consider whether the social benefits to be gained by optimizing the allocation of resources in a particular decision is greater than the social costs of overriding a concerned public. A pessimistic view on "going with the public" might argue that "it only encourages the forces of irrationality (indirectly giving credence to astrology, superstition, and the like)." An optimistic view might be that risk questions are going to be with us for a long time. For a society to deal with them wisely, it must learn about their subtleties, including how appearances can be deceiving. One way of learning is by trial and error. Often, the experts will be able to say "we told you so. It would have been better to listen to us." In other cases, they may be surprised. Learning is possible as long as some basic respect remains between teacher and pupil. That respect may be one of a society's greatest assets.

REFERENCES

Bazelon, D. L., Risk and responsibility, Science, 205:277-280 (1979).
Bickerstaffe, J., and Peace, D., Can there be a consensus on nuclear power?, Social Studies of Science, 10:309-344 (1980).
Brokensha, D. W., Warren, D. M., and Werner, O., Indigenous knowledge: Systems and development, Lanham, M.D., University Press of America (1980).
Commoner, B., The politics of energy, New York, Knopf (1979).
Crouch, E. A. C., and Wilson, R., Risk analysis, Cambridge, Massachusetts, Ballinger (1981).
Doern, G. B., Science and technology in the nuclear regulatory process: The case of Canadian uranium miners, Canadian Public Administration, 21:51-82 (1978).

Fischhoff, B., Lichtenstein, S., Slovic, P., Derby, S., and Keeney, R., Acceptable risk, N.Y., Cambridge University Press, (1981).
Fischhoff, B., Slovic, P., and Lichtenstein, S., Knowing what you want: Measuring labile values, in: Cognitive processes in choice and decision behavior, T. Wallsten, ed., Hillsdale, New Jersey, Erlbaum (1980).
Fischhoff, B., Slovic, P., and Lichtenstein, S., Lay foibles and expert fables in judgments about risk, American Statistician, 36:240-255 (1982).
Fischhoff, B., Slovic, P., Lichtenstein, S., Read, S., and Combs, B., How safe is safe enough? A psychometric study of attitudes towards technological risks and benefits, Policy Sciences, 8: 127-152 (1978).
Frankel, C., The rights of nature, in: When values conflict, C. Schelling, J. Voss, and L. Tribe, eds., Cambridge, Massachusetts, Ballinger (1974).
Friedman, S. M. Blueprint for breakdown: Three Mile Island and the media before the accident, Journal of Communication, 31:116-129 (1981).
Gamble, D. J., The Berger inquiry: An impact assessment process, Science, 199:946-951 (1978).
Green, C. H., Risk: Attitudes and beliefs, in: Behavior in fires, D. V. Canter, ed., Chichester, Wiley, (1982).
Hammond, K. R., and Adelman, L., Science, values, and human judgment, Science, 194:389-396 (1976).
Handler, P., Public doubts about science, Science, 208:1093 (1980).
Hanley, J., The silence of scientists, Chemical and Engineering News, 58:(12), 5 (1980).
Harris, L., Risk in a complex society, Public opinion survey conducted for Marsh and McLennan Companies, Inc. (1980).
Mazur, A., The dynamics of technical controversy, Washington, D.C., Communications Press (1981).
Nelkin, D., Technological decisions and democracy, Beverly Hills, California, Sage (1977).
Rothschild, N. M., Rothschild: An antidote to panic, Nature, 276: 555 (1978).
Schnaiburg, A., The environment: From surplus to scarcity, New York, Oxford University Press (1980).
Seligman, M. E. P., Helplessness, San Francisco, Freeman (1975).
Sheridan, T. B., Human error in nuclear power plants, Technology Review, 82:(4), 23-33 (1980).
Sjöberg, L., Strength of belief and risk, Policy Sciences, 11:539-573 (1979).
Slovic, P., Fischhoff, B., and Lichtenstein, S., Rating the risks, Environment, 21:(30, 14-20, 36-39 (1979).
Slovic, P., Fischhoff, B., and Lichtenstein, S., Facts vs. fears: Understanding perceived risk, in: Societal risk assessment: How safe is safe enought?, R. Schwing and W. A. Albers, Jr., eds., New York, Plenum (1980).

Slovic, P., Lichtenstein, S., and Fischhoff, B., Characterizing perceived risk, in: Technological hazard management, R. W. Kates and C. Hohenemser, eds., Cambridge, Massachusetts, Oelgeschlager, Gunn & Hain, in press.

Stallen, P. J., Risk of science or science of risk?, in: Society, Technology, and Risk Assessment, J. Conrad, ed., London, Academic Press (1980).

Starr, C., Societal benefit versus technological risk, Science, 165:1232-1238 (1969).

Thompson, M., Aesthetics of risk: Culture or context, in: Societal Risk Assessment, R. C. Schwing and W. A. Albers, Jr., eds., New York, Plenum (1980).

Vlek, C., and Stallen, P. J., Rational and personal aspects of risk, Acta Psychologica, 45:273-300 (1980).

Warner, F., and Slater, D.H., The assessment and perception of risk, London, The Royal Society (1981).

Wynne, B., Technology, risk and participation, in: Society, Technology, and Risk Assessment, J. Conrad, ed., London, Academic Press (1980).

Zentner, R. D., Hazards in the chemical industry, Chemical and Engineering News, 57:(45), 25-27, 30-34 (1979).

COPING WITH NUCLEAR POWER RISKS:

A NATIONAL STRATEGY*

> Chauncey Starr
>
> Electric Power Research Institute
>
> Palo Alto, California

In the energy field, national policy centers about two basic issues that require societal balancing. The first is the determination of the size of the societal investment allocated to providing a future energy supply, as compared with the potential social value of anticipated future energy uses.

Societal investment here includes both the usual tangible economic resources and such typical intangible social costs as those of environmental improvement, risk avoidance, and increased national security. Difficult as it is to evaluate such intangibles, and to explicitly include them as part of a societal investment, such evaluations are always an implicit part of the national decision process.

The second issue is planning the mix of supply alternatives, each of which has a different set of benefits and disbenefits. These two issues are not wholly independent of each other, but traditionally have been treated so.

Balancing these current and long-term benefits and disbenefits is, then, the problem of all national decision making in the energy field. Our criterion for this balance must not be solely to achieve a low economic cost energy supply, traditional as this goal had been. Rather, it is to achieve our social objectives at the lowest social costs (including rists), as measured by our social values and taking

---

*Condensed from "Risk Criteria for Nuclear Power Plants: A Pragmatic Proposal," C. Starr; ANS/ENS International Conference, Washington, D.C., Nov. 16/21, 1980 (to be published).

into account the future as well as the present. Thus, for example, a currently higher economic cost indigenous supply, such as shale oil, would provide increased national security and help conserve other fossil fuels for a future need; and thus may have a favorable net social cost.

But the real world is not kind to the analyst. We do not understand all the important physical aspects of any of the energy sysstems. And any society is a contradictory mix of group values and objectives. Further, we understand only vaguely the feedbacks and interactions in our societal network, particularly those parts dependent on our energy systems.

And, finally, the important societal decisions are rarely left to technocrats. Final decisions are usually in the hands of our nontechnical leadership - those who control our major social institutions, whether political, industrial, or intellectual. We must, therefore, consider social benefits and disbenefits in their terms, rather than from a technical professional perspective.

In this context, the question of risk policy for nuclear power plants must be addressed in the framework of a total social benefit and cost perspective that is not fully defined, and which may always lack a unanimous consensus in our pluralistic and politically varying society. We can, nevertheless, emphasize those aspects that may be important in our final balance, particularly where the social values involved are culturally self-evident.

What are such important social values? To illustrate these, let us start with a global view and proceed to the individual nuclear plant. The growing worldwide need for energy is such that the future of many nations and the very subsistence of millions of people depend crucially on an assured energy supply. Nuclear power is one of the new, foreseeable, and sure energy sources available to many nations, and with the breeder, it could eventually be fuel supplyindependent. So, a good part of the world is clearly choosing to go nuclear now, because it represents the optimum choice among the long-range alternatives available to them.

The benefits of nuclear are clear. For the world, prime values are avoidance of fuel resource wars between the military powers, maintaining the stability of the developed countries, and increasing the stability of developing nations. To the energy-hungry nations, the values are even clearer. Electricity is being successfully produced from about three hundred worldwide nuclear plants (180 GWe); its operational reliability is about the same as that of other available alternatives; its environmental impact is observably much less than that of the fossil fuels - no air pollution or CO_2 increement -; and there has been no observable physical harm to the public. And finally, where full-scale operation has provided an economic comparison, it is clearly a lower-cost source of electricity than coal and very much less than electricity based on imported oil and gas.

What is its observable disbenefits? First, nuclear power is high technology, very capital-intensive, and demanding on the quality of operation and infrastructure support. It requires a large front-end investment of capital and professional skills. Both are in short supply worldwide, and, therefore, their selective allocation to nuclear power constrains other economic development. Second, nuclear operations involve highly radioactive substances, so that occupational or long-term public hazards can result from improper, unskilled, or careless handling. Third, reliability and economics of performance require the highest level of continuous maintenance. The Three Mile Island accident was an extreme example of this last risk.

I have not included in this listing of observable disbenefits the issue of public risk, because it belongs in the category of hypothetical risks for which no significant empirical information exists. To this hypothetical list we should add the risks of permanent waste storage, the potential for a marginal increase in nuclear weapons proliferation, and the possibility of acts of terrorism that might be byproducts of the power system. Coping with these risks is a generic issue of preparing for projected hypothetical events likely to be unpredictable and very rare, but also with the potential for extensive consequences. We do not have a national philosophy to quide us in the management of such risks.

These hypothetical risks are the most troublesome to evaluate. On the one hand, we have had about 6000 research and power reactor-years and thirty time-years of experience in handling all portions of the nuclear reactor cycle, with no verifiable occurrence of a public risk from nuclear stations, or uncontrollable high-level wastes, or the clandestine diversion of nuclear power plant fuel to weapons purposes, or of the use of nuclear installations for terrorist purposes. On the other hand, a variety of scenarios can be imagined which might lead to such outcomes. The issue, then, becomes one of estimating their plausibility, and this has been done on other occasions. To most of us matured in the nuclear arts, these hypothetical risks appear quite small, based on our own multi-decade cumulative professional experience and analyses. Nevertheless, these are the areas in which the public anxiety has been greatly stimulated by presentations of the frightening possibilities of the extreme "what if" scenarios.

Given a worldwide and a national policy to deploy nuclear power, the question still remains, "How safe is safe enough?", for those impacted by their proximity to nuclear plants. Let us return to the balance of benefits, costs, and risks. If we accept the international and national benefits of additional energy sources, including nuclear power, then each U.S. citizen shares in these benefits. The fact that the whole nation receives these benefits, but only a small group in the locality of the nuclear station has a potential risk exposure, does raise an issue of equity. In several countries this has been addressed by providing the residents around nuclear plants with reduced electricity rates or reduced local taxes as a balancing financial benefit. The empirical social evidence is that such compensation does go a long way to creating local acceptance.

Given such a local compensating benefit, is it possible to find a risk level that would be accepted locally? I believe the answer is yes, because the benefits to the group potentially at risk can be made credible to them - reduced potential of a major resource war; reduction in international instabilities (for example, Middle East conflicts, revolutions in Chile, Africa, etc.); an improved economy in the U.S.; increased public health and welfare; and financial compensation for their proximity to a station. The benefits are credible, and common sense tells us they are worth some risk. This, then, leaves us with the question of determining an appropriate upper bound to this risk, specifically for those regionally adjacent to nuclear facilities and in accord with the above benefits.

Reducing the public risk by reasonable technical means has been the subject of most professional studies and government agency reviews. Their major purpose has been assurance that the events and sequences studied in detail have not overlooked some rare circumstance that would significantly increase the probability of a major accident and consequent risk. Such evaluations are inherently uncertain, because in the absence of empirical information, the feasibility and quantification of any suggested sequence must be based on professional judgment. As a result, such evaluations have been very conservative. There are recent analyses of the experience of the past decades which suggest that even in an extreme case, the public exposure would be limited to the region close to the plant and is likely to be much less than previously estimated. Overall, the preponderance of professional technical opinion is that the nuclear risk is low compared to the public risks of major alternatives, even if there is a wide range of uncertainty in its quantification.

Assuming our ability to project a performance domain, how, then, should we establish a criterion for an acceptable risk level? There are several approaches. The least conservative is to set the acceptable upper bound of the range of potential nuclear risks just below the aggregate risks from all other normal activities, so that the proximity of nuclear power would not significantly alter the group's statistical risks. Thus, for example, the actuarial insurance premiums for life and health coverage would not be higher for the residents near a nuclear plant.

The most conservative is to reduce nuclear risk to a theoretically negligible level. In may early papers on risk, it was estimated that a worldwide minimum risk to anyone from uncontrollable natural accidents was about one death per year per million population. It was suggested then that an acceptable upper bound of potential nuclear risk is arbitrarily set at one-tenth of this level (10^{-7} per person per year). This would place nuclear risks at the "negligible" level. The size of compensating benefits plays no role in arriving at this decision.

A broader public health approach is to seek the most cost-effective mixture of expenditures on improving the health of the group involved. Each cause of accident or health reduction has some funds spent on its reduction. If these sums, including those spent on the nuclear sector for this purpose, are totaled, then the national pool of health-oriented funds might be reallocated among all the causes to achieve their most effective use in improving public health. Thus we may find it worthwhile to spend less on nuclear plant risks and, for example, more on emergency ambulance service, prenatal care, water system purification, or smog reduction. This equalization of the marginal effectiveness of each dollar spent on health would be the most effective way to use our total resources. Evaluations of aggregated regional risks suggest that such an approach would raise the acceptable risk level about 1000-10,000 times higher than the negligible level suggested in the first method above. This approach presumes that nuclear risks would be acceptable at the present "medium" to "low" level. From an overall public health view, a policy to equalize the marginal cost-effectiveness of safety investment is undoubtedly the most beneficial for the society. Further, because this approach is free of ideology implications, it could be rationalized on a public health basis alone.

These suggested approaches have as their objective the establishment of a performance target with an upper limit of public risk, presumably to be used as a criterion for regulatory purposes. Inherent in the question of a risk criterion is the implied premise that the nuclear utilities and their vendors might not sufficiently minimize public risk without such a regulatory overview and measure. The validity of this premise merits some examination.

As was shown at TMI, a plant can be seriously disabled without any public risk. The spectrum of financially serious disabling events without public risk will always be such that their integrated cost to the utility may be roughly 10-100 times that of the cost of a publicly significant core melt.*

We have, then, a situation in which an informed utility industry should be seeking reliability and safety levels better than the

---

*This is discussed in the companion paper,"Coping with Nuclear Power Risks: The Electric Utility Incentives," Starr and Whipple, presented at this meeting.

"negligible" level for the public. Certainly, a self-regulating system is more desirable than an imposed surveillance, so we should find a way to utilize this financial incentive effectively. A full awareness in the nuclear power industry of the economic value of achieving a high level of performance reliability and safety would, of course, do this.

CONCLUSIONS

Combining these concepts of nuclear risk, the following is suggested as a means of coping with the public perceptions, professional projections, and utility performance needs. First, the design goals for the plants should be sufficiently low in public risk that any individual potential risk exposure should be a small fraction of the minimum statistical risk due to uncontrollable natural hazards (storm, lightning, etc.). Such a "negligible" risk design target would simultaneously meet the utility need for financial reliability, and it would be in the utility's interest to operate so as to achieve or better this goal.

Second, the resident population near a nuclear station should be given a compensatory benefit sufficient to justify accepting a risk equivalent to the normal risk of living. This is likely to be a 1000-10,000 times greater risk than the design target. This range is much greater than we believe is needed to cover the uncertainties of engineering design predictions.

This approach would appear to satisfy the national policy objectives of both targeting design for a negligible risk and providing a basis for local acceptance. It also meets the professional need for a margin of uncertainty. If we are ever to guide the Nation to a sensible course, we should provide a professional consensus on the guidelines. This may be such a pragmatic course. The need for the Nuclear Regulatory Commission (NRC) to establish such a sound and understandable risk policy should not be underrated. In the past, the NRC has been less then reassuring to an anxious public. Its response to such events as the TMI accident has created an image of panic and bewilderment. It has reacted to TMI like a bureaucratic urban police department asking for more regulations and policemen to contain crime. The difference, of course, is that crime pays off for the criminal, but unsafe nuclear operation benefits no one.

It is important that the NRC undertakes to reassure the public that nuclear risks are manageable and acceptable, and that it is cooperating with the utilities to make it so. This is especially needed when the expressed fears of the public are likely to be based on the highly publicized imagery created by hypothetical "what if" scenarios, rather than on a body of observable experiences or professional evaluations. The case studies comparing such public views with available professional analyses do not instill confidence in the soundness of the public's perceptions of new and complex risks. Considering the long lifetime of energy systems, and the relatively short lifetime of political credos and popular concepts, we should provide a more enduring and professionally credible basis for the balance of benefits, costs, and risks needed to guide national policy.

HEALTH IMPACT OF TOXIC WASTES:

ESTIMATION OF RISK

>Renate D. Kimbrough
>
>Centers for Disease Control
>Public Health Service
>U.S. Department of Health and Human Services
>Atlanta, Georgia

Toxic waste disposal has been very haphazard in the past. Only recently has it been recognized that certain chemicals may persist for many years, that they may migrate, and that drums containing them eventually corrode. Toxic wastes have been and are still being handled in a number of ways. They may be temporarily or permanently stored in controlled and uncontrolled landfills, salt mines, in warehouses, mixed into salvage oil, dumped at night on private property or along road sides, dumped into rivers, lakes and the ocean, treated in special lagoons or ponds with chemicals, bacteria and/or ultraviolet light to degrade them (Johnson, 1977; Dunphy and Hall, 1978). They may be discharged into sewage treatment plants or they may be retained by the toxic waste generator on his property, in which case they are either burned or stored in drums above or below the ground.

In the more recent past, toxic waste treatment plants have been built where chemical waste is incinerated (Dunphy and Hall, 1978).

These different situations present different potential problems and different risk situations.

Occupational exposure may occur when such wastes are handled, remedial actions are taken or warehouses containing chemical wastes catch on fire.

The general public may be exposed directly or indirectly. For instance, in 1971, a salvage oil dealer picked up extremely toxic chemical waste, mixed it with salvage oil and sprayed it on three riding arenas for dust control. Many birds, cats, dogs, and horses died and several people suffered illness (Carter et al., 1975).

If persistent chemicals are discharged into rivers, they will be biomagnified in the food chain and contaminate human food sources, mainly wildlife and fish. For many of our poor, particularly in the southeastern part of the United States, fish and wildlife are a major source of protein (Kreiss, et al., in press). Since fish and wildlife are caught by individuals, it is not known how much polluted fish and game are actually consumed in the United States or any other country. In many instances, it is not possible to determine whether people living around chemical dumps have actually been exposed. Love Canal is a case in point. Most of the chemicals that were determined in the air of the houses and in the soil to which people might have been exposed were solvents which are rapidly metabolized in the body and excreted. They can, therefore, only be detected in people immediately after exposure. The only persistent chemicals at Love Canal that would accumulate in humans were Lindane (γ-hexachlorocyclohexane) and chlorinated dibenzodioxins. These were found in a few soil samples, basement sumps and remote areas like the storm sewer system. Analysis for chlorinated dibenzodioxins are very difficult and Lindane has not been found in elevated levels in human cases where monitoring was done. However, human monitoring was very limited. When the Centers for Disease Control was asked to conduct a study in the summer of 1980, it was not possible to determine what the exposures might have been in the past. The canal had been capped with a clay cap and a trench had been built around the canal. Drainage from the canal is now collected in the trench and treated in a waste treatment plant built on the site. The people who had lived in the first two rings of houses and presumably had had the highest exposure had been moved out in 1978. Thus, it was not possible to reconstruct exposure and make any risk assessment. Since weather may greatly influence chemical movements from dumps, environmental measurements made in 1978, for instance, can give no information about past exposures. In addition, little is presently known about the effects of multiple chemical exposures, nor is it known whether chemicals measured in soil at Love Canal are actually bioavailable to humans.

In handling chemical wastes, it must be remembered that they may be corrosive, flammable, explosive, and radioactive in addition to being acutely or chronically toxic. Furthermore, a multitude of chemicals are usually involved. To estimate the real and potential risk of stored wastes is a very complex problem. The information needed to attempt any risk assessment is given in Table 1.

As the questions in Table 1 illustrate, each toxic waste situation will have to be evaluated on its own merits. If the amount of toxic waste is limited (a few drums), this can usually be taken care of fairly easily unless supertoxic chemicals are involved. If a warehouse contains primarily nitroglycerin and nitro-cellulose, for example, or many flammable solvents, the danger of a fire or an explosion is very real, and should be of foremost concern. In some cases, acute toxicity may be of utmost concern. However, many chem-

Table 1. Information Needed for Assessment of Risk to Human Health
         by Toxic Chemical Wastes

---

1. How much waste is there?
2. How is the waste presently contained?
3. What is the waste composed of?
4. How likely is it that people will get exposed?
5. By what route and how would people get exposed?
6. Is there evidence of migration or spread of the toxic wastes?
7. What toxicity information is available on the chemicals present in the waste?
8. At what concentrations would people most likely be exposed to these chemicals?
9. Do we know whether the conglomeration of chemicals in the toxic waste would have additive, synergistic, potentiating or antagonistic effects?

---

icals may not be very toxic on an acute basis, but may have chronic effects, such as permanent paralysis or cancer. Other concerns are fetotoxicity, teratogenesis, and mutagenesis. If information about the effects of the chemicals in toxic waste is available from animal studies, such data can be utilized to estimate what the risk might be if humans get exposed to particular chemicals. If human data are available, this should, of course, also be considered. For animal toxicity data, extrapolations as they are made for food additives and pesticides for which tolerance levels are set (Complicance Policy Guides Manual, 1973, and Code of Fed. Reg. 21, 1980) could be used. If drinking water is contaminated, water standards can be used. In such cases, the no-effect level of a lifetime study is taken (NOEL) of the most sensitive species if that is available and a 100-fold safety factor built in.

Time weighted averages or threshold level values (A.C.G.I.H., 1980) could be used as a baseline to decide what a dangerous air level might be. It must be remembered that these air levels are for workers who are on 7-8 hour shifts and are not exposed for 24 hours every day. Levels would have to be lowered to protect the infirm, infants and those who have special susceptibility because of their genetic makeup (Beutler, 1972; Cleaver, 1968).

The situation is quite different if we are dealing with mutagens or carcinogens. There are now many chemicals which have been shown to either be carcinogenic or mutagenic in experimental systems

(Christensen and Fairchild, eds., 1976). For only a few of these chemicals does sufficient evidence exist that they have also caused cancer in humans (Wolff and Oehme, 1974) and even less information is available for human mutagens. The reason that many of these chemicals have not been shown to cause cancer in humans may be that it is difficult to demonstrate carcinogenic properties of chemicals in humans. First of all, the latency period between exposure and development of cancer may be 20 years or longer. If a chemical produces a tumor which occurs frequently anyway, an increase might not be detected unless it is overwhelming. Groups of people exposed to higher concentrations of a chemical in occupational exposure may be too small and their exposure so varied that it may not be possible to demonstrate an increase in cancer. At the present state of our knowledge, in order to protect the public health, chemicals that produce cancer in animals must be considered to be potential carcinogens in humans. The same would be true for mutagens. Presently, most experimental data suggest that the number of cancers produced in a particular group of animals decreases as the dose of the carcinogen given to the animals decreases. Thus, a dose response curve can be developed. There are some scientists who feel that eventually a threshold is reached below which no cancers will develop (Brown, 1976). However, this is not generally accepted (Crump and Masterman, 1979). Most likely, if lower doses are given, less tumors develop and eventually none will be detected because the background incidence of tumors in a particular animal species or in humans cannot be separated from the tumors produced by the chemical. Another reason would be that the animal groups are too small to detect statistically significant differences between dosed animals and controls. For these reasons, in animal studies, relatively high doses are usually used and risk assessments are then based on extrapolations from these animal studies. A dose response curve is developed and much has been argued about the lower end of the response curve both for radiation, as well as chemical exposure. It is beyond this presentation to discuss this in detail. The reader is referred to a number of recent publications (Hoel et al., 1975; Guess et al., 1977). None of the calculations take species differences into account, or differences in metabolism at lower doses. These calculations make risk assessments for rats or for mice or for whatever animal species the cancer study was conducted in rather than for humans.

In the last few years, it has been established that there are different types of carcinogens, the initiators, the promoters, direct-acting carcinogens and those that have to be metabolized to be active. Whether all of them present the same risk is debatable. It could be argued, for instance, that promoters are less dangerous than initiators. But what about promoters that are persistent in the body and are only very poorly excreted?

Be this as it may, it would be theoretically possible to establish cancer risks or other health risks for each individual chemical present in a chemical dump. Unfortunately, it is not that simple. First of all, the mere fact that a chemical is in a dump does not mean that humans have been exposed. For many reasons, exposure data are very difficult to develop. If a dump is well contained, then there is probably no exposure. If it has existed for many years and if contamination of homes from the dump has occurred, a single measurement of air levels, for instance, may be very misleading. Such levels may fluctuate a great deal and may change over the years. In addition, it is not known what the combined effect of a multitude of chemicals even at low concentration might be or do (Schmähl, 1976). Unless exposure has been to very high levels of the chemicals involved as is sometimes seen in occupational exposure (Cannon et al., 1978), it may not be possible to get these answers by studying people living around dump sites. The reasons for this are as follows: Usually, little or no information is available about exposure. The number of people living around a dump is limited. They have not all had the same length of exposure which stratifies them. They are not all of the same age or sex. By the time all of this has been considered, very small numbers are left in the different groups. For all of these reasons and the fact that people are very heterogeneous and lead different life styles, most of such studies will not be able to demonstrate an effect. This is not to say that toxic waste does not present a public health risk. It may also have an emotional impact on people. They become concerned, and this concern may become a fixation which can completely dominate peoples' personal lives. Because of this preoccupation, it is conceivable, for instance, that children would get less attention and that peoples' productivity would suffer.

The dumping of toxic wastes will also over time increasingly contaminate our environment. A build up of those nonbiodegradable chemicals will occur in the food chain.

As the world's population increases, land use will increase and as this happens, the potential for human exposure will increase. To prevent this situation, a proper toxic waste management on a national and international basis needs to be insituted. Many chemical wastes could be reclaimed, others should be incinerated, and landfills should only be used as a last resort. Standards in chemical waste disposal plants need to be improved so that safe work practices exist, so that chemicals are not stored in large quantities for long periods of time, and so that their disposal is done correctly and adequately. Small chemical companies may not be able to pay for the disposal of their toxic wastes. States attracting their business should also provide assistance in disposal of their toxic wastes.

REFERENCES

1. American Conference of Governmental Industrial Hygienists (ACGIH). Threshold limit values for chemical substances and physical agents in the workroom environment with intended changes for 1980. Publication Office, ACGIH, P. O. Box 1937, Cincinnati, Ohio 45201.
2. Beutler, E., Glucose-6-phosphate dehydrogenase deficiency. In the metabolic basis of inherited disease, 3rd ed., J. B Stanbury, J. B. Wyngaarden, and D. S. Frederiskson, eds., McGraw-Hill, New York, pp. 1358-1388 (1972).
3. Brown, C. C., Mathematical aspects of dose response studies id carcinogenesis. The concept of thresholds. Oncology, 33, 62-65 (1976).
4. Cannon, S. B., Veazey, J. M., Jackson, R. S., Burse, V. W., Hayes, C., Straub, W. E., Landrigan, P. J., and Liddle, J. A., Epidemic kepone poisoning in chemical workers, Am. J. Epid., 107, 529-537 (1978).
5. Carter, C. D., Kimbrough, R. D., Liddle, J. A., Cline, R. E., Zack, M. M., Jr., Barthel, W. F., Koehler, R. E., and Phillips, P. E., Tetrachlorodibenzo-dioxin: An accidental poisoning episode in horse arenas. Science 188:738-740 (1975).
6. Christensen, H. E., and Fairchild, E. J., Suspected carcinogens, 2nd ed., U.S. Dept. HEW, NIOSH, Government Printing Office, Supt. of Doc., Washington, D.C. 20402.
7. Cleaver, J. E., Defective repair replication of DNA in xeroderna pigmentosum. Nature 218:652-656 (1968).
8. Code of Federal Regulations 21, part 192, 365-381, April 1, 1980. U.S. Government Printing Office, Washington, D.C. 20402. Also available NTIS UB/C/220.
9. Compliance Policy Guides Manual. FDA Executive Director for Regional Operations, Division of Field Regulatory Guidance, Field Compliance Branch, December 3, 1973.
10. Crump, K. S., and Masterman, M. D., Review and evaluation of methods of determining risks from chronic low level carcinogenic insult in Environmental contaminants in food, Congress of the United States, 1979, Library Congr. Cat. No. 79-600207. Sup. of Doc., U.S. Printing Office, Washington, D.C. 20402. Stock No. 052-00200724-0.
11. Dunphy, J. H., and Hall, A., Waste disposal: its a dirty business. Chemical Week, pp. 25-29, March 1, 1978.
12. Dunphy, J. H., and Hall, A., Waste disposal, settling on safer solution for chemicals. Chemical Week, pp. 28-32, March 8, 1978.
13. Guess, H., Crump, K., and Peto, R., Uncertainty estimates for low-dose-rate extrapolations of animal carcinogenicity data. Cancer Research 37:3475-3483 (1977).
14. Hoel, D. G., Gaylor, D. W., Kirschstein, R. L., Saffiotti, U., and Schneiderman, M. A., Estimation of risks of irreversible delayed toxicity. J. of Toxicology and Environmental Health, 1:133-151 (1975).

15. Johnson, C. J., Toxic soluble waste disposal in a sanitary landfill site draining to an urban water supply. A.J.P. H., 67:468-469 (1977).
16. Kreiss, K., Zack, M., Kimbrough, R. D., et al., Cross-sectional study of a community with exceptional exposure to DDT. JAMA, in press.
17. Schmähl, D., Combination effects in chemical carcinogenesis (experimental results), Oncology, 33:73-76 (1976).
18. Wolff, A. H., and Oehme, F. W., Carcinogenic chemicals in food as an environmental issue. JAVMA, 164:623-629 (1974).

PERCEPTION OF RISK:

A JOURNALIST'S PERSPECTIVE

>Joanne Omang
>
>Environment and Energy Reporter
>The Washington Post
>Washington, D.C.

It clearly goes without saying that actual risks have little to do with perception of risk. I would take it further and say that neither actual nor perceived risk have much to do with fear.

I lived in Turkey for two years and we had a maid who washed the clothes by filling the kitchen floor with sudsy water and beating on the clothes with a stick. She came every week and washed all the clothes every week, and she used to shake her head and chide us about washing too often. Of course we clucked about the poor, dirty peasants.

One day I was putting on a housecoat and the entire sleeve fell off. The same day my roommate came home with her skirt shredded up the back. It turned out that the fabric on almost everything we had was completely rotted through. We looked at the box the maid had been using to make the suds and it was pure lye.

The maid had hands and feet that were calloused and rough, and she said of course the lye was strong - that's why she used a stick. We told her the stuff is dangerous, and she said well, at least she was smart enough not to wash her clothes in it every week.

The common chemicals of everyday life have been more or less harmful forever. People have always known that and used them anyway, and do so even now when we know more about the long-term risks. My father is a skilled mechanic, and he still washes his greasy hands in gasoline no matter what I tell him about cancer. It's fast and it does the job, and he says he washes it right off so it won't soak in.

The actual risks from toxic chemicals probably peaked in the late 1950s and early 1960s when there were few controls on industry, but public awareness of the risk is a product of the last 15 years. I would argue now that public FEAR has also peaked, even though we are learning more every day at the scientific level about new risks - or rather, old risks newly discovered.

Paper mill workers have always had skin trouble, just as miners have always coughed. Some kind of sickness is associated with just about every kind of job, just part of the job description - like heart disease goes with being a white collar executive now. If your husband worked at the steel mill, you put up with the soot on your laundry and the constant sore throat.

But then unions and scientists and doctors and environmental groups began saying two things, both equally important: first, that the chemicals were dangerous, real problems; and second, that the problems could be fixed. Only then did the risks from toxic chemicals begin to make people angry and afraid.

The first of those announcements - that some chemicals were real and dangerous problems - upset some people more than others. Information on risk, even assuming it is precise and convincing, which it often is not, doesn't seem very well related to fear. People ride in cars knowning very well that 55,000 people die in accidents every year. But we would never allow a new machine, or a new chemical, to be produced, no matter how useful, if its environmental impact statement predicted that many deaths.

We fear any new risk if we perceive it to be optional, but we often shrug off massive risks we assumed long ago. If we were rational about risks, nobody who could read would ever smoke cigarettes. Everyone would wear seat belts and nobody would eat sugar-coated yum-yums for breakfast. But people clearly are not rational about risk.

This is considered normal. In fact, there are some people who have thought about all this and are afraid to leave the house, terrified by automobiles, streets full of criminals and air full of chemicals that we all know kill lots of people every year. We call these people irrational agoraphobics and send them to doctors.

We make it normal not to be afraid of some very real risks, because the price of being afraid is that you have to do something about it. Often that price is too high to pay. The woman whose husband works in the steel mill is more afraid of unemployment than the soot, so she ignores the risks.

We simply cannot do without automobiles any more, no will we allow a police state to halt crime completely, nor can we do without chemicals, plastics, synthetic fabrics and most of the rest of modern life - although you will note that some people advocate these solutions.

This was the second important change of the last 15 years: after science and unions and environmental groups said the risks were real, they then said that something could be done about it at a reasonable price.

I lived as a teenager in central Florida, which produces citrus fruit and phosphate fertilizers. We used to swim in the old phosphate quarry pits, but the town closed them down when articles appeared on the chemical risk. We would just close the windows briefly when the spray planes came over the groves, but the town passed ordinances about spraying over houses when we began learning about DDT. The cost of those changes was not very high.

We used to water ski on the lakes, avoiding the green patch near the sewage plant and the beaches with the dirty brown foam. The town took one of the frozen fruit juice plants to court over dumping foul-smelling waste into the lakes, I remember, and that cost quite a bit. But the green stuff spread and the foam was on all the beaches, so finally the town closed the lakes to swimming and built a new sewage system. That cost a lot.

It used to be that everything science did was good. DDT saved thousands of lives, if not millions; frozen orange juice has probably saved all of us from scurvy. The transformation of petroleum into clothing, telephones, steering wheels and lettuce crispers has transformed the planet, and very few people would say it should never have happened.

But scientists now are in a real bind. Society has discovered side effects. As serious scholars, scientists invent things, and at the same time they must tell us both the benefits and the dangers. But what does that mean, danger? Whose definition do you use? How dangerous is it? What's the price of avoiding that danger, and is it worth the price?

This entire conference is focused on these questions, yet very few of those are scientific questions. Most are political questions, yet science has somehow gotten involved in answering them.

As scholars, scientists can say at a minimum that so many mice given such-and-such a dose of something over so long a time developed X number of cancers, rashes, deformities, and so on. This is where we are now - and always were - in terms of certainty: no one doubts that such things may be said with confidence.

At the moment, science is trying hard to say at least that much about all the known chemicals and waste products and to keep up with new ones, and that alone is not easy. But as a journalist I feel qualified to tell you that the mere collection of data has never been enough.

Someone has to impose a pattern on raw information and say what it means, or the information is useless. A dictionary has all the raw material for War and Peace, a piano for a Beethoven sonata, but both are only tools with which to build a pattern. People want to know how afraid they ought to be of something that causes 2.4 cancers among 1000 rats dosed in milligrams per day over six months, and somebody ought to tell them. What does it mean in English?

You have always known this, for scientists have always told us when the data meant something good. They said that when X numbers of mice were free of scurvy, it meant Vitamin C could cure it. That one was true. I know that responsible scientists have never claimed more for a discovery than they could prove, and drugs have always had contraindication warnings. Many scientists blame the media for blowing the promises of sciencies out of proportion, but somehow we heard less about that than we have been hearing about the dangers of blowing the risks out of proportion.

Somehow, with benefits, nobody seemed to mind blue-sky's-the-limit descriptions. DDT was going to save the world from bugs and penicillin would rid us of infections, plastics would bring consumer goods to the world and nuclear power would be too cheap to meter.

Somehow it's different with risks: anyone who talks apocalypse is completely irresponsible. Risk is measured in jargon of ten to the minus five chances in zillions and the public is told not to worry. Science clearly must share some blame for the fact that the public tends to be suspicious about such talk. We always used to be told to assume the best for a benefit, so why this reluctance to consider the worst for a risk?

It must be in part that science stands accused of having duped us all these years, not talking about the hidden price attached to its wonders. The public is feeling somewhat betrayed and leery of new promises, especially the notion that a little risk is good for you. We don't live in or want a risk-free society, but we want to know what we're getting into.

What do the raw numbers of cancers or deformities or other damage mean in English? Well, it depends.

It depends first on who's talking. If scientists could agree on how much risk was worrisome, the public would be delighted. But six PhDs will advise six different levels of concern for the same finding. There are some parallels between the risks and benefits of nuclear power and the risks and benefits of a chemical society.

Both were developed by scientists who knew very well that they were taming dangerous items for a larger good. But as technology taught us more about the risks of low levels of exposure, a lot of people soured on the whole production.

The black humor of nuclear power and toxic chemical exposure is all sick and very similar. "Pleased to meet you; I'm Mrs. Jones and this is my husband and that's our daughter over there in the aquarium." You don't know whether to laugh or throw up. One cartoon showed a window overlooking Love Canal's toxic dump in New York with a baby in a cradle nearby. The baby had one hand up and the fingers were webbed. The fears are the same for nuclear power: cancer, deformed babies, and death.

Since, as we have seen, fear has no relation to reality, even very tiny amounts of evidence can produce terror. In Memphis, Tenn., the whole neighborhood of Frayser now lives in fear and sickness, convinced their homes are built over an old toxic waste dump. The Environmental Protection Agency has sifted and tested and measured the place to death and cannot find what it calls significant traces of any chemical wastes. They say the whole thing is in these people's heads.

But the people disagree that what was found is insignificant. They say chlordane levels are too high, and they say evidence has been suppressed. They are also undeniably sick - with cancers, benign lumps, skin troubles and breathing problems. What's going on there? Are these people making themselves sick? Are we wrong about what levels of chemicals cause damage? Who knows? How do we find out?

With all this, I think the public fear of toxic chemicals may have peaked. Word of yet another old waste dump brings only a shrug. People know an agency full of bureaucrats is assigned to take care of it. Every time some familiar substance like coffee is found to be related to cancer, most people just sigh and say well, you've got to die of something.

In some ways this is ironic since regulators and experts in the field tell me they are just beginning to get into the worst problem, which is groundwater contamination. But at least it's no longer fashionable to be cynical and say that everything causes cancer and it's all a racket. Once people are convinced that scientists and government agencies are responsive to the issues, and once we have a few case studies to prove it, I think toxic chemicals will take a much lower place on the list of things people spend their time worrying about.

DEPLETION OF STRATOSPHERIC OZONE AS A RESULT

OF HUMAN ACTIVITIES

R. D. Hudson

NASA/Goddard Space Flight Center
Laboratory for Planetary Atmospheres
Greenbelt, Maryland 20771

INTRODUCTION

The purpose of this paper is to introduce the reader to the earth's stratosphere and to discuss the role that man-made perturbations may have on its stability. The phenomena observed in the stratosphere are, as in the troposphere, the result of complex interactions between radiative, chemical and dynamic forces. In order to make this paper of reasonable length, I will concentrate on the photochemistry.

Stratospheric Chemistry

There are three major atmospheric species below an altitude of 90 km: molecular nitrogen, which comprises 78% of the atmosphere; molecular oxygen which comprises 21% of the atmosphere; and argon which comprises 1%. Molecular nitrogen and argon are inert below 90 km. Molecular oxygen, on the other hand, is chemically active; that is, ultraviolet radiation that reaches the stratosphere and mesosphere can dissociate molecular oxygen into two atomic oxygen atoms:

$$O_2 + \text{Solar Ultraviolet (170-240 nm)} \rightarrow O + O \tag{1}$$

The atomic oxygen that is released will combine with a molecular oxygen molecule to form the ozone molecule:

$$O_2 + O + M \rightarrow O_3 + M \tag{2}$$

The ozone in its turn can be dissociated by ultraviolet radiation into its two initial components:

$$O_3 + \text{Solar Ultraviolet } (200-320 \text{ nm}) \rightarrow O_2 + O \qquad (3)$$

However, the rate at which atomic oxygen combines with molecular oxygen is extremely fast. Thus that cycle of formation and destruction is rapid, and the net effect of reactions 2 and 3 is only to determine the ratio of atomic oxygen to ozone in the stratosphere. To the stratospheric scientist, ozone and atomic oxygen are essentially indistinguishable and are referred to as "odd oxygen." Odd oxygen can be destroyed in only two ways:

$$O + O_3 \rightarrow O_2 + O_2 \qquad (4)$$

$$O_3 + O_3 \rightarrow O_2 + O_2 + O_2 \qquad (5)$$

In both of these reactions, odd oxygen has been converted to molecular oxygen. The reactions shown above are those first put forward by Chapman (1930). When detailed measurements were made of the ozone content of the stratosphere and of the solar flux, and when better determinations of the reaction rates were obtained in the laboratory, it became obvious that the amount of ozone in the stratosphere was considerably less than would have been predicted by the Chapman reactions alone.

In order to understand what other loss mechanism there might be, we must now begin to look at molecules that are present in trace quantities (trace species). One of these is nitrous oxide, which arises principally from denitrification in the soil and is present in the atmosphere in 1 part in 10^7. Other species are water vapor, found in the stratosphere in parts per million, and methane, also present in parts per million. Finally there are the organic halogens such as methyl chloride, the CFM's, and methylchloroform, which are present in the range of 1 part per billion. These substances, known as source molecules, do not in themselves react with ozone or atomic oxygen to control the odd oxygen in the stratosphere. In order to be capable of such reactions, they must first be broken down into chemically active species known as radicals. There are two basic mechanisms by which this is done. The first is through the photodissociation of the molecules:

$$CF_2Cl_2 + \text{ultraviolet radiation} \rightarrow Cl + CF_2Cl \qquad (6)$$

The second is through the dissociation of ozone itself:

$$O_3 + \text{ultraviolet radiation} \rightarrow O_2 + O^*$$

The star indicates that the atomic oxygen atom is left in a highly excited state. This highly excited atom has enough energy to break apart both nitrous oxide and water vapor as described by the following reactions:

$$O^* + N_2O \rightarrow NO + NO \tag{7}$$

$$O^* + H_2O \rightarrow OH + OH \tag{8}$$

The three species, NO, OH, and the chlorine atom, are the fundamental radicals which start the complex photochemistry that leads to the recombination of odd oxygen in the stratosphere.

This recombination is achieved through chemical cycles, known as catalytic cycles. For example, nitric oxide, i.e., NO, goes through the following chemical cycle:

$$NO + O_3 \rightarrow NO_2 + O_2 \tag{9}$$

$$\frac{NO_2 + O \rightarrow NO + O_2}{O + O_3 \rightarrow O_2 + O_2} \text{ net reaction.} \tag{10}$$

There are two things to note about this chemical cycle. First, the net result is to recombine ozone and atomic oxygen and produce molecular oxygen. Second, the nitric oxide molecule is unchanged at the end of this cycle and is, therefore, free again to repeat the cycle until it is removed from the stratosphere. A similar catalytic cycle exists for the hydroxyl radical OH.

$$OH + O_3 \rightarrow HO_2 + O_2 \tag{11}$$

$$\frac{HO_2 + O_3 \rightarrow OH + O_2 + O_2}{O_3 + O_3 \rightarrow O_2 + O_2 + O_2} \text{ net reaction.} \tag{12}$$

In this case, the net result is the recombination of two ozone molecules giving three molecules of oxygen. This reaction is especially important at low altitudes in the stratosphere, where the ultraviolet radiation which dissociates molecular oxygen can no longer penetrate. Atomic chlorine also has its cycle which is analogous to that for nitric oxide:

$$Cl + O_3 \rightarrow ClO + O_2 \tag{13}$$

$$\frac{ClO + O \rightarrow Cl + O_2}{O + O_3 \rightarrow O_2 + O_2} \text{ net reaction.} \tag{14}$$

If we could deal with the perturbations produced in the stratosphere as separate entities (i.e., that due to chlorine, that due to nitrogen, or that due to hydration), then the calculation of the effects of perturbations on the column content of ozone would be relatively simple. However, considerable complexity is introduced because the chemical families interact with one another, and perturbations of one of the families can effect the efficiency of the other families in the recombination of odd oxygen. This is illustrated in Fig. 1. The nitrogen and the hydrogen families interact with one

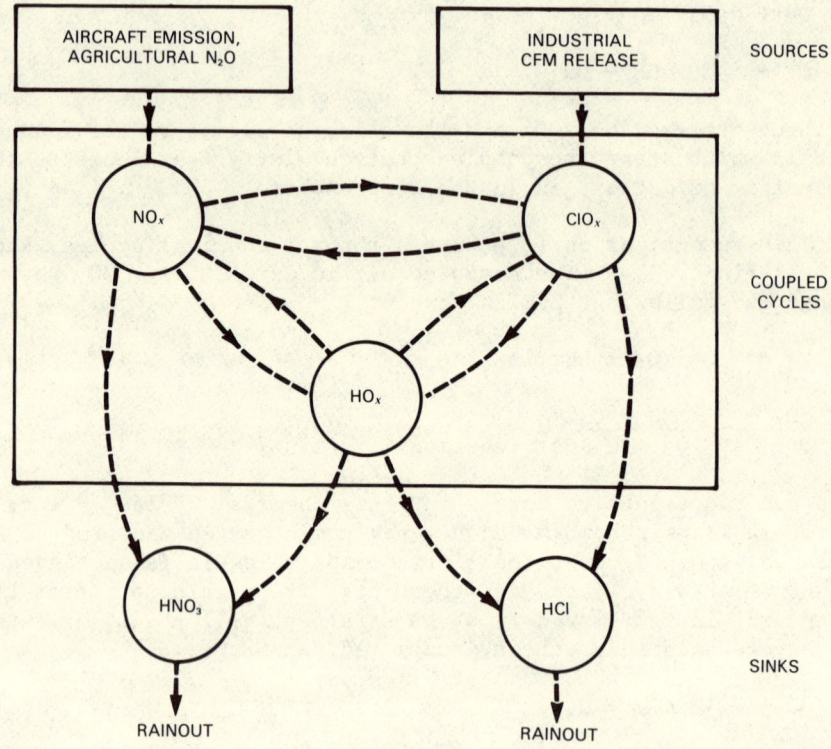

Fig. 1. Schematic of interaction between chemical families.

another to produce nitric acid which is a relatively inert sink for both of these families. The nitric acid is eventually transported to the troposphere where it is rained out. In a similar way the chlorine family and the hydrogen family interact to produce hydrogen chloride which is also eventually rained out. It will be appreciated that the processes shown here are simplified and that the true interactions between these families can be more complex. The overall chemical cycle that goes on in the stratosphere is illustrated in Fig. 2. The tropospheric sources are carried out to the stratosphere; they are broken down to form the radicals; the radicals control the amount of odd oxygen in the stratosphere; they interact amongst themselves to form the reservoir molecules which are eventually transported down to the troposphere where they are rained out.

The absorption of ultraviolet radiation by ozone, which results in its dissociation, also produces heat; and as a consequence, the stratosphere has a temperature profile that increases with altitude. The decrease with altitude in the troposphere stops at the tropopause; and at about 20 km altitude the temperature begins to rise

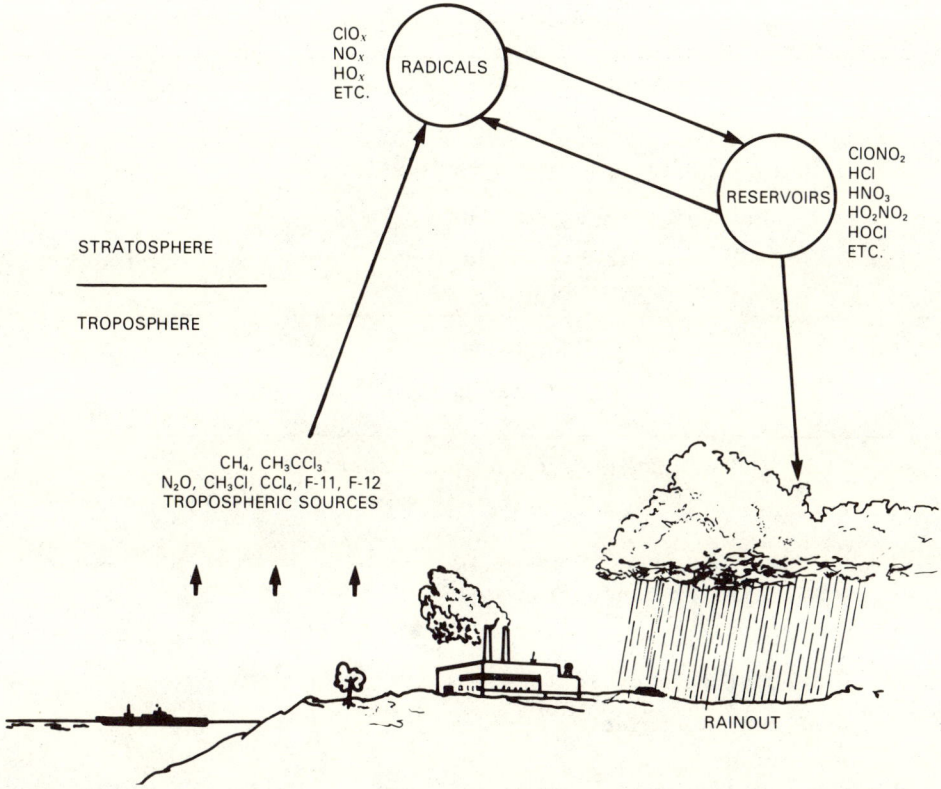

Fig. 2. Overall chemical cycle in the troposphere and stratosphere.

rapidly to a peak of about 280°K at 45 to 50 km, a point that is known as the stratopause. An atmosphere whose temperature is falling with altitude such as the troposphere is essentially unstable. The opposite is true, of course, for an atmosphere whose temperature is increasing with altitude (a temperature inversion is a good analogy in the troposphere), and thus the stratosphere is inherently a stable region of the atmosphere. This stability leads to long lifetimes for molecules placed in the stratosphere, an average residence time being between 1 and 2 years. Unlike the troposphere where pollutants tend to be flushed out fairly rapidly (within days), in the stratosphere this process is much longer; and, therefore, radical species can continue in their catalytic cycles for long periods of time before they eventually interact to form source molecules and are eliminated from the stratosphere. A typical catalytic cycle takes between one-half and one day to complete. However, because the radical species have a long lifetime in the stratosphere, they can have a significant effect on the odd oxygen in concentratins of the order of 1 part per billion.

Table 1

Natural ozone perturbations

o Volcanoes	Cl_x
o Solar particle events (PCS, REP)	NO_x
o Solar radiation variations	ALL
o Galactic cosmic ray variations	NO_x

Table 2

Man-made ozone perturbations

o CFM's	Cl_x
o SST's	NO_x, HO_x
o Fertilizer production	NO_x
o Carbon dioxide	ALL
o Methyl chloroform (CH_3CCl_3)	Cl_x
o Carbon monoxide	Cl_x
o Forest removal	NO_x
o Nuclear explosions	NO_x
o Solid-fueled rockets	Cl_x
o Subsonic aircraft	NO_x, HO_x

Perturbations of the Stratosphere

Tables 1 and 2 list the perturbations that have been identified as having an impact on the ozone in the stratosphere. Table 1 lists the natural ozone perturbations while Table 2 lists those perturbations due to man's activities.

The first two perturbations, listed in Table 1, arise from sudden events. Analysis of the ozone perturbations following a solar particle event have provided proof that the catalytic cycle for the nitrogen oxides does take place in the upper stratosphere. Analysis of the radiative effects following a volcanic eruption have also given us confidence in the detailed radiative transfer calculation in the lower stratosphere. Perturbations due to solar radiation and

Table 3

Total ozone changes (steady state)

Perturber	1979	1981
1. CFM's	−15 to −19%	−5 to −7%
2. Aircraft (NO_x) at 17 km	+1 to +3%	−2.2%
3. Aircraft at 9 km	—	+0.5%
4. Doubling NO_2	−2 to +3%	−8 to −12%
5. Doubling CO_2	—	3 to +6%
Composite scenarios		
6. 2 × CO_2 + CFM	—	−3 to −5%
7. 2 × N_2O + CFM	−9 to −12%	−8 to −12%

galactic cosmic ray radiation are of a much longer period. Unfortunately data have not been collected with any precision for a long enough time to elucidate all of these variations, but it is obvious that changes in the ultraviolet radiation from the sun will have a marked effect on the ozone balance in the stratosphere.

There are many man-made perturbations on the ozone in the stratosphere that have been identified. The two most widely known are supersonic transports and the chlorofluoromethanes. These two have been actively discussed in the literature and have been the subject of several national reports (i.e., National Academy of Sciences, 1975, National Academy of Sciences, 1979, Hudson and Reed, 1979). The supersonic transports release both nitrogen and the hydrogen oxides into the middle stratosphere. However, other perturbations also release nitrogen oxides into the stratosphere. Increased fertilizer production leading to an increase in the nitrous oxide emitted into the troposphere will also increase the amount of nitrogen oxides in the stratosphere as will forest removal, nuclear explosions, and the increased flights in the lower stratosphere of the subsonic aircraft. The chlorofluoromethanes release chlorine atoms into the stratosphere as does the release of methyl chloroform and the solid fuel from rockets.

Other man-made perturbations have a secondary but nevertheless significant effect on the ozone balance. Carbon monoxide can remove the hydroxyl radicals from the troposphere and in so doing removes a natural cleansing agent for the higher hydrocarbons and chlorocarbons. Thus increased carbon monoxide leads to an increase of the chlorocarbons in the stratosphere where they can remove odd oxygen. Carbon dioxide controls the temperature profile in the stratosphere; and since most of the reaction rates that determine ozone are strongly temperature dependent, an increase in carbon dioxide will also have an effect on the ozone balance.

Table 3 lists calculations made in 1979 and 1981 of the effects of some man-made perturbation on the total column of ozone in the stratosphere. The most significant change in this 2-year period has been in the estimate of the effect of supersonic aircraft at 17 km. The estimate has gone from a positive increase in the total column ozone to a decrease. The calculations made in 1979 were based upon some new measurements of reaction rates involving the hydrogen family. Since 1979, the complete set of hydrogen reactions has been remeasured; and the latest estimates are the results of using these improved values. As mentioned earlier, the chemical families are strongly linked. This can be immediately seen when the estimates of the CFM perturbations made in 1979 and 1981 are examined. The 1981 numbers are about 40% of those calculated in 1979, and the decrease can be largely assigned to the changes in the nitrogen oxide chemistry brought about by changes in the hydrogen chemistry.

Also shown in Table 3 are the effects of doubling the nitrous oxide and the carbon dioxide in the troposphere. Doubling of the carbon dioxide in the atmosphere could be considered a realistic scenario on which to base calculations; however, the doubling of the nitrous oxide is probably not realistic as the present rate of increase of nitrogen oxide is of the order of half a percent per year. Doubling the carbon dioxide alone would lead to an increase in the column content of ozone from 3 to 6%. If the same emission rate for the chlorofluoromethanes is maintained as in 1975, the total column content ozone would be decreased from 5 to 7%. The effect of these two scenarios combined is to decrease the column content of ozone from 3 to 5%. It is obvious that the net effect on the stratosphere is not the simple addition of these two perturbations and indicates again the strong interaction that exists between the chemical families. Thus the net effect of any perturber must be considered in light of other perturbations, and the release rates of all perturbing elements must be known in order to produce a realistic assessment of the impact of man's activities.

REFERENCES

Chapman, S., "Environmental Impact of Stratospheric Flight," Phil. Mag., 10, 363-393, 1930, Washington, D.C. 1975.
Hudson, R. D., and Reed, E., "The Stratosphere: Present and Future," National Aeronautics and Space Administration Reference Publication 1049, 1979.
National Academy of Sciences, "Stratospheric Ozone Depletion by Halocarbons: Chemistry and Transport," Washington, D.C. 1979.

RELATIONSHIP OF STRATOSPHERIC OZONE DEPLETION

TO RISK OF HUMAN SKIN CANCER

Frederick Urbach

The Center for Photobiology
Skin and Cancer Hospital
Temple University School of Medicine
Philadelphia, Pennsylvania

The high energy, short wavelength portion of the solar electromagnetic spectrum (wavelengths shorter than 320 nm) is potentially very detrimental to living cells and tissues. A low concentration of ozone formed in the stratosphere absorbs photons of ultraviolet radiation (UVR) and thus prevents most of them from reaching earth. However, even in the presence of this ozone layer, which varies in thickness in various latitudes and at various seasons, a biologically significant amount of UVR reaches the surface of the earth.

It is a working assumption that byproducts of human activity in recent years and in the foreseeable future may penetrate to the level of ozone formation in the stratosphere and could result in a dep'.- tion of this important "ozone shield."

Model calculations performed in the past decade have result in the suggestion of a parametric range for the potential decre; in stratospheric ozone due to various causes of 1% to 50% (medi 15%) (CIAP, 1975). A recent report of the National Academy of ences (NAS) entitled, "Protection against Depletion of Stratos｟　｠Ozone by Chlorofluorocarbons" (NAS, 1979) estimated a median decrease in stratospheric ozone at equilibrium of 16.5%, assuming continuing discharge of chlorofluorocarbons (CFC) into the stratosphere (Dickinson et al., 1978; Crutzen et al., 1978).

The major effects on humans of UVR in the UVB (280 to 320 nm) range are on the skin and the eyes. Acute effects consist of "sunburn," an inflammatory response of the tissues which may be no more than mild redness or slight stinging of the eyes or may develop into the equivalent of second degree (blistering) "burns." The acute effects of single overdoses of UVB are transient, heal without scarring, and in the skin lead to adaptive changes of skin thickening and pigmentation which afford some degree of protection (Giese, 1965). The only established beneficial effect of UVB in humans is the production in the skin of vitamin D precursors which are absorbed into the blood stream and prevent rickets, a serious vitamin deficiency disease (DeLuca, 1971). It should be recognized that most work has been on the harmful effects of UVB, and relatively little attention has been given to possible beneficial effects.

Repeated UVB exposure, prolonged over many years, can result in chronic degenerative changes in skin, characterized by skin "ageing" and by the development of premalignant and malignant skin lesions. The skin cancers of man can be broadly divided into two types: non-melanoma skin cancer (NMSC) and malignant melanoma (MM).

There is excellent, although circumstantial, evidence that NMSC is primarily due to repeated exposure of the skin to UVB (Blum, 1959; Urbach et al., 1972). The major arguments in favor of such a causal role of UVR in NMSC are:

o The most frequent location of NMSC is on the most exposed skin sites (head, neck, arms, hands);

o Pigmented races, who sunburn much less readily than people with white skin, have much less NMSC; and when it does occur, it does not affect the sun-exposed sites;

o Among white skinned people, those who spend more time outdoors and live in areas of greater UVR exposure (near the equator and in tropical and semitropical areas) have much greater risk for NMSC;

o Genetic diseases resulting in greater sensitivity to solar UVR are associated with premature NMSC development (albinism); and

o Skin cancers of the NMSC type can readily be induced in the skin of experimental animals; and the upper wavelength limit for this is 320 nm, similar to the spectral range producing sunburn in man.

Recent epidemiologic research has resulted in the development of a preliminary dose-response relationship of NMSC to UVB. Reasonable assumptions, based on tissue culture and animal experiments, for the relationship of various wavelengths of UVR to NMSC induction have been proposed (Setlow, 1974). While there is still discussion about the exact relationships of various parts of the UVR to carcinogenesis, existing data appear reasonably sufficient for the development of model systems that allow for some predictions as to the effect of reductions of stratospheric ozone to the probable increases in the incidence of NMSC secondary to increase in UVB (Forbes et al., 1978).

Because of uncertainties in the amount of UVB actually reaching various populations, and because parts of any population are significantly more sensitive (for inherited reasons) to the chronic effects of UVR, estimates of possible NMSC increases have ranged from a few thousand to many hundred thousand of additional cases in the U.S.A. alone. Projected estimates are based on available data (which is improving) (Scotto et al., 1980) and best estimates of their meaning (where considerable differences of opinion still exist) (Vitaliano and Urbach, 1980).

In summary, there is reason to believe that most NMSC (but not all, for about one-third of basal cell cancer occurs in areas not receiving much UVB and not showing other sunlight damage [Urbach et al., 1972]) is causally related to chronic, repeated UVB exposure, and that a dose-response function exists that can, with considerable future refinement, be used to predict changes in NMSC incidence given alterations of the UVB climate.

The situation is somewhat different where MM is concerned. The NAS report (NAS, 1979) uncritically suggests that the relationship of MM to solar UVR exposure is basically similar to that of NMSC and that the same model systems apply equally to NMSC and MM for predictive purposes.

There is considerable evidence that such assumptions are not tenable. The matter is of great importance, since MM is a serious, often fatal, cancer. Although much less frequent in incidence, its potential ability to cause death makes it of greatest importance to attempt to develop a model for its causation, which may allow predictions as to future incidence to be made.

Because of its seriousness, recording of cases of MM is usually very good, so that epidemiologic studies that are quite reliable have been carried out in many countries in the past three decades. Very briefly, the following are the salient findings:

- There has been a consistent, worldwide increase in MM, incidence rates increasing 3 to 7% per year, leading to a doubling of incidence every 10 to 15 years (Magnus, 1981). This trend for increasing incidence began with people born at or before the turn of the century, and has been progressively accelerating, so that younger persons have a greater risk for development of MM than older ones. This is very unusual behavior for a malignant tumor (Lee et al., 1979).

- There are striking differences in anatomic distribution by sex - MM is much more frequent on the backs of men and the lower legs of women. The incidence in those areas has been increasing rapidly, in contrast to MM of the head and neck area, and on the feet, where incidence rose only slightly. This distribution is entirely different from that of NMSC (Malec and Eklund, 1978).

- Most MM is a disease of young adults. In contrast to NMSC, risk of MM rises sharply during adolescence, plateaus in middle age, and rises again in old age. NMSC begins to appear in late middle age and rises sharply in old age (Houghton et al., 1980; Sober et al., (1980).

- In contrast to NMSC, only about 10% of MM (a subtype that closely resembles squamous cell cancer in its biologic behavior) show significant evidence of chronic solar skin damage.

- While there are latitude gradients for MM within some countries, the normal gradient of increasing incidence from higher to lower latitudes is reversed in central Europe, and, particularly in the Scandinavian countries, show incidence rates greatly in excess of those expected based on relative UVB intensity found at their latitude (Cutchis, 1978; Crombie, 1979).

- From available data, it appears that the latent period for MM development is short, perhaps even 3 to 5 years. The latent period for NMSC is certainly in excess of 20 years (Lee, 1972).

- In contrast to NMSC, which affects mainly those chronically exposed to sunlight (i.e., outdoor workers), MM is much more frequent in white collar, educated, more affluent city dwellers (Lee and Strickland, 1980).

- As yet, no animal model for experimental production of MM exists using UVR. Only recently has it been possible to produce MM experimentally in guinea pigs, using a chemical carcinogen (Pawlowski et al., 1980).

From the foregoing it is obvious that there are major differences in most features of NMSC and MM.

It appears that, if UVR is causally related to the development of MM, the mechanisms are very different. Certainly, except for the Lentigo Maligna Melanoma that appears on the head and neck of old persons and makes up no more than 10% of all MM, most MM can not reasonably be caused by the cumulative effect of repeated exposures to UVR, as is the case with NMSC. The various theories that have been proposed for its etiology are:

Related to UVR:

- Intermittent, high dose UVR exposure, presumably due to social changes, i.e., sunbathing, weekend gardening, golf, tennis, etc., and vacation trips to sunny places (Spain, North Africa, the Caribbean, etc.) (Fears et al., 1976; Eklund and Malec, 1978; Magnus, 1981).

- Production by UVR of a "circulating factor" which causes distant effects on precursor lesions (Lee and Merrill, 1971).

- "Initiation" of melanocytes, either by chemicals or UVR, with either UVR or chemicals acting as a "promotor" (i.e., two stage carcinogenesis) (Clark et al., 1978).

- "Promotion" of preexisting, abnormal precursor cells or lesions by UVR (precursor lesions perhaps genetically determined) (Clark et al., 1978).

In any case, it is obvious that at this time, there is no possibility of assuming a reasonable dose-response relationship of UVR to MM development. In the absence of such information, attempts to predict what changes may take place in MM incidence if the intensity or spectrum of UVR changes are not appreciable.

Risk Analysis - NMSC

From a few detailed epidemiologic investigations and the data available from a world compilation presented in Cancer Incidence in Five Continents some figures for incidence of basal cell and squamous cell carcinoma can be obtained which show a geographic variation of annual skin cancer incidences from as low as 5/100,000 to over 200/100,000. From these data it is also possible to estimate that the incidence of skin cancer approximately doubles for every 10° of latitude, provided that the population is of reasonable similar genetic stock.

Various models have been proposed, relating a decrease in stratospheric ozone to potential increases in NMSC. Most of these are based on two steps: a calculation of the expected increase of "biologically effective" ultraviolet radiation, and then a number relation UVR to increase in NMSC.

All available evidence suggests that virtually all human skin squamous cell cancer but only $^2/_3$ of basal cell cancer are related to total accumulated lifetime dose of biologically effective solar ultraviolet radiation (Urbach, 1969; O'Beirn et al., 1968; Gallin et al., 1965; Scotto et al., 1974; Vitaliano and Urbach, 1980). It is also clear that pigmentation of the skin confers a marked protection and that there is considerable variation in the sensitivity to UVR among "white" people, and that much of this variation is genetic (Urbach, 1969). It is highly associated with "ease of sunburning" and ability to tan (Vitaliano and Urbach, 1980). However, given sufficient UVR exposure for sufficient time, most white people can develop NMSC. The great difference in susceptibility, which is by no means regularly distributed geographically, is one of the major confounding factors for any dose-response calculation in humans.

All investigators who have attempted to model dose-response effects for NMSC agree that there are also a number of latitude-related variables which modify the relation of actually received UVR dose to incidence of NMSC. This conclusion is based in part on the indication of nonlinearity in incidence-latitude figures, and in part on the intuitive concept that latitude-related climatic and behavioral effects must modify the UVR dose actually reaching the skin of a target population.

There is obviously a relationship between latitude and solar energy intensity, and this gradient is exaggerated in the UV portion of the spectrum. Although the shape of this relationship is relatively complex and dependent on a number of variables (altitude, albedo, cloud cover, aerosols, etc.) for a range of northern mid-latitude values (30° to 50° N), the form closely approximates a straight line for calculated (Green, 1978) and integrated skin erythema action spectrum weighted values (Scotto et al., 1980).

The relative effects of the "lifestyle" factors are largely speculative. MacDonald (1971) suggested that it contributes about 50% of the gradient, and van der Leun and Daniels (1975) considered this not an unreasonable estimate. To accept such an estimate in conjunction with an exponential dose-incidence relationship, however, implies that these "other factors" are also exponentially related to latitude. If the assumed magnitude of these factors is speculative, an assumed exponential relationship is considerably more so.

If the input of potentially effective UVR is linearly related to latitude, and if the factors relating this input to NMSC are nonlinearly latitude dependent, it follows that the relationship of skin cancer incidence to latitude must be nonlinear. Indeed, the epidemiologic evidence available (Gordon, 1976; Scotto et al., 1980) shows this to be the case.

In order to obviate the unknown "other factors," various modelers have used a two step procedure: in the first step, the amount of average stratospheric ozone is related to the biologically effective UVR dose (physical amplification factor); in the second step, the annual UVR dose is related to estimates of skin cancer incidence (biological amplification factor) (Green and Mo, 1974; Urbach et al., 1975; Scott and Straf, 1977).

The advantage of this approach is that one can explicitly account for dose differences which are not related to the amount of stratospheric ozone. No attempt was made to account for possible genetic or behavioral differences between the various populations.

A fundamentally different approach is to evaluate the effect of an ozone reduction indirectly by using a dose-response theory to describe the effect of an increment in the annual UVR dose (Rundel and Nachtwey, 1978). This approach resembles and uses the aforementioned "two step approach." The difference is that the step in which the annual UVR dose is related to the incidence is not based on epidemiologic data, but on a dose-response model inferred from animal experiments. The advantage of this approach is that one deals with one population (hairless mice) and does not have the problem of genetic, environmental and behavioral differences at different locations (Rundel and Machtwey, 1978).

Finally and most recently, De Gruijl and van der Leun (1980) have developed a dose-response model based on animal experiments. Using this model, they derived a mathematical formula which gives the relationship between a fractional increment in the animal UVR dose and the corresponding fractional increment in skin cancer incidence.

For such model calculations, estimates of the "physical amplification" factor are necessary. Extensive calculations have been made by Green et al. (1974), Mo and Green (1974), Johnson et al. (1976), Scott and Straf (1977). These calculations are based mostly on UVR measurement performed by Bener (1972), estimates of local average O_3 concentrations, estimates for effect of cloud cover, etc. With minor variations, the optical amplification factor has been found to be approximately 1.7 (Green and Mo, 1974) based on an action spectrum similar to UVR absorption by DNA.

There is still extensive discussion among modelers whether the biological dose-response relationship follows an exponential or a power law relationship. Nevertheless, irrespective of the model used, the "biological amplification factor" ranges in the vicinity of 2× (Urbach et al., 1975; Green et al., 1976; Green, 1978; Scotto et al., 1980; de Gruijl and van der Leun, 1980). It should, however, be pointed out that, particularly if the power law applies, the "biological amplification factor" (AF_B) will be less away from the equator and more near the equator (e.g., Seattle, 47.5° N, AF_B 1.47; New Orleans, 30° N, AF_B 2.57) (Scotto et al., 1980).

Thus, the general estimate for the overall amplification factor obtained from the most recent models remains at 4×, i.e., a 1% reduction in ozone may result in a 4% increase in nonmelanoma skin cancer. Preliminary results from two dimensional modeling suggest the nonuniform distribution of O_3 depletion could result in a lower factor (Pyle and Derwent, 1980).

One other point needs to be made before the above described dose-resonse relationships are put to general use. That is that there is a significant difference in risk for developing basal cell or squamous cell cancer of the skin. It has been pointed out before that of basal cell carcinomas, about $1/3$ seem to have no relationship to UV exposure (Urbach et al., 1972; Gellin et al., 1965). Also, the most recent epidemiologic survey performed by NCI (Scotto et al., 1980) shows a much shallower north-south incidence slope for basal cell carcinoma than squamous cell carcinoma. Finally, Vitaliano and Urbach (1980) have found that the increased risk for the most susceptible population is very much greater for development of squamous cell carcinoma than basal cell carcinoma, and that, given the same level of cumulative lifetime solar exposure, subjects over 60 years of age had a higher risk for development of NMSC than those at younger ages.

Clearly, the existing models need considerable refining before realistic risk estimates relating reduction in ozone to increase in NMSC can be made, and present models certainly overestimate the risk.

Finally, as far as changing incidence of NMSC with time is concerned, there is a small amount of data that the incidence has been increasing by 2-3% per year in the past decade (Scotto et al., 1980). This has been happening although there has been no measurable decrease in ozone in that time period (Berger, personal communication; London, 1980). It may be the reflection of a continuing increase in exposure for sociologic reasons - i.e., "a tan is beautiful."

Risk Analysis - MM

In the most recent publication on relationship of UVR exposure to risk of developing MM (NAS, 1979), the assumption has been made that the models developed for NMSC are applicable to MM. As will be shown below, this is a naive and untenable assumption.

The two major findings that support the contention that NMSC is primarily caused by chronic exposure to solar UVR are the anatomic distribution of the cancers (on sun exposed areas) and the striking latitude gradient of their incidence.

It has been pointed out in a previous section that the anatomic distribution of MM does not clearly parallel the presumed sites of maximum solar exposure. In the case of one type of MM, Lentigo Maligna Melanoma, the characteristics of age distribution, anatomic distribution, and histologic evidence of solar connective tissue damage all seem to be virtually identical to the features of squamous cell carcinoma of the skin. Thus it must be assumed that solar UVR exposure is the primary cause of LMM. However, this then applies to only about 10% of MM (Larsen and Grude, 1979).

Cutchis (1978) was one of the first to point out that the geographic distribution of MM, reported in Vol. III of Cancer Incidence in Five Continents (Doll et al., 1976), shows significant anomalies if the assumption is made that exposure to solar UVR is the most significant factor in the causation of MM.

The major discrepancies found at that time were (a) the relatively large incidence of MM in the Scandinavian countries, particularly Norway (Magnus, 1975, 1977) and Sweden (Malec and Eklund, 1978): (b) the lack of any evidence for a latitude gradient (or even a reversal from the usual north-south increase) for central European countries (Crombie, 1979); and (c) the peculiar male-female ratio of MM, which usually remains below unity, in sharp contrast to NMSC, where the M/F ratio exceeds 4:1 or more (Urbach, 1972).

The data described above have been accumulated by the International Agency for Research in Cancer (Doll et al., 1976) from worldwide cancer registries. They certainly represent information based on various time periods and acquired by different means. Since the incidence of MM has been rising sharply world wide, it may not be too surprising if these data did not represent strictly comparable enumerations.

If solar UVR is the primary cause of MM, then, as in NMSC, there should exist recognizable latitude gradients for the incidence and mortality of MM, with rates increasing toward the tropics. The existence of such latitude gradients has been reported from Norway (Magnus, 1975), Sweden (Magnus, 1977; Eklund and Malec, 1978), Finland (Teppo et al., 1978), and the U.S.A. (Cutler and Young, 1975; Fears et al., 1976).

There are, however, inconsistencies in a number of these studies which have recently been noticed. Of major importance have been the observations that major cities show a disproportionately high incidence of MM, which could not be explained on latitude alone, and that most Nordic countries show incidence rates for MM greater than can be expected based on their latitude (Lee and Issenberg, 1972; Eklund and Malec, 1978; Crombie, 1979; Viola and Houghton, 1978; Houghton et al., 1980; Jensen and Bolander, 1980).

For instance, Teppo et al. (1978) found a distinct north-south gradient for MM incidence along Finland, with the incidence rates 30% or more greater in the south of that country. However, when the incidence rates were adjusted to urban/rural population ratios, the latitude gradient markedly diminished for the 1953-59 time period, and disappeared for 1961-73. They concluded that there was a marked effect of urbanization and that factors other than latitude operate. Eklund and Malec (1978) performed an epidemiologic study similar to that of Teppo (1978) and Magnus (1975, 1977). Their major findings were: Sweden has a much higher incidence rate for MM than central Europe. There was a distinct north-south gradient along Sweden, but the major cities had a disproportionately high incidence of MM. The difference in the cities was primarily due to increases of MM of the trunk and arms of men and the arms and legs of women. There was no disparity in incidence of MM of the head and neck between rural and city areas. The differences could not be explained on latitude alone.

In contrast to the above, Beardmore (1972) and Little et al. (1980) noted that in Queensland, Australia, MM was more frequent in subtropical than in tropical areas, and Holman noted a reverse gradient (i.e., away from the equator) in Western Australia. This finding was ascribed to the above noted increased incidence of MM in large cities.

Most recently, Crombie (1979) examined data from a large series of tumor registries for Europe (Doll et al., 1976) and concluded that the incidence of MM increased with increasing latitude in central Europe from Italy to Germany, the reverse of that found in England and Wales, the Nordic countries and the U.S.A. He investigated possible confounding effects and concluded that the result was highly significant and not due to confounding factors. He also noted the apparently excessively high incidence of MM in Scandinavia and the excess found in large cities. Crombie concluded that this latitude reversal may be due to darker pigmentation of southern Europeans' skin, but pointed out that the effect of change in susceptibility would have to be very large to overcome a latitude effect.

Jensen and Bolander (1980) reviewed again the IARC and newer incidence data. They comment on great variation in incidence of MM by country, the low incidence in countries with predominantly pigmented inhabitants and point out that there are latitude gradients within countries (England and Wales, Sweden, Canada, U.S., Norway) which increase as latitude decreases and that there are exceptions to this phenomenon (Western Australia, Texas, Finland).

Lee (1977) was the first to point out a peculiar association of incidence of MM with certain, unexpected, occupational groups. He observed, first in England and Wales, that the occupational groups that suffered the highest mortality from melanomas of the skin was clerical and professional workers and their wives. This initial finding has been confirmed by Viola and Houghton (1978) for Connecticut, by Holman et al. (1980) for Western Australia, and summarized by Lee and Strickland (1980).

The size and consistency of the relationship of MM risks to some factors associated with better education, high social status, or more money, and the presence of the relationship in both employed men and among their wives suggest that the effect is real. It suggests that the relationship is a biological one between MM incidence and some feature of life associated with education or economic status. It may also explain the above noted preponderance of MM incidence to large cities, since the more affluent are more likely to live in cities and their suburbs.

Teppo, Pukkala, et al. (1980) in a remarkable study of the

relationship of "Way of Life and Cancer Incidence in Finland" confirm the predominance of MM in the towns and low rates in rural areas. They show a consistently positive association with parameters describing the socioeconomic level of the municipality. A negative association was seen between farming and forestry and MM. In contrast, for cancer of the lip (the best reported form of NMSC) a strong inverse association was seen between the incidence of lip cancer and that describing the standard of living of the municipality. No other primary cancer site displayed such strong negative associations than that found for lip cancer. The risk for lip cancer was highest in the farming, forestry and fishing occupations, and in rural areas. Thus a striking difference between MM and lip cancer was found related to income, city dwelling, and outdoor occupations.

It has been pointed out previously that the incidence of MM has been rising rapidly in white populations for many years. For example, in Norway a continuous rise in incidence of about 7% per year has been observed since 1955 (Magnus, 1980). A major component of the causation of this rising incidence is a systematic increase in risk of successively later born cohorts (Lee and Carter, 1970; Elwood and Lee, 1975; Magnus, 1975; Eklund and Malec, 1978; Lee et al., 1979; Holman et al., 1980). This cohort increase began as early as the last quarter of the nineteenth century in Australia (Holman et al., 1980) and about 1900 in Norway (Magnus, 1975). It is of interest that cohorts born since 1925-30 show a stabilization of this increased risk but at a higher rate (Elwood and Lee, 1975; Holman et al., 1980). The increase and cohort variations seem to be much greater for MM of trunk and lower limbs than for MM of the face-neck area (Magnus, 1980). This indicates that the trend in carcinogenic exposure through life may be different for face-neck and trunk-leg areas. Why people born 5 years later should go through life with a substantial increase in risk of dying from MM compared with their elder peers in the same population is not known (Lee et al., 1979).

In addition to the above described geographic variations of MM incidence and mortality, several other observations have been proposed so as to relate the risk of MM development to sunlight exposure.

Anaise et al. (1977) and Movshovitz and Modal (1973) reported that MM was found more frequently in European Jews than in African or Asian Jews in Israel; that the incidence was higher in Israeli born, European descended Jews than recent immigrants and attributed this to greater UV exposure in a sunny country. However, Hinds and Kolonel (1980) and Holman et al. (1980) noted reverse conditions, i.e., more MM in recent immigrants to Hawaii or Western Australia than in the native born, European descendants. Houghton et al.

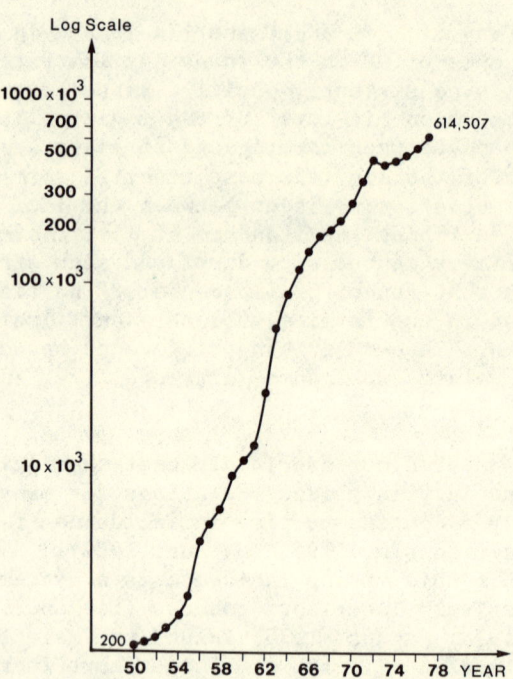

Fig. 1. Number of persons, by year, on charter tours from Denmark 3 days or longer, 1950 to 1978. Notice the huge increase in travel, particularly after 1965-70.

(1978) noted a cyclic increase in MM superimposed on the previously reported continuing rise in MM incidence two to three years after sunspot maxima in Connecticut, New York, and Finland, but not in Norway. Wigle (1978) confirmed this observation in Canada. Houghton et al. ascribed this effect to a decrease in stratospheric ozone associated with increases in galactic cosmic rays. They suggested ozone decreases of the order of 1 to 3% secondary to sunspot influence. However, no such evidence of ozone decrease has been observed in the long term Dobson measurements reported by London (1980), nor have such changes in ozone been noted in association with sunspot activity (London and Reber, 1979).

Scotto and Nam (1980) observed a strong seasonal pattern with summertime peaks for MM of females (particularly legs) and a similar peak for MM of upper extremities of men. This was also noted by Malec and Eklund (1978).

On the other hand, Leach et al. (1978) observed a consistent increase in stratospheric ozone over England (in keeping with the results of London, 1980), an accompanying fall in NMSC in Bristol,

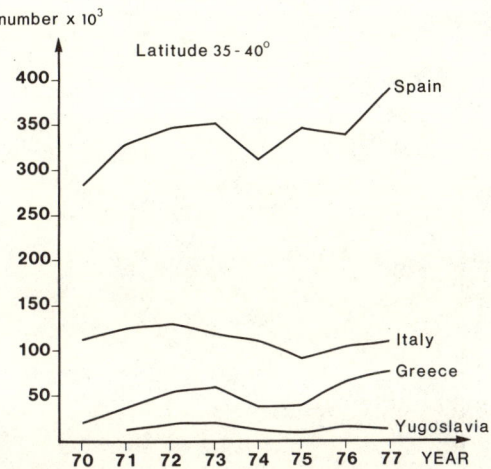

Fig. 2. Number of persons traveling from Denmark to sunny areas (latitude 35° to 40°). Note almost no increase in the past decade.

but a steadily rising incidence of MM opposite to the ozone concentration in the stratosphere.

It has been suggested by a number of observers (Lee, 1972; Fears et al., 1976; Magnus, 1975, 1980) that the increase in the incidence of MM, particularly among the affluent population in the cities, may be due to unaccustomed, intermittent overexposure to sunlight on vacations, weekends, etc. Furthermore, the changes in clothing and outdoor recreation habits in the decades since World War II would allow more exposure (men going shirtless, women with shorter skirts and nylon stockings, more sun exposure for cosmetic tanning purposes, etc.).

To test this hypothesis, Eklund and Malec (1978) investigated the issuance of passports in Sweden, reasoning that to get to a sunny area one had to leave Sweden and go south. Indeed, it was found that more passports were issued in cities than in rural areas, but the difference in passport issue could at most explain 25% of the increase of MM in Swedish cities. Furthermore, Brodthagen (personal communication) investigated the frequency of Danes participating in charter tours since 1950 (Fig. 1), the number of persons in organized tours of more than three days' duration traveling to areas 35° to 40° latitude by year since 1970 (Fig. 2), and the number of persons traveling to various areas on these tours (Fig. 3). As can be seen, the frequency of Danes leaving the country has increased dramatically (Fig. 1), but there has been little change in the past decade in those going to sunny areas, the greatest increase had been in travelers to England (i.e., the same latitude), probably

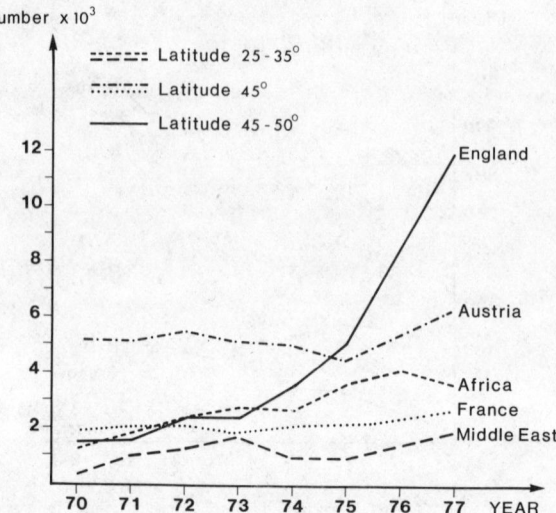

Fig. 3. Number of persons traveling from Denmark on charter tours. Note that only travel to England has increased significantly.

because of favorable foreign money exchange. It thus appears that greater exposure to sunlight in the south, at least for Danes, is not likely to be the major cause of the increase in MM incidence.

In summary, it appears that the pathogenesis of NMSC and MM may be quite different as far as the role of UVR is concerned. The predominance of NMSC on the head and neck of older persons strongly suggests that cumulative, total UVR dose is involved. The relatively early onset and different distribution of MM, as well as the secular changes described above (cohort effects), strongly suggest that the latent period for MM is very much shorter than that for NMSC (Lee, 1972; Magnus, 1980).

Lee (1972) in an elegant analysis of the potential mechanisms of MM induction suggested:

1) that the causative mechanism for MM is complex and specific, and certainly not "chronic irritation" (i.e., cumulative UVR damage);

2) that exposure to the eliciting agent operates with a short latent period, so that MM develops in the susceptible persons reasonably soon after they have a possibility of developing the tumor;

3) that either a "systemic agent" which operates at a distance is involved, or that the skin of susceptible persons contains "dominant precursors," i.e., something akin to premalignant clones of cells that are stimulated into fully developed cancers by an environmental agent, of which UVR may be one.

In summary, the geographic variation of the incidence of MM is not nearly as clearcut as is that of NMSC. While there are respectable latitude gradients within some countries, these are affected by the very real increase in MM incidence in big cities, which, at least in Scandinavia, are located in the southern portions of the countries. In contrast to this is a reverse latitude gradient along central Europe, which appears to be real. Further studies will be needed to sort out the degree to which exposure to solar radiation is responsible for the development of MM.

At this time there is no realistically usable model for estimating the risk of increase of incidence of MM with increase in UVR secondary to depletion of stratospheric ozone.

Finally, a seriously confounding factor in all such risk analyses involving sun exposure of humans is the varying attitude of people regarding sunbathing. Greiter et al. (1981) have performed a most enlightening study in this regard.

The attitude towards sunlight and the motivations responsible for a certain form of behavior were determined by means of a questionnaire addressed to 20,000 persons of both sexes, between 20 and 55 years of age, living in Australia, the Federal Republic of Germany (FRG), France and Austria. Multivariance analyses carried out in a randomized sample of 913 test subjects provided a clear two-factor structure as expression of the fact that behavior of exposure to sunlight is determined by two BIPOLAR ATTITUDES.

Factor I is a "pleasure dimension," ranging between the two extremes "pleasant" and "unpleasant." Persons very conscious of this dimension would regard sunbathing as extremely pleasant, restful and relaxing - or the opposite.

A second dimension of attitude, which has no bearing on the first one, is the "health factor" which can be illustrated by the two extremes "beneficial to one's health" and "detrimental to one's health." Persons whose behavior is mainly determined by this factor are convinced that sunbathing is either good or, in the other extreme, bad for their health.

Since these two factors are independent of each other, any combination of attitudes can be observed - in particular (pleasant and good or unpleasant and bad) - which should result in a marked tendency either in favor of (or against) sunbathing. Strong, but

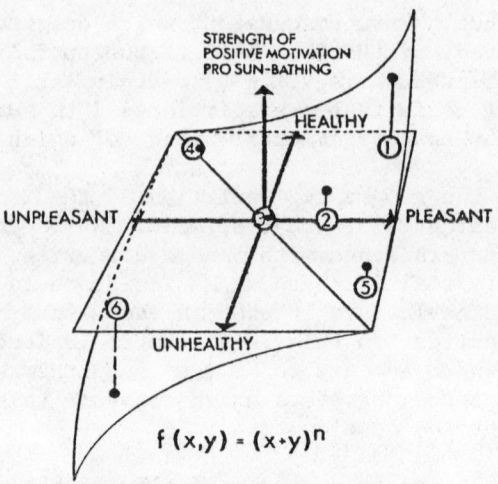

Fig. 4. This figure illustrates behavior to sun exposure. The arrows in the third dimension show the actual intensity of attitude (upwards – positive, toward the sun; downwards – negative, against the sun).

contradictory attitudes (pleasant and unhealthy) might result in the same weak motivation as the lack of any distinct attitudes.

Since a number of theoretical studies on attitudes have shown that the weight of an attitude increases with the degree of its intensity, an exponent was introduced which can be determined empirically from the data. The result is the following expression for the actual attitude intensity in favor of (or against) sunbathing:

BEHAVIOR VIS-A-VIS SUNLIGHT = (FACTOR I + FACTOR II)^n =

$$\left(\begin{array}{c} \text{attitude intensity} \\ \text{PLEASURE} \end{array} + \begin{array}{c} \text{attitude intensity} \\ \text{HEALTH} \end{array} \right)^n$$

An analysis of the individual factor scores permits the placement of each individual on the attitude scale illustrated in Fig. 4. The third dimension shows the actual intensity of attitude (upwards – positive, towards the sun).

Thus subject 1 is an enthusiastic "sun worshipper" whose positive attitude in both dimensions makes him a person at risk who tends to underestimate the harmful influence of an exaggerated exposure to sunshine.

Subject 6 on the other hand shows an equally exaggerated "excape behavior" with regard to sunshine. The empirical analysis of the data does, however, also confirm that neither intensive but opposing attitudes (subject 4) nor the actual absence of attitudes (subject 3) induce confidence - such an absence of attitude was found only in an extremely small number of subjects. A slight predominance of one of the two partial attitudes again results in a positive attitude (subject 5) or in an avoidance behavior. With only one of the two attitudes being intense, the actual readiness of behavior will also be markedly reduced (subject 2).

In the analyses, the exponent n was always near 1.00 (highest value 1.10) which enabled one to use the simple linear additive model for the prediction of behavior. With regard to attitude intensity, direction, and attitude consistency, one can distinguish the following five groups:

Group 1: Weak intensity in the two dimensions. Neutral and indifferent attitude towards sunbathing.

Group 2: Positive attitude in both directions. Extreme "sun worshippers."

Group 3: Negative attitude in both directions - exposure to sun is avoided.

Group 4: Sunbathing is pleasant but harmful. "Sun worshippers" who are convinced that exposure to sunlight is detrimental to their health.

Group 5: Beneficial, but unpleasant. Persons belonging to this group take a sunbath as they would take a drug: or they stay in the shade and are worried because they deny themselves something that is good for their health.

There are clear-cut distinctions between these five groups. Group 1 was separated by limits parallel to the coordinate axis in a distance of 1 sigma. The remaining four groups are formed by the cases remaining in the respective quadrants. Because of the evident accumulations, a narrowing or extension of the separating lines hardly changes the individual groups.

The following table shows the distribution over the different classes of behavior both of the entire group and of the inhabitants of four selected countries with different sun experience due to different climatic conditions:

While the percentage of persons who were completely indifferent is very low, a nationwide comparison of the distribution over the other four quadrants reveals remarkable differences:

While the values obtained in the Federal Republic of Germany

	Attitude groups				
	1	2	3	4	5
Total Group	5%	20%	19%	36%	20%
FRG	5%	23%	18%	36%	18%
Austria	5%	16%	17%	35%	27%
France	3%	21%	30%	30%	15%
Australia	3%	28%	19%	46%	4%

approximate the total group distribution, in Austria a marked shift toward negative attitudes can be observed (4 and 5).

In France there was a marked trend towards consistent avoidance behavior, while in Australia one sees a striking trend towards the dimension "pleasant": about $^3/_4$ of the persons asked showed this basic attitude, but at the same time 461 persons were convinced that sunbathing is bad for their health!

It is thus clear that, while the actual risk for the development of skin cancer in the most susceptible population (e.g., Australia) is high, the perceived risk varies with people, and the known ill effect of sun exposure is greatly outweighed by the pleasant aspect, in spite of warnings against such behavior.

REFERENCES

Anaise, D., Steinitz, R., and Ben Hur, N., Solar radiation: A possible etiological factor in malignant melanoma in Israel, Cancer 42:299-304 (1978).

Beardmore, G. L., The epidemiology of malignant melanoma in Australia, in: Melanoma and Skin Cancer, New South Wales Government Printer, Sydney (1972), pp. 39-64.

Bener, P., Approximate Values of Intensity of Natural Ultraviolet Radiation for Different Amounts of Atmospheric Ozone, European Res. Office, U. S. Army (Contract DA & A 36-68-C-1017) (1972).

Blum, H. F., Carcinogenesis by Ultraviolet Light, Princeton, Princeton University Press (1959).

CIAP Monograph No. 5, Impacts of Climatic Change on the Biosphere, Climatic Impact Assessment Program, Department of Transportation, (September 1975).

Clark, W. H., Jr., Reiner, R. R., Greene, M., et al., Origin of
familiar malignant melanoma from heritable melanocytic lesions,
Arch. Dermatol., 114:732-738 (1978).

Crombie, I. K., Variation of melanoma with latitude in North America
and Europe, Brit. J. Cancer, 40:774-781 (1979).

Crutzen, P. J., Isaksen, I. S. A., and McAfee, J. R., The impact of
the chlorocarbon industry on the ozone layer, J. Geophys. Res.,
83:345-363 (1978).

Cutchis, P., On the Linkage of Solar Ultraviolet Radiation to Skin
Cancer, Institute for Defense Analysis, Washington, D.C., U.S.
Department of Transportation (1978), pp. 105-106.

Cutler, S. J., and Young, J. L., Third National Cancer Survey: Incidence Data, NCI Monograph 41, DHEW Publication No. (NIH)75-787,
Bethesda, National Cancer Institute (1975).

DeGruijl, F. R., and van der Leun, J. C., A dose-response model for
skin cancer induction by chronic UV exposure of a human population, J. Theor. Biol., 83:487-504 (1980).

DeLuca, H. F., Vitamin D: A new look at an old vitamin, Nutr. Rev.,
29:179-181 (1971).

Dickinson, R. E. Liu, S. C., and Donohue, J. M., Effect of chloro-
fluoromethane infrared radiation on zonal atmospheric temperature, J. Atmos. Sci., 35:2142-2151 (1978).

Doll, R., Muir, C., and Waterhouse, J., Cancer Incidence in Five
Continents, Vol. III, International Union Against Cancer, IARC
No. 15 (1976).

Eklund, G., and Malec, B., Sunlight and incidence of cutaneous
malignant melanoma, Scand. J. Plast. Reconstr. Surg., 12:231-241
(1978).

Elwood, J. J., and Lee, J. A. H., Recent data on the epidemiology
of malignant melanoma, Semin. Oncol., 2:149 (1975).

Fears, T. R., Scotto, J., and Schneiderman, M. A., Skin cancer,
melanoma and sunlight, Am. J. Publ. Health, 66:461-464 (1976).

Forbes, P. D., Davies, R. E., and Urbach, F., Experimental ultraviolet photocarcinogenesis: Wavelength interactions and time-
dose relationships, in: NCI Monograph No. 50 (1978).

Gellin, G. A., Kopf, A. W., and Andrade, R., Basal cell epithelio-
mas: A controlled study of associated factors, Arch. Dermatol.,
91:38-45 (1965).

Giese, A. C., Photophysiology, Vol. 4, New York, Academic Press
(1965), pp. 139-202.

Gordon, D., and Silverstone, H., Worldwide epidemiology of premalignant and malignant cutaneous lesions, in: Cancer of the
Skin, R. A. Andrade, ed., Philadelphia, W. B. Saunders (1976),
pp. 405-455.

Green, A. E. S., Ultraviolet exposure and skin cancer response, Am.
J. Epidemiol., 107:277-280 (1978).

Green, A. E. S., Findley, G. B., Klenk, K. F., Wilson, W. M., and
Mo, T., The ultraviolet dose dependence on nonmelanoma skin
cancer incidence, Photochem. Photobiol., 24:353-362 (1976).

Green, A. E. S., and Mo. T., Proc. Third Conf. on CIAP, DOT-TSC-OST-74-15, pp. 518-522 (1974).
Green, A. E. S., Mo. T., and Miller, J. H., A study of solar erythema radiation doses, Photochem. Photobiol., 20:473-482 (1974).
Greiter, F., Bilek, P., Bachl, N., et al., The effect of artificial and natural sunlight upon some psychosomatic parameters of the human organism, Submitted for publication (1981).
Hinds, M. W., and Kolonel, L. N., Malignant melanoma of the skin in Hawaii, 1960-1977, Cancer, 45:811-817 (1980).
Holman, C. D. J., Mulroney, C. D., and Armstrong, P. K., Epidemiology of pre-invasive and invasive malignant melanoma in Australia, Int. J. Cancer, 25:317-323 (1980).
Houghton, A., Flannery, J., and Viola, M. V., Malignant melanoma in Connecticut and Denmark, Int. J. Cancer, 25:95-104 (1980).
Houghton, A., Munster, E. W., and Viola, M. V., Increased incidence of malignant melanoma after peaks of sunspot activity, Lancet I:759-760 (1978).
Jensen, O. M., and Bolander, A. M., Trends in malignant melanoma of the skin, World Health Statistics Quart., 33:2-26 (1980).
Larsen, T., and Grude, T. H., A retrospective histological study of 669 cases of primary malignant melanoma in clinical stage I. VI, The relation of dermal solar elastosis to sex, age, and survival of the patient, and to localization, histological type and level of invasion of the tumor, Acta Path. Microbiol. Scand. Sect. A. 87:361-366 (1979).
Leach, J. F., Beadle, P. C., and Pingstone, A. R., Effect of ozone variation on disease in Great Britain. I. Skin cancer, Aviation, Space and Environ. Med., 49:512-516 (1978).
Lee, J. A. H., Sunlight and the etiology of melanoma, in: Melanoma and Skin Cancer, Sydney, New South Wales Government Printer (1972), pp. 83-94.
Lee, J. A. H., Current evidence about the causes of malignant melanoma, in: Clinical Cancer, Ariel, ed., New York, Grune and Stratton (1977), p. 151.
Lee, J. A. H. and Carter, A. P., Secular trends in mortality from malignant melanoma, J. Natl. Cancer Inst., 45:91-97 (1970).
Lee, J. A. H., and Issenberg, H. J., A comparison between England and Wales and Sweden in the incidence and mortality of malignant skin tumors, Brit. J. Cancer, 26:59-66 (1972).
Lee, J. A. H. Petersen, G. R., Stevens, R. G., and Vesanen, K., The influence of age, year of birth and date on mortality from malignant melanoma in the populations of England and Wales, Canada and the white population of the U.S., Am. J. Epidemiol., 110:734-739 (1979).
Lee, J. A. H., and Merrill, J. M., Sunlight and melanoma, Lancet I:550 (1971).
Lee, J. A. H., and Strickland, D., Malignant melanoma: Social status and outdoor work, Brit. J. Cancer, 41:757-763 (1980).
Little, J. H., Holt, I., and Davis, N., Changing epidemiology of malignant melanoma in Queensland, Med. J. Austral., 1:66-69 (1980).

London, J., The observed distribution and variation of total ozone, in: Proc. NATO Advanced Study Institute on Atmospheric Ozone, U.S. Department of Transportation, FAA-HAPP, Washington, D.C. (1980), pp. 31-44.

London, J., and Reber, C. A., Solar activity and total atmospheric ozone, Geophys. Res. Lett., 6:869-872 (1979).

MacDonald, E. J., Assessment of possible SST effects on the incidence of skin cancer, Tech. Report, Department of Atmospheric Physics, University of Arizona (1971).

Malec, E., and Eklund, G., The changing incidence of malignant melanoma of the skin in Sweden 1959-1958, Scand. J. Plastic Reconstr. Surg., 12:19-27 (1978).

Magnus, K., Epidemiology of malignant melanoma of the skin in Norway with special reference to the effect of solar radiation, in: Biological Characterization of Human Tumors, Excerpta Medica Congress Series, No. 375 (1975).

Magnus, K., Incidence of malignant melanoma of the skin in the five Nordic countries, Int. J. Cancer, 20:477-485 (1977).

Mo, T., and Green, A. E. S., A climatology of solar erythema dose, Photochem. Photobiol., 20:483:496 (1974).

Movshovitz, M., and Modan, B., Role of sun exposure in the etiology of malignant melanoma, Epidemiological Conference, J. Natl. Cancer Inst., 51:77 (1973).

NAS (National Academy of Sciences), Report of the Comittee on Impacts of Stratospheric Change (CISC): Protection against Depletion of Stratospheric Ozone by Chlorofluorocarbons, Washington, D.C. (1979).

O'Beirn, S. F., Judge, P., Urbach, F., et al., Skin Cancer in County Galway, Ireland, Proc. 10th Intl. Cancer Conference, Am. Cancer Soc. (1968).

Pawlowski, A., Habermann, H. F., and Menon, I. A., Skin melanoma induced by DMBA in albino guinea pigs and its similarity to skin melanoma in humans, Cancer Res., 40:3652-3660 (1980).

Pyle, J. A., and Derwent, R. G., Possible ozone reductions and UV changes at the earth's surface, Nature, 286:373-375 (1980).

Rundel, R. D., and Nachtwey, D. S., Skin cancer and UV radiation Photochem. Photobiol., 28:345-356 (1978).

Scott, E. L., and Straf, M. L., Ultraviolet radiation as a cause of cancer, in: Origins of Human Cancer, Cold Springs Harbor Lab. (1977), pp. 529-546.

Scotto, J., Fears, T. R., et al., Incidence of nonmelanoma skin cancer in the United States 1977-78, DHEW Publication No. (NIH)80-2154, U.S. DHEW, National Cancer Institute (April 1980).

Scotto, J., Kopf, A. W., and Urbach, F., Nonmelanoma skin cancer in four areas of the U.S., Cancer 34:1333-1338 (1974).

Scotto, J., and Nam, J. M., Skin melanoma and seasonal patterns, Am. J. Epidemiol., 111:309-314 (1980).

Setlow, R. B., The wavelengths in sunlight effective in producing skin cancer: A theoretical analysis, Proc. Natl. Acad. Sci., 71:3363-3366 (1974).

Sober, A. J., Lew, R. A., Fitzpatrick, T. B., and Marvell, R., Solar exposure patterns in patients with cutaneous melanoma, Clin. Res., 28:561A (1980).

Teppo, L., Pakkanen, M., and Hakulinen, T., Sunlight as a risk factor of malignant melanoma of the skin, Cancer, 41:2018-2027 (1978).

Teppo, L., Pukkala, E., Hakama, M., Hakulinen, T., Herva, A., and Saxen, E., Way of life and cancer incidence in Finland: A municipality based ecological analysis, Scand. J. Social Med. Suppl., 19:50-54 (1980).

Urbach, F., Geographic pathology of skin cancer, in: The Biologic Effects of Ultraviolet Radiation, F. Urbach, ed., Oxford, Pergamon Press (1969).

Urbach, F., and Davies, R. E., Estimate of ozone reduction in the stratosphere on the incidence of skin cancer in man, in: CIAP Monograph No. 5, DOT-TST-75-55, National Technical Information Service, Springfield, Virginia (1975).

Urbach, F., Rose, D. B., and Bonnem, M., Genetic and environmental interactions in skin carcinogenesis, in: Environment and Cancer, Williams and Wilkins, Baltimore (1972), pp. 355-371.

van der Leun, J. C., and Daniels, F., Jr., Biologic effects of stratospheric ozone decrease, A critical review of assessment, in: CIAP Monograph No. 5, Appendix B, pp. 7-105, DOT TST-75-55, National Technical Information Service, Springfield, Virginia (1975).

Viola, M. V., and Houghton, A., Melanoma in Connecticut, Conn. Med., 42:268-269 (1978).

Vitaliano, P. P., and Urbach, F., The relative importance of risk factors in nonmelanoma carcinoma, Arch. Dermatol., 116:454-456 (1980).

Wigle, D. T., Malignant melanoma of skin and sunspot activity, Lancet II:38 (1978).

DEPLETION OF STRATOSPHERIC OZONE:

IMPACT OF UV-B RADIATION UPON NONHUMAN ORGANISMS

Robert C. Worrest

U.S. Environmental Protection Agency
Corvallis Envirnomental Research Laboratory
Corvallis, Oregon, U.S.A.

For years, scientists and laymen alike have casually noted the impact of solar ultraviolet radiation upon the nonhuman component of the biosphere. It was not until recently, when human activities were thought to threaten the protective stratospheric ozone shield, that researchers undertook intensive studies into the biological stress caused by the previously slightly short-wavelength edge of the global solar spectrum. It is stratospheric ozone that functions effectively as an ultraviolet screen, filtering out solar radiation in the 220-320 nm waveband as it penetrates through the atmosphere, allowing only small amounts of the longer wavelengths of radiation in this waveband to leak through to the surface of the earth. Although this radiation (UV-B radiation, 290-320 nm) comprises only a minute fraction (less than 2%) of the total solar spectrum, it can have a major impact on biological systems due to its actinic nature. Many organic molecules, most notably DNA, absorb UV-B radiation which can initiate photochemical reactions. It is Life's ability, or lack thereof, to cope with enhanced levels of solar UV-B radiation that has generated the concern over the potential depletion of stratospheric ozone. The defense mechanisms that serve to protect both plants and animals from current levels of UV-B radiation are quite varied. Whether these mechanisms will suffice under conditions of enhanced levels of UV-B radiation is the subject of this paper.

Agricultural Crops, Native Terrestrial Vegetation, and Marine Phytoplankton

Although some forms of plant life are motile and exhibit negative phototaxis in response to ultraviolet (UV) radiation, the avoidance of UV radiation for most plants does not include escape movements [1]. Unlike animals, which can reduce exposure of their tissues to solar radiation through protective coverings and behavioral responses, terrestrial plants have evolved to expose much of their tissue to sunlight to utilize its energy for photosynthesis.

Many studies have demonstrated that photosynthesis is inhibited by UV-B radiation [2-9]. In addition, UV-B radiation can affect leaf expansion and abscisic acid content [3, 4, 10-16], pigment concentrations [12, 17-20], plant growth [8, 10, 17, 21-23], carbohydrate metabolism [24], fruit growth and yield [15, 25], and pollen germination and pollen tube growth [26].

In marine systems, UV-B radiation (e.g., 310 nm) penetrates approximately the upper 10% of the coastal marine euphotic zone before it is reduced to 1% of its surface irradiance [27]. There is good evidence that current levels of UV radiation depress near-surface primary production in marine waters. Steemann Nielsen [28] noted that the UV component of sunlight was capable of depressing both the rate of light-saturated photosynthesis and the initial slope of the Photosynthesis-versus-Irradiance curves for water samples from Friday Harbor, Washington. Jitts et al. [29] also found that surface samples from marine waters underwent greater primary production when covered with a UV-B filter than when exposed to full sunlight. The authors speculated that most reports for primary production in surface waters are overestimated by about 40% due to the common use of thick-walled glass bottles in the studies. Several other authors have also described the impact of UV-B radiation upon marine primary production [30-36]. Furthermore UV-B radiation impacts survival [37] and growth rate [33, 38].

Terrestrial and Marine Animals

Very little informatin has been published regarding the impact of UV-B radiation on nonhuman, terrestrial animals. Solar UV-B radiation is the probable cause of ocular squamous cell carcinoma (cancer eye) [39-41]. Cancer eye is found in most breeds of cattle, but it is most prevalent in the Hereford breed. Certain strains of the Hereford breed lack any pigment around the eye. UV-B radiation also acts as a significant enhancing factor in the etiology of keratitis in cattle [42] and reindeer [43]. The two-spotted spider mite can discriminate and respond to the presence of UV-B radiation [44]. In the study, the egg-laying capacity of the female mite was reduced in a linear fashion as the UV-B dose was increased. Radiation in the 320-400 nm waveband did not elicit either result.

Information exists regarding the impact of UV radiation on finfish raised in hatcheries [45, 46]; however, until recently, little documentation has appeared in the literature concerning UV stress upon pelagic fishes. Marinaro and Bernard [47] concluded from their study that solar UV radiation may be a stressful agent for pelagic fish eggs, which are often located near the surface of the water, especially those eggs spawned during the winter. Hunter et al. [48, 49] have noted an impact of UV-B radiation upon anchovy and mackerel eggs and yolk-sac larvae. Investigators have also described the impact of UV-B radiation upon marine copepods [50, 51], shrimp larvae, crab larvae, and euphausids [52, 53]. Results of the reports by Karansas et al. [48, 49] indicate that UV-B radiation increased mortality of the copepod populations and decreased fecundity of the survivors. Damkaer et al. [52, 53] have concluded that beyond threshold levels, activity, developmental rats, and survival rates of marine invertebrates are depressed by UV-B radiation. In a report of Jokiel [54], the author found that "shade-loving" sponges, bryozoans, and tunicates were killed by one to two days of exposure to solar UV radiation.

Field Validation of Laboratory Experiments

It is obvious from the references cited in the previous sections of this paper that UV-B radiation imposes a stress upon many living organisms. The question arises as to whether valid conclusions regarding the potential photochemical impact of stratospheric ozone depletion can be based on the laboratory experiments. For example, tests of more than 100 species or varieties of species in laboratory environmental growth chambers indicate that about 20% are sensitive to daily levels of UV-B exposure approximating levels currently found in Florida. Twenty percent were resistant to exposure levels four times greater than this, and the remaining 60% showed some intermediate sensitivity [55]. These tests would suggest that a significant fraction of the present agricultural varieties are currently under UV stress and would suffer decreased production with a 16% ozone reduction. However, tests of 15 species and varieties, irradiated in the field under conditions of high visible light and UV-A radiation, appeared to be more UV resistant than plants grown in the laboratory environmental growth chambers.

Where it is possible to make dose-response comparisons for various growth parameters for the same type of plants, the difference in resistance, comparing results from laboratory and field studies, is on the order of four-fold. If this result turns out to be general, the higher sensitivity indicated in the growth chambers does not represent what would happen in the field and the expected consequences of ozone depletion become considerably reduced. At this time, results from laboratory environmental growth chambers cannot be used to extrapolate to field conditions. However, they do provide a data base for identification of symptoms that might be encountered under stress from UV-B radiation [56].

There is evidence that some plant species acclimatize to enhanced levels of UV-B radiation. The occurrence of Temperate-Zone exotics at high elevations in the tropics, where the UV-B irradiance is significantly greater than that which would occur with a 16%

ozone reduction at temperate latitudes, attests to the ability of
many species to acclimatize [57, 58]. If acclimatization to environments with high UV fluence is merely a phenotypic response, the rate
at which atmospheric ozone depletion takes place would not matter.
If, on the other hand, acclimatization involves genotypic changes
for some species, the rate at which the atmospheric ozone layer
changes must be considered. About half of the equilibrium ozone reduction is now predicted to occur within the next 30 years [55].
This may be insufficient time for genotypic changes to occur for
long-lived plant species or for those that reproduce asexually.

Hunter et al. [59] evaluated the uncertainties in making an assessment of UV effects on natural anchovy populations. They concluded that a realistic assessment of the effects on the anchovy
larval population requires knowledge of seasonal changes in larval
abundance, incident UV-B radiation, spatial and temporal variation
in the depth of penetration of UV-B radiation in the habitat, vertical distribution of larvae in the sea, rates of vertical mixing of
eggs and larvae, and natural rates of mortality. At present, many
of these parameters can only be approximated or are unknown for
anchovy. Much less is known for most other species.

Hunter et al. [59] did not consider larval abundance a major
uncertainty in their calculations for anchovy, but did consider it
a major uncertainty in species where the seasonality of annual larval
production is poorly defined. One of the principal reasons that they
considered the relative effect of increased UV radiation due to an
ozone decrease to be minor is that anchovy larvae are abundant at
other depths as well as in the upper few meters. A 25% reduction in
ozone in moderately productive waters is equivalent to about a one-meter increase in the depth of penetration of the biologically effective UV-B radiation for anchovy larvae. An increase in depth of
penetration of one meter would not have a major effect on a population distributed over about 60 meters. Calkins and Thordardottir
[37] have pointed out that reproduction by surviving organisms could
replace those killed by UV-B radiation. As long as the dose of UV-B
radiation does not exceed a certain "replacement limiting dose" for
each species of organism then, according to the model, populations
could survive.

Large uncertainties for the impact of UV-B radiation upon anchovy larvae are associated with vertial distribution and mixing,
and it is unrealistic to establish upper and lower limits for these
variables. The number of larvae present in tows taken at a 2-m depth
can vary from 0.25% to 27% of the population and may be dependent on
the depth of the mixed layer [59]. Thus, seasonal changes in vertical distribution, or possible biases in calculating the average
distribution, could either increase, or reduce to nearly zero, estimated UV effects. Similarly, vertical mixing could reduce estimated
UV effects to zero at current or reduced ozone concentrations.

Hunter et al. [59] concluded that the parameters of vertical distribution and mixing could account for the full range of possible UV-B effects on the population independent of all other variables.

Simulation of Solar UV-B Radiation

Experimental enhancement of UV-B radiation for ozone depletion studies occurs through use of artificial sources - sources whose spectral output cannot conform exactly to the solar spectrum. Due to this lack of conformity and because not all wavelengths of UV radiation are equally effective in producing biological effects, the relative biological effectiveness of radiation in the UV waveband must be determined in order to compare the results of the experiments with what might occur in nature. Photobiological action spectra demonstrate the relationship between biological effectiveness and wavelength of incident radiation. Analytic representations of action spectra can then be used to weight the spectral irradiance of interest. Weighting the irradiance over all wavelengths yields the biologically effective irradiance.

This approach presupposes knowledge of the action spectrum of interest. Usually the selection of a weighting function occurs "after the fact" by determining which of the published action spectra best "fits" the data in question, not an unrealistic approach if there are no other practical alternatives. For example, Smith and Baker [60, 61] and Smith et al. [34] have demonstrated that results of acute radiocarbon uptake studies with marine algae are consistent with a photoinhibition action spectrum [62], but results of chronic exposures [33] appear to be best described by an analytic representation [63] of a DNA action spectrum [64]. The results of a study by Worrest et al. [35] were evaluated utilizing the absolute dose as well as three different biologically effective doses: one weighted by a generalized plant action spectrum [65], another by a DNA action spectrum [64], and the third by a photoinhibition action spectrum [62]. There was a trend favoring the biologically sensitivity functions weighted more heavily at the shorter wavelengths (plant and DNA) rather than the photoinhibition action spectrum which has a "flatter" response or a response more closely resembling the absolute irradiance spectrum.

The major uncertainty associated with the calculation of the biologically effective irradiance is introduced by the choice of weighting function. As an example, in the report by Worrest et al. [35], a level of irradiance selected in an experiment performed during the fall of the year was determined by weighting the irradiance from the artificial UV-B source by a generalized plant action spectrum [65]. Two of the treatments were designed to simulate a current level of plant-effective irradiance for October and a plant-effective irradiance simulating a 15% ozone reduction [66]. The components of the marine ecosystem under study were naturally recruited from existing populations in Yaquina Bay (Oregon). If the sensitivity of responses for the components of the experimental ecosystems was consistent with a generalized plant action spectrum, the lack of conformity between the artificial UV-B sources and the existing solar radiation would be inconsequential. However, if the impact of the UV-B waveband upon the ecosystem was more consistent, for example, with a DNA action spectrum [64], the experimental design would have been simulating the solar conditions in August rather than October, and a 19% ozone reduction rather than a 15% ozone reduction.

An additional complication in experimental studies is that, in most situations, the artificial UV-B sources are turned on at a single irradiance level for from 4-6 hours to obtain a given daily dose comparable to that from the sun. This is obviously unnatural and subjects organisms to much higher exposures thatn the sun at some times and possibly lower doses than from the sun at noon.

Traditionally, biological action spectra have been determined by exposing organisms to monochromatic radiation at the same irradiance and scoring the biological response, or by supplying the level of monochromatic fluence necessary to elicit a certain biological endpoint. In either case, the emphasis has been to determine the fine structure of the action spectrum with as much spectral resolution as possible in order that potential chromophores might be identified. Seldom have the tails of these action spectra been elucidated, because usually they contribute little to identification of chromophores. Since solar spectral irradiance changes by orders of magnitude within the UV spectrum, tails of action spectra can be quite important even though these may represent greatly diminished biological effectiveness compared to the maxima. Thus, the spectra to be used as weighting functions should ideally include any tails that lie within the waveband where there is solar flux.

Since biological action spectra are normally determined with monochromatic radiation, the assumption must be made that radiation at different wavelengths does not interact when presented simultaneously. There is evidence that synergisms may be involved when organisms are exposed to polychromatic UV radiation. For example, there was a close correspondence between the action spectrum for a particular type of DNA lesion, pyrimidine dimers, and death of cultured Chinese Hamster cells over a broad waveband from 254 to 313 nm when the action spectrum was determined with monochromatic radiation [67]. However, in a comparison of responses of the same type of cell cultures to monochromatic 254 nm radiation and polychromatic UV-B from a sunlamp, the survival curves and DNA lesion frequencies did not support the hypothesis that the same DNA photolesion was primarily responsible for cell death [68, 69]. Thus, polychromatic UV-B radiation may produce different kinds and proportions of DNA photolesions than monochromatic UV radiation. These differences in turn may result in different physiological responses. For cultured mammalian cells, this was expressed in different characteristic survivorship curves.

Such evidence should encourage caution in the application of action spectra determined with monochromatic radiation. Various repair systems activated by light may also complicate spectral assessments. Photoreactivation is a process that repairs a certain type of DNA damage. This mechanism is driven by long-wavelength UV and short-wavelength visibile radiation and can greatly diminish the resultant damage caused by UV-B radiation. Plant photosynthesis also appears to be much less inhibited by UV-B radiation in the presence of strong visibile radiation than if leaves are subjected only to UV-B radiation or to UV-B radiation in conjunction with dim visible radiation. Although it is unlikely that photoreactivation per se is effective in repairing photosynthetic damage, some mechanism is operating to alleviate the inhibition. Thus, repair systems or other interactions that might alter the expression of UV damage in plants diminishes the usefulness of biological action spectra determined with monochromatic radiation. To be useful in an ecological context, weighting functions should be verified by exposing organisms to different spectral distributions of polychromatic radiation to determine if the response is consistent with a proposed action spectrum. Such an approach has seldom been undertaken.

Epilog

Although an assessment of the impact of enhanced levels of UV-B radiation upon the biosphere is fraught with complications and caveats, the evidence that solar UV-B radiation is a stressful agent is growing. It has been demonstrated that exclusion of solar UV-B radiation from the current solar spectrum enhances primary production [28, 29, 70]. Although it does not necessarily follow that enhanced levels of solar UV-B radiation will further decrease productivity from current levels, one might assume that organisms do not possess repair mechanisms which function to "pre-adapt" the organism to levels of radiation not previously encountered [71].

As determined from the work of Van Dyke and Thomson [30], alteration of a marine algal community structure may result from exposure of microcosms to enhanced levels of UV-B radiation. A similar impact of UV-B radiation on the community structure of marine microcosms was noted by Worrest et al. [31]. Also noted in the latter study was depressed chlorophyll a concentrations, reduced biomass, and increased autotrophic indexes. Both of these studies utilized artificial (relatively low-level) light sources in the laboratory, whereas in an investigation by Worrest et al. [35], marine microcosms were exposed to about 90% of the natural, visible fluence. The results of the study parallel those found by Worrest et al. [31] and Van Dyke and Thomson [30] which, in part, corroborates the previous laboratory results.

There are, also, subtle effects of solar UV-B radiation mentioned in the literature which may have a significant impact upon ecosystems. In some cases plant growth [17] and shellfish and finfish reproduction [53, 59] are restricted to certain times of the year, usually spring. The environmental factors influencing this period of development involve temperature, moisture, day length, nutrients, and possibly other factors. The amount of UV-B radiation early in the growing season, when young UV-B-radiation-sensitibe plants and animals are developing, may not exceed the threshold under unperturbed conditions. However, given a reduction of stratospheric ozone, the UV-B irradiance may exceed threshold levels for some days, thus shortening the reproduction/development period.

Karanas et al. [51] have noted that the fecundity of a marine copepod is reduced by nonlethal levels of UV-B radiation. This impact upon reproductive rates was also noted by Worrest et al. [36] and utilized to determine the impact of enchanced UV-B radiation upon a model ecosystem. Shifts in community composition occurred in the model, a result corroborated by a subsequent study involving naturally recruited marine ecosystems [35]. Fox and Caldwell [72] noted a similar shift resulting from UV-B exposure in an investigation into the competitive interaction between pairs of plant populations.

The interpretations of changes in community diversity is subject to speculation; but one can say that, if the species composition of the primary producers is altered, the quality and possibly the quantity of food present for primary consumption could also be altered. Therefore, organic carbon exchange between trophic levels could be affected. The impact would be significant if the organisms selected for by enhanced UV-B radiation were of lower nutritional value (i.e., if they impacted growth and fecundity of the consumers). Another effect might be to alter the size distribution of the component producers in the ecosystem. For example, in a marine ecosystem, decreasing the size of the representative producers upon which consumers graze can significantly increase the energy allotment required for consumption, thereby reducing the feeding efficiency of the consumer. As stated by Worrest et al. [35], the actual impact of altering the community composition of an ecosystem requires further investigation.

REFERENCES

1. Halldall, P., Photoinactivations and their reversals in growth and motility of the green alga Playmonas (Volvocales), Physiol. Plant., 14:558-575 (1961).
2. Van, T. K., and Garrard, L. A., Effect of UV-B radiation on net photosynthesis of some C_3 and C_4 crop plants, Proc. Soil Crop Sci., Florica, 35:1-3 (1975).

3. Sisson, W. B., and Caldwell, M. M., Photosynthesis, dark respiration, and growth of Rumex patientia L. exposed to ultraviolet irradiance (288 to 315 namometers) simulating a reduced atmospheric ozone column, Plant Physiol., 58:563-568 (1976).
4. Sisson, W. B., and Caldwell, M. M., Atmospheric ozone depletion: Reduction of photosynthesis and growth of a sensitive higher plant exposed to enhanced UV-B radiation, J. Exp. Bot., 28: 691-705 (1977).
5. Van, T. K., Garrad, L. A., and West, S. H., Effects of UV-B radiation on net photosynthesis of some crop plants, Crop. Sci., 16:715-718 (1976).
6. Van, T. K., Garrard, L. A., and West, S. H., Effects of 298-nm radiation on photosynthetic reactions of leaf discs and chlorplast preparations of some crop species, Environ. Exp. Bot., 17:107-112 (1977).
7. Brandle, J. R., Campbell, W. F., Sisson, W. B., and Caldwell, M. M., Net photosynthesis, electron transport capacity, and ultrastructure of Pisum sativum L. exposed to ultraviolet-B radiation, Plant Physiol., 60:165-169 (1977).
8. Basiouny, F. M., Van, T. K., and Biggs, R. H., Some morphological and biochemical characteristics of C_3 and C_4 plants irradiated with UV-B, Physiol. Plant, 42:29-32 (1978).
9. Teramura, A. H., Biggs, R. H., and Kossuth, S., Effects of ultraviolet-B irradiances on soybean, Plant Physiol., 65:483-488 (1980).
10. Berg, R. H., and Garrard, L. A., The production of tobacco haploid plants: Utilization for UV-B-stree ultrastructural studies, Proc. Soil Crop Soc., Florida, 36:160-164 (1976).
11. Dickson, J. G., and Caldwell, M. M., Leaf development of Rumex patentia L. (Polygonaceae) exposed to UV irradiation (280-320 nm), Amer. J. Bot., 65:857-863 (1978).
12. Lindoo, S. J., and Caldwell, M. M., Ultraviolet-B radiation-induced inhibition of leaf expansion and promotion of anthocyanin production, Plant Physiol., 61:278-282 (1978).
13. Lindoo, S. J., Seeley, S. D., and Caldwell, M. M., Effects of ultraviolet-B radiation stress on the abscisic acid status of Rumex patentia leaves, Physiol. Plant, 45:67-72 (1979).
14. Teramura, A. H., Effects of ultraviolet-B irradiances on soybean: I. Importance of photosynthetically active radiation in evaluating ultraviolet-B irradiance effects on soybean and wheat growth, Physiol. Plant, 48:333-339 (1980).
15. Kossuth, S. V., and Biggs, R. H., Ultraviolet radiation affects blueberry fruit quality, Sci. Hort., 14:145-150 (1981).
16. Brabham, D. E., and Biggs, R. H., Cis-trans photoisomerization of abscisic acid, Photochem. Photobiol. (1981), in press.
17. Krizek, D. T., Influence of ultraviolet radiation on germination and early seedling growth, Physiol. Plant, 34:182-186 (1975).
18. Robberecht, R., and Caldwell, M. M., Leaf epidermis transmittance of ultraviolet radiation and its implications for plant sensitivity to ultraviolet-radiation induced injury, Oecologia (Berl.), 32:277-287 (1978).

19. Semeniuk, P., and Stewart, R. N., Seasonal effect of UV-B radiation on poinsettia cultivars, J. Amer. Soc. Hort. Sci., 104: 246-248 (1979).
20. Semeniuk, P., and Stewart, R. N., Comparative sensitivity of cultivars of Coleus to increased UV-B radiation, J. Amer. Soc. Hort. Sci., 104:471-474 (1979).
21. Ambler, J. E., Krizek, D. T., and Semeniuk, R., Influence of UV-B radiation on early seedling growth and translocation of ^{65}Zn from cotyledons in cotton, Physiol. Plant, 34:177-181 (1975).
22. Vu, C. V., Allen, L. H., and Garrard, L. A., Effects of supplemental ultraviolet radiation (UV-B) on growth of some agronomic crop plants, Proc. Soil Crop Sci. Soc., Florida, 38:59-63 (1978).
23. Biggs, R. H., Kossuth, S. V., and Teramura, A. H., Response of 19 cultivars of soybeans to ultraviolet-B irradiances, Physiol. Plant. (1981), in press.
24. Garrard, L. A., Van, T. K., and West, S. H., Plant response to middle ultraviolet (UV-B) radiation: Carbohydrate levels and chloroplast reactions, Proc. Soil Crop Sci. Soc., Florida, 36: 184-188 (1976).
25. Kossuth, S. V., and Biggs, R. H., Sunburned blueberries, Proc. Fla. State Hort. Soc., 91:173-175 (1978).
26. Chang, D. C., and Campbell, W. F., Responses of Tradescantia stamen hairs and pollen to UV-B irradiation, Environ. Exp. Bot., 16:195-199 (1976).
27. Jerlov, N. G., Irradiance, in: Marine Optics, pp. 127-150, Elsevier Scientific, Amsterdam (1976).
28. Steemann Nielsen, E., On a complication in marine productivity work due to the influence of ultraviolet light, J. Cons. Cons. Int. Explor. Mer., 29:130-135 (1964).
29. Jitts, H. R., Morel, A., and Saijo, Y., The relation of oceanic primary production to available photosynthetic irradiance, Aust. J. Mar. Freshwater Res., 27:441-454 (1976).
30. Van Dyke, H., and Thomson, B. E., Response of a simulated estuarine community to UV irradiation, in: Impacts of Climatic Change on the Biosphere, Part I: Ultraviolet Radiation Effects, D. S. Natchtwey, eds., pp. 5-95 to 5-113, U.S. Dept. Transportation, DOT-TST-75-55, Washington, D.C. (1975).
31. Worrest, R. C., Van Dyke, H., and Thomson, B. E., Impact of enhanced simulated solar ultraviolet radiation upon a marine community, Photochem. Photobiol., 27:471-478 (1978).
32. Worrest, R. C., Brooker, D. L., and Van Dyke, H., Results of a primary productivity study was affected by the type of glass in the culture bottles, Limnol. Oceanogr., 25:360-364 (1980).
33. Wolniakowski, K. U., The physiological response of a marine phytoplankton species, Dunaliella tertiolecta, to mid-wavelength ultraviolet radiation, M.S. Thesis, Oregon State University, 101 pp. (1980).

34. Smith, R. C., Baker, K. S., Holm-Hansen, O., and Olson, R., Photoinhibition of photosynthesis in natural waters, Photochem. Photobiol., 31:585-592 (1980).
35. Worrest, R. C., Thomson, B. E., and Van Dyke, H., Impact of UV-B radiation upon estuarines microcosms, Photochem. Photobiol. (1981), in press.
36. Worrest, R. C., Wolniakowski, K. U., Scott, J. D., Brooker, D. L., Thomson, B. E., and Van Dyke, H., Sensitivity of marine phytoplankton to UV-B radiation: Impact upon a model ecosystem, Photochem. Photobiol., 33:223-227 (1981).
37. Calkins, J., and Thordardottir, The ecological significance of solar UV radiation on aquatic organisms, Nature, 283:563-566 (1980).
38. Thomson, B. E., Worrest, R. C., and Van Dyke, H., The growth response of an estuarine diatom (Melosira nummuloides [Dillw.] ag.) to UV-B (290-320 nm) radiation, Estuaries, 3:69-72 (1980).
39. Anderson, D. E., Cancer eye in cattle, Mod. Vet. Pract., 51:43-47 (1970).
40. Kopecky, K. E., Ozone depletion: Implications for the veterinarian, J. Amer. Vet. Med. Assoc., 173:729-733 (1978).
41. Kopecky, K. E., Pugh, G. W., Jr., Hughes, P. E., Booth, G. D., and Cheville, N. F., Biological effects of ultraviolet radiation on cattle: Bovine ocular squamous cell carcinoma, Amer. J. Vet. Res., 40:1783-1788 (1979).
42. Hughes, D. E., and Pugh, G. W., A five year study of infectious bovine kerato-conjunctivities, J. Amer. Vet. Med. Assoc., 157:443-451 (1970).
43. Rehbinder, C., Keratitis in reindeer, Acta Vet. Scand., 18:75-85 (1977).
44. Barcelo, J., Photoeffects of visible and ultraviolet radiation on the two-splotted spider mite, Tetranychus urticae, Photochem. Photobiol., 33:703-706 (1980).
45. Bell, G. M., and Hoar, W. S., Some effects of ultraviolet radiation on sockeye salmon eggs and alevins, Can. J. Res., 28:35-43 (1950).
46. Dunbar, C. E., Sunburn in fingerling rainbow trout, Prog. Fish Cult., 21:74 (1959).
47. Marinaro, J. Y., and Bernard, M., Contribution à l'etude des oeufs et larves pélagiques de Poissons méditerranéens, I. Note preliminarie sur l'influence léthale du rayonnement solaire sur les oeufs. Pelagos, 6:49-55 (1966).
48. Hunter, J. R., Taylor, J. H., and Moser, H. G., Effects of ultraviolet irradiation on eggs and larvae of the northern anchovy, Engraulis mordax, on the Pacific mackerel, Scomber japonicus, during the embryonic stage, Photochem. Photobiol., 29:325-338 (1979).
49. Hunter, J. R., and Kaupp, S. E., Effects of solar and artificial ultraviolet-B radiation on larval northern anchovy, Engraulis mordax, Photochem. Photobiol. (1981), in press.

50. Karanas, J. J., Van Dyke, H., and Worrest, R. C., Midultraviolet (UV-B) sensitivity of Acartia clausii Giesbrech (Copepoda), Limnol. Oceanogr., 24:1104-1116 (1979).
51. Karanas, J. J., Worrest, R. C., and Van Dyke, H., Impact of UV-B radiation (290-320 nm) on the fecundity of Acartia clausii (Copepoda), Mar. Biol. (1981), in press.
52. Damkaer, D. M. Dey, D. B., Heron, G. A., and Prentice, E. F., Effects of UV-B radiation on near-surface zooplankton of Puget Sound, Oecologia (Berl.), 44:149-158 (1980).
53. Damkaer, D. M., Dey, D. B., and Heron, G. A., Dose/dose-rate responses of shrimp larvae to UV-B radiation, Oecologia (Berl.) (1981), in press.
54. Jokiel, P. L., Solar ultraviolet radiation and coral reed epifauna, Science, 207:1069-1071 (1980).
55. NAS, Committee on Impacts of Stratospheric Changes, in: Protection Against Depletion of Stratospheric Ozone by Chlorofluorocarbons, National Academy of Sciences, Washington, D.C., 392 pp. (1979).
56. E. I. Du Pont de Nemours and Company, Comments on the advance notice of proposed rulemaking - Ozone-depleting chlorofluorocarbons: proposed production restriction, Wilmington, Delaware (1981).
57. Caldwell, M. M., Robberecht, R., and Billings, W. D., A steep latitudinal gradient of solar ultraviolet-B radiation in the arctic-alpine life zone, Ecology, 61:600-611 (1980).
58. Robberecht, R., Caldwell, M. M., and Billings, W. D., Leaf ultraviolet optical properties along a latitudinal gradient in the arctic-alpine life zone, Ecology, 61:612-619 (1980).
59. Hunter, J. R., Kaupp, S. E., and Taylor, J. H., Assessment of effects of UV radiation on marine fish larvae, Proceedings of the NATO Advanced Research Institute on the Role of Solar Ultraviolet Radiation in Marine Ecosytems, Copenhagen, Denmark, 1980 (1981).
60. Smith, R. C., and Baker, K. S., Stratospheric ozone, middle ultraviolet radiation, and carbon-14 measurements of marine productivity, Science, 208:592-593 (1980).
61. Smith, R. C., and Baker, K. S., Biologically effective dose transmitted by culture bottles in ^{14}C productivity experiments, Limnol. Oceanogr., 25:364-366 (1980).
62. Jones, L. W., and Kok, B., Photoinhibition of chloroplast reactions, 1. Kinetics and action spectra, Plant Physiol., 41: 1037-1043 (1966).
63. Green, A. E. S., and Miller, J. H., Measures of biologically effective radiation in the 280-340 nm region, in: Impacts of Climatic Change in the Biosphere, Part I: Ultraviolet Radiation Effects, D. S. Nachtwey et al., eds., pp. 2-60 to 2-70, U.S. Dept. Transportation, DOT-TST-75-55, Washington, D.C. (1975).
64. Setlow, R. B., The wavelengths in sunlight effective in producing skin cancer: A theoretical analysis, Proc. Nat. Acad. Sci. USA, 71:3363-3366 (1974).

65. Caldwell, M. M., Solar ultraviolet radiation as an ecological factor for alpine plants, Ecol. Monogr., 38:243-268 (1968).
66. Green, A. E. S., Cross, K. R., and Smith, L. A., Improved analytic characterization of ultraviolet skylight, Photochem. Photobiol., 31:59-65 (1980).
67. Rothman, R. H., and Setlow, R. B., An action spectrum for cell killing and pyrimidine dimer formation in Chinese hamster V-79 cells, Photochem. Photobiol., 29:57-62 (1979).
68. Elkind, M. M., Han, A., and Chang-Liu, C.-M., "Sunlight"-induced mammalian cell killing: A comparative study of ultraviolet and near-ultraviolet inactivation, Photochem. Photobiol., 27:709-715 (1978).
69. Elind, M. M., and Han, A., DNA single-strand lesions due to "sunlight" and UV light: A comparison of their induction in Chineses hamster and human cells, and their fate in Chinese hamster cells, Photochem. Photobiol., 27:717-724 (1978).
70. Lorenzen, C. J., Ultraviolet radiation and phytoplankton photosynthesis, Limnol. Oceanogr., 24:1117-1120 (1979).
71. Conrad, M., Biological adaptability: The statistical state model, BioScience, 26:319-324 (1976).
72. Fox, F. M., and Caldwell, M. M., Competitive interaction in plant populations exposed to supplementary ultraviolet-B radiation, Oecologia (Berl.), 36:173-190 (1978).

CHAIRMAN'S SUMMARY

Claud S. Rupert

The risks considered in this final session bear some resemblance to those associated with nuclear radiations and toxic wastes, discussed earlier in the meeting. Like those other risks, they arise from unintended side effects of large scale human endeavors toward desirable ends. But their estimation involves also several layers of uncertainty concerning both scientific facts and future human actions. These uncertainties raise additional problems about the meaning and evaluation of "actual" risks in relation to those "perceived."

By the early 1970's a growing understanding of the earth's atmosphere made it evident that man-made materials released into the air might alter processes in the stratosphere, 10 to 50 km above the surface of the earth, and change the concentrations of critical constituents there. While 10 to 50 km is a substantial distance up and the constituents affected are present in amounts ranging from only parts per million to parts per billion, such changes might not be inconsequential. Ozone, for example, is a major absorber of solar radiation; and its concentration - although always small - affects both the heat budget of the stratosphere, and the amount of short wavelength ultraviolet radiation reaching the surface of the earth. Changes in both these quantities could be matters of concern. The changed heating should have some effect on atmospheric dynamics, and therefore possibly on climate - although the details would be hard to forsee - and the changes in solar ultraviolet radiation (UV) had an obvious point of impact in biological organisms exposed to daylight.

Some knowledge exists about the biological action of UV. This radiation is strongly absorbed in constituents of all living cells, where it can induce abnormal photochemical reactions. Particularly in nuclear acids, but also in proteins and other materials, most effects of these reactions are detrimental to organisms. Fortunately normal concentrations of stratospheric ozone absorb most of this UV and only a minor "fringe" leaks past the long wavelength edge of the main ozone absorption to reach the surface of the earth. Sunburning, germicidal and other damaging effects of sunlight are essentially due to this so-called "UV-B" radiation; and its worldwide increase with any ozone depletion would increase the magnitude of all such effects on plants and animals.

The first concerns about stratospheric alteration were focussed on effects of nitrogen oxides released by supersonic aircraft flying in the mid-stratosphere, but as the economic attractiveness of these aircraft faded, attention turned more toward materials released at the surface: nitrous oxide (N_2O) arising from natural microbial degradation of nitrogen fertilizers (particularly when they are heavily applied), and chlorofluorocarbons widely used in industrially developed countries. Both are important substances, and neither could be lightly curtailed.

The problem is: 1) Given some reasonable scenario for emission of these materials into the air, what depletion of stratospheric ozone would be expected? 2) Considering the ultraviolet radiation increase which would accompany such a depletion, what biological effects on humans, plants, and animals would be expected? The uncertainties in anticipated consequences of various material emissions are compounded of the uncertainties in answers to both these questions.

As Dr. Hudson has indicated, ozone is continually made and destroyed in a maze of chemical reactions (about 100 in all) taking place in the stratosphere, so that the effect of a particular stratospheric pollutant depends on the rates of a number of different reactions. Consequently, the predictions change where advances in knowledge change the values assigned to reaction rate constants. The effect of a particular pollutant also depends on the concentrations of many other reacting materials including other pollutants; and the predictions, therefore, change with changes in the scenario adopted for release of such pollutants into the air. This multiplicity of governing factors adds a certain volatility to the esimates of possible ozone depletion. While the calculated numerical values have not changed greatly in the last 5 years (still lying within earlier estimates of their uncertainty), the relative importance of nitrogen oxides and chlorofluorocarbons has been shifted, changing the relative impact of different human activities on the stratosphere. Such a shift is awkward in communicating the situation accurately to non-specialists and raises questions about the meaning of the risk estimate itself.

CHAIRMAN'S SUMMARY 319

When one is trying to foresee a problem in order to prevent it, when the chain of reasoning is long and complex, and when successive improvements in understanding of the system have qualitatively altered the prediction about which human activities are most closely linked to the anticipated problem, what is the meaning of the "actual" risk associated with these activities? What is the risk that the risk estimates themselves are flawed? By what criteria should one decide when the predictions are reliable?

In these circumstances, a natural reaction might be to "wait-and-see," deferring any corrective action until it is clear that something detrimental is happening. But this course has its own risks. The slow transport of materials from the troposphere into the stratosphere means that waiting for signs of visible trouble will load up the troposphere with a larger quantity of material to leak into the stratosphere afterward, thereby increasing the difficulty should it then be desired to stop the pollution. In the case of long-lived materials, like chlorofluorocarbons, one would be committed to a final ozone depletion somewhat larger than that existing when corrective action was initiated, followed by very leisurely recovery. The entire excursion back to the ozone levels before counter measures were taken would take a major part of a human lifetime. Clearly one must make some choice of risks and then live with them.

There is no question that sufficient UV-B radiation will produce serious damage, leading to strong physiological responses in the cells of plants, animals, and humans; but the increase in damage accompanying moderate ozone depletions, such as might arise from forseeable stratospheric pollution are not easily estabished. It has been determined in laboratory tests that different agricultural plants have different UV-sensitivities and that about 20% of the cultivars tested show damage from daily UV-B exposures similar to those of Southern U.S. sunlight. However, it has also been determined that sensitivities in controlled growth chambers under artificial illumination are higher than in the easily controlled (but agriculturally normal) conditions of the open field - a fact which hampers testing considerably.

Similarly it is known that UV-B at current sunlight levels can injure organisms of aquatic ecosystems living near the water surface. This damage depresses the light-saturated rate of photosynthesis in primary food producers, and diminishes the survival of eggs and larval forms of at least some animals valuable for fishery purposes. However, the impact of UV-B on production of the overall system is not yet known. Not only can there be some protection from UV-B by movement deeper into the water column, but also only a small proportion of larvae survive to adult life anyway, such that the effect of additional UV killing on the size of final population is not clear.

As Dr. Worrest has outlined, a number of technical problems exist in testing organisms with supplementary UV-B radiation. While it should be possible to overcome these difficulties and to finally determine any possible economic impact of a specified ozone depletion in agricultural output and fishery yields, this cannot be done presently. It is neither possible to show with assurance that there would or would not be significant losses in these important food sources. How one reacts to this uncertainty depends on whether the individual is more concerned with the economic consequences of interfering with ongoing industrial or agricultural operations or with a worldwide environmental change of unknown dimensions.

Concern over the effects of UV-B on humans, where the welfare of the individual is a prime consideration, centers on skin cancer. As Dr. Urbach outlined for us, skin cancers fall into two different groups: malignant melanomas and all others (the latter, therefore, frequently referred to as the "non-melanoma skin cancers"). While melanomas are deadly, usually with rapid development and metastasis after onset so that treatment is unlikely to be successful, non-melanoma cancers are usually removable by surgery with subsequent full recovery. (The removal does entail cost and discomfort, however, as well as occasional disfigurement; and the cancers would be dangerous if untreated).

The cause of malignant melanoma is not known, but non-melanoma skin cancers are clearly linked by the circumstances of their occurrence and, by parallel animal experiments, to chronic ultraviolet radiation exposure. Although only rough estimates can be made of the increase expected with plausible ozone depletions, it seems clear that this would be modest compared with current incidence and that a very considerable portion of existing skin cancers could be avoided by more prudent sunlight exposure in the population. Indeed, it might be possible to reduce skin cancer incidence in the face of an appreciable ozone depletion if the population could be persuaded to control its sunlight exposure more carefully. Here we have another curious example in the sociology of risk acceptance: people are willing to undertake some demonstrable risks voluntarily but are deeply concerned about others where no certain risks can be shown. The factors involved in these risk perceptions certainly need to be better understood.

If the rates at which man-made pollutants contribute to stratospheric ozone destruction were an order of magnitude less than they are, the prospective effects on human welfare would be considered trivial; and we would not be discussing them here. If these rates were an order of magnitude greater, there would be obvious environmental hazards which we would attempt to control. Instead, they fall awkwardly in between; and effects of the pollutants depend on many details, some of them not well understood. This is also true of other substances that are released in the operation of a technological civilization and that produce changes in the environment. Ozone depletion is, therefore, only one example of a more general problem that we must learn to handle satisfactorily.

PANEL DISCUSSION

DR. NELSON

...Now we will ask each of the panel chairmen to present 10 or 15 minute distillation of their experience and their perceptions of the proceedings of the past few days. Then there will be some opportunity for questions from the floor. Then I will make some very brief remarks. First, Dr. Hohenemser.

DR. HOHENEMSER

On the first day, I was the chairman of the auto panel with the unpronouncible name; and I thought that the panel's presentation was so clear and the issues so well debated that it seems a shame to try to improve on it or discuss it again. I think there were good opinions expressed on both sides. Therefore, I'd like to sum up by discussing another topic.

This is a conference about perceived vs. actual risk; and by "actual" what is meant is scientifically or objectively defined risk as opposed to subjectively defined risk - that is what most of us meant. In general, we found that this actual risk really dominated the conference and furthermore that this actual risk was defined almost always in terms of mortality risk, that is, a conditional probability of dying in some way or another. And the conditions of dying were sometimes very complicated, as the last session showed. We have had many mortality risk numbers, whose complexity was exemplified by Hans Joksch's opening talk. Beyond that, Jeff Harris spoke at length about various ways to express the objective risk of cigarette smoking in the hopes that one of the ways would be understandable to the public or would get them to form their perceptions correctly. Leonard Hamilton spoke about the objective risk of nuclear power in terms of mortality. The whole cancer symposium was about mortality; and Chauncey Starr's new target criterion for risk acceptability was expressed in terms of mortality. Throughout this, perceived risk was thought of as the failure to comprehend the objective risk correctly.

We will know there is something peculiar going on with perceived risk; and I would like to remind you that insofar as any evidence about perceived risk has been presented here, it is mainly from the work of the Decision Research Group consisting of Paul Slovic and Baruch Fischhoff and Sarah Lichtenstein. Based on this work there's really no reason to believe that perceived risk has much to do with mortality rate. Therefore, the lay public and we many be talking past each other, if we continuously quote mortality rates or mortality experience when we refer to risk.

I think there is a solution to the problem: I think we can develop ways of defining the risk of technology in a more sophisticated and perhaps a more complex way than mortality numbers. Our group at Clark University (namely Bob Kates, Roger Kasperson and several graduate students) has been thinking about an extended definition of "risk" in our recent work. We think that as a substitute for "risk" as mortality one should define something which we call "hazardousness." We use "hazardousness" to denote scientifically defineable characteristics of hazards that can be referred to objectively just as are those mortaility numbers. Recently we have written down and coded the "hazardousness" of 93 different technological hazards! Let me show you very quickly what we have done. (This is really to entice you to write us for more information and then I'll close up with the moral of that story.)

We thought of hazards as a causal sequence of events which end with some kind of human or biological consequence, maybe death. It begins with human needs and wants; and in between there is a choice of technology which may then lead to some harmful consequences. What we have tried to do is to measure characteristics of these causal sequences in various way, actually in terms of 12 or 16 variables. The names of the variable are written down below each stage of causal evolution. At the far right, there are elements that have to do with exposure-consequence relations: for example, the delay of consequences after exposure. Further up the line, we characterized the releases that are involved in the hazardous sequence in terms of reasonably well defined physical variables such as spacial extent, concentration, persistence, reoccurrence time, etc. And way up the line, where choice of technology is involved, we have introduced some variables which are more or less economic in nature such as market value, substitutability and soforth. We have taken these variables and concocted out of them a more general definition of "hazardousness." Our general definition of hazardousness is essentially the profile of scores which we obtained on each 12-16 descriptors of causal sequences. All of these variables or descriptors enter the discussion and are of concern to people who are trying to judge the hazard, even though they sometimes can't determine what the mortality numbers are.

The next slide shows our list of hazards. We scored all of these hazards on the scales that were shown on the previous slide. For example, the spacial extent of various hazards varies from one square meter to 10^{14} square meters and so you need factor of 10 scales in order to span such a range. Furthermore, we realized that there is so much argument about the details of the hazard chain that it doesn't make much sense to define variable scores closer than by a factor of 10. Using these factors of 10 scales, we scored all of our 93 hazards and came up with a set of scale ratings for each of them.

I don't want to tell you what all of those scale ratings are - that's a large statistical matrix; but let me say that we referred, as much as possible, to the scientific literature. Now for 93 hazards in 16 dimensions, it takes quite a bite of work, and our judgments are surely preliminary; but we think they are right within a factor of 10 or maybe a hundred. If that kind of rough scoring was anything at all to do with the problem, then our scores would mean something.

As one test of our broader definition of risk (that is, what we called "hazardousness") we compared our results to lay perception, by getting Slovic and Company to ask people questions about those same hazard dimensions using our list of hazards. Thus, Decision Research asked lay people to judge, for example, the spacial extent, the delay of consequences, the reoccurrence time and a variety of other things that we had scientifically scored. Our scientific judgments are compared to lay judgments in the next slide. If you are interested in correlation coefficients, those correlations coefficients are all rather high, 0.9 or so. If you are interested in scatter, the graph shows you that there are many cases of hazards that are "misjudged" by the lay public by a factor of a 1000 or more. So you can see it either way - the glass is half full or half empty depending upon your viewpoint.

There is a particular feature here: If lay judgment were in perfect agreement with our scientific estimates the lines would have a 45° slope. None of them have this slope; and all judgments are "compressed." Thus, lay people in general don't perceive the very large ranges that scientists do. We conclude that in a very broad sense people can understand the dimensions of hazards that we define, but that there are important differences between lay and scientific judgments of our scales.

The next two slides show a comparison of our scientifically judged risk dimensions with lay judgments of "perceived risk." Those of you who are familiar with the Decision Research work will know that they have for a long time measured something called "perceived risk" - which is basically an answer to the question: on that scale of 1 to 100, tell us how risky this activity or that hazard is. And this global variable, "perceived risk," does not involve just mortality but involves a variety of qualities beyond mortality. You can see from our table that about the only dimension of hazard that shows no correlation with the global variable "perceived risk" is "annual mortality." So people know about all those qualities of risk which we have "objectively judged;" and they produce correlation coefficients of 0.4 or so for "perceived risk" and these qualities, but the correlation coefficient for annual mortality and for a couple other things such as reoccurrence and concentration time is nonexistent in a statistical sense.

In the last slide, we have done a related thing, using a reduced set of dimensions (these are five factors which were gotten by factor analysis from the 11 dimensions). The factors are defined to be orthogonal. Again, factor 4, which involves mortality, doesn't correlate very well at all with "perceived risk." So the bottom line of this comparison (using factors) is that we can explain 50 to 70% of the variance in "perceived risk" across 81 hazards.

Now let me draw the moral from this story. I think that as risk analysts we really need to ask ourselves not whether we should adopt our way of looking at the problem, but whether it is reasonable to think of "risk" as equal to mortality. It seems to me that the lesson of our exercise is that there is an alternative. We know already from many sources that the public is concerned about a broader set of issues than mortality numbers; they frequently misunderstand mortality numbers. We should try to find out the concerns of the public, and then try to define those concerns scientifically and objectively. In this way we can agree that we used correct scientific methods to get at the answer, once the right questions has been asked. In this way, we might also be on the way of bringing the public into a central role of our business. Thus we might find out whether the public is worried about delayed hazards as opposed to immediate hazards; whether they are worried about transgenerational effects or not, and soforth. Given answers to such questions we can objectively define hazards using the best tools of science. I proposed that this is an alternative to simply continuing to quote elegantly derived mortality numbers, which then are misunderstood by the public because they don't tell the public about things they are interested in.

DR. O'RIORDIAN

Thank you very much. Like Dr. Hohenemser, I wish to raise new items on the agenda rather than simply to repeat what was said, though I also wish to make a few comments about the session that I chaired. Let us look at the diagram. The idea is to indicate there isn't such a thing as a lay public, but a large number of different publics that are called "lay" and that, at probably different times of the day certainly different times of the year, we may well take up many different views of the world making it difficult for those who are trying to interact with the lay public in terms of perceptions or risk. Now I won't go into this diagram in detail, since I have discussed this elsewhere (O'Riordan 1980).

Broadly speaking you can look at the way in which the environmental movement operates in two very general categories. One we may call <u>technocentricism</u>, basically a view of the world which holds that it can be contained and managed and that growth can continue. The other is called <u>ecocentricism</u> which days that the world can't be contained and managed and that growth can't continue. Ecocentrists argue, therefore, that major alterations have to be made to human institutions in accordance with biological constraints. This is not a new philosophy: its roots were nurtured by the transcendentalists in this country. But nowadays it has new meanings.

If one subdivides technocentrism, one finds that there are two inputs coming in from the right (and the right is politically the way one should see it as well). We have those who may be termed cornucopians who believe there is no limit to man's ingenuity to cope with his problems, and then there are those who can be described as accommodators. Accommodation means that one adjusts but one never really fundamentally changes the way in which power is distributed in society. That is, democratic institutions are not changed, nor is the political economy of the Western world. Accommodation is an ever varying and constantly dynamic phenomenon, but it involves no fundamental shift in political power, a factor to bear in mind. If one goes farther to the left into the ecocentric camp, we are speaking about two groups of people: those who believe in a small decentralized society, the so called small-is-possible camp who I will call federalists. They prefer to rely on a decentralized economy based upon the so-called "soft-technology" based on reusable energy, durable materials and collective maintenance. Finally, there are those who are known in the American literature as the "deep ecologists." These people are basically an offshoot of the transcendentalist movement reincarnated 150 years later and who firmly believe that society ought to constrain its whole operations in relation to biological imperatives.

Each of their major world views are but modes in the pattern of human beliefs and there is no doubt that groups of people at different times of their lives align firmly with these groups. When one therefore looks at the reactions to risk and the contexts of those three or four modes, one will find very, very different responses. It is my firm belief that they also differ from one person to another at different times in their lives. Society is generally shifting leftward from cornucopianism to accommodation currently the fashionable mode. It includes such devices as environmental impact assessment in risk analysis. Risk analysis is really a form of accommodation at the end of the day in terms of its operation in the technical system. But we haven't yet gotten to the point of a major shift into the ecocentric mode, which a lot of people would like to see. In order to understand the distinction between real vs. perceived risk, one must do so in the context of these belief patterns for they are in turn influenced by the way in which people see the world in the context of their social and political position. The weaker members of society, politically speaking, tend to be found in the left hand side of these columns. They are the ones who want to get a toehold on power, and they are using the environmental motif as a way of altering the world or changing the system in which they find themselves, a system which they find distasteful. There is certainly an element of this feeling about the anti-nuclear movement. Therefore, criticisms of technological risk assessments in the nuclear area are likely to be coming particularly from those who espouse ecocentric philosophy.

Now when one sees how groups and society politically orientate themselves, we find that they are divided into three categories. The first is known as the private environmental group. They are really interested in protecting their own backyards, they are not particularly concerned about changing the mode of life in society. They tend to fall into the technocentric mode, but they divide between accommodations and cornucopianism. They are often very powerful groups for they have a unity of purpose often with a lot to fight for (for example, they are the ones that try to stop motorways and power plants - anything that might affect their own environmental amenity and well being).

The second group is the public interest environmental group espoused by frieds of the Sierra Club and who believe in causes rather than parochial issues but take these local affairs and treat them as principles upon which to join battle in the environmental cause. They generally straddle uneasily the accommodation motif and federated soft-technology motif. These groups are actually divided intellectually between different modes of thought. This is often an important reason why they fit uneasily into the political spectrum, not being sure whether to work from within or campaign from the outside. Then finally in the far left, one tends to find the so-called "green" parties which politically have never been very powerful, but which should never be ignored. In Europe particularly they continue to show up at general elections. At the recent French election, although the French Ecology Party didn't have a very large vote overall, they did hold the balance of power in terms of the choice of a president.

Thus, the way in which the public introduces debate into the political arena depends upon the way in which they mobilize their political ideologies and how they in turn relate to the belief patterns described above. Therefore I am suggesting that when we look at "lay" versus "expert" perception of risk, it would be very useful to look at these in the context of these wider political belief patterns in society. Also it may be helpful to bear in mind that if you simply take a group of students or a group of League of Women voters or a group of Sierra Club members, it would be unwise to assume that they espouse any single philosophy because they are in fact a complex bundle of different rationalities which vary from time to time. Therefore, I am not sure if there is such a thing as lay perception of risk. I think there is something much more confused than this, which emerges through the political debate rather than in the minds of people. I sense that I am being slightly provocative, for I am sure that many people would disagree.

Where does all this lead to in terms of the nuclear issue and what I have learned from this conference? Well again, let me be slightly controversial. What struck me particularly about the first day was the difference in societal response between automobile safety and nuclear safety. Now for automobile safety, we seem to have very clear information about the nature of the risk and we also have very clear information about what can be done to stop it, yet somehow society is still not doing enough to reduce the risks when the opportunities are there. On the other hand, we have a nuclear risk, a far less demonstrable risk as far as the public is concerned (this was stressed repeatedly in the session particularly by Chauncey Starr the following afternoon). Yet government and particularly the nuclear industry are bending over backwards to try to control that risk beyond what is regarded by technical people as a reasonable level. Chauncey Starr made it quite clear that the acceptable technical level of societal risk would involve a probability of 10^{-6}. Yet the nuclear industry is prepared to go to 10^{-7}. just to be on the safe side. And to protect themselves further they are prepared to spend hard cash to "buy out" those who feel they might still be suffering from residual risk. In my view, a utility may be making itself even more vulnerable because it is making a confession of guilt which could be easily exploited by ecocentrists.

Why is it that we have this major difference in social response between the two kinds of risk? Various psychologists point to the fear of the unknown and unfamiliar and the potentially catastrophic character of the nuclear-related risk. Within the automobile field, risk is rather well known, it is almost part of the public culture, and it is largely regarded as voluntary. Where social action is accepted is in the area of involuntary risk - for example by establishing laws to safeguard babies and young children who are clearly the ones who are most likely to suffer because they cannot understand what the risk probability is. In the nuclear area there is a major difference btween what the technologist regards as an acceptable level of risk and what the public and the politicians regard as safe enough. In addition there is the ethical issue of social equity. In automobile accidents, it may well be that the lower income members of society are more vulnerable than the higher income members of society partly because they are more likely to be in risky situations when driving and partly because they are less likely to put on any kind of seat belt or any other restraining apparatus. (Indeed they may not even have working seat belts.) Likewise we find a similar problem in cigarette smoking, namely that the poorer members of society tend to smoke more and to be more exposed to active smokers. They may not always know about the risk associated with passive smoking. In the nuclear field the distributional issue is not nearly so clear cut. It may well be that relatively well off members of society are equally at risk compared with those who are less well off. Perhaps there is a lesson to learn from that comparison. I suspect the same kind of situation happens with the ozone issue, but that is not quite so clear cut.

Turning now to the session which I chaired on Monday afternood, let me remind you of its purpose. This was to look at the role of major generic risk studies in the nuclear field, studies done by technically competent people, to see how these influenced the technical debate, influenced public opinion, altered both regulatory standards and political judgments and whether they led to any improvements in levels of safety in the design of new nuclear plants.

The papers that were presented were largely at an exploratory stage, but a number of introductory observations can be made.

1. The style of government generally plays a crucial part in determining how risk studies will be used. The US and Sweden enjoy open, adversarial governmental styles where information is legally available and various waves of reviewing committees can be set up to scrutinize the work of each other. In the US, no less than nine studies followed in the ake of the Three Mile Island (TMI) incident. The Swedes and the West germans also produced "post TMI" risk studies while the British concentrated on how the major faults could not be reproduced in the UK because of the different style of regulatory activity. In Britain and Canada no specific "post-TMI" study was published. In Sweden there have been five nuclear risk studies of one sort or another between 1974 and 1978. In both the US and Sweden, information shortage is not the problem, how to grapple with it and make sense for practical decisions is the major task.

In Britain, Canada, and West Germany information is not provided by official bodies unless authorized by ministerial mandate. This means that major risk studies are produced on official terms; and responses to them are usually kept confidential - or at least what is publicized is what is wanted to be publicized. It follows that the press plays a more limited role in commentating on independent assessments of official reports, and have to relay on investigative journalism (an activity which is in itself curtailed by laws of secrecy and libel) to obtain the "real" views of officials and experts. In these countries, policy making and standard setting are consultative activities involving discussions between various interested parties but excluding the general public and, normally, those who are officially described as "trouble makers."

It is too early to tell how national political style influences nuclear risk analyses, but one can suggest that events may be bringing about an interesting convergence. In the US, for example, there are moves to incorporate an element of confidential discussion between interested parties before the "set price" hearings are conducted and to allow the nuclear industry greater say in the enforcement of its own safety practices. Meanwhile the British and Canadians, long noted for official secrecy, have begun to open up their regulatory procedures to much wider public scrutiny (thus moving toward the Swedish and American approach) while continuing to place the burden of risk regulation and safety upon the industry itself.

2. The locus of risk regulation varies remarkably from country to country. In the US the prime responsibility rests with the federal regulatory agency which must set and enforce specific standards and is responsible if these standards, even when followed, prove to be inadequate. The British argue that this locus of responsibility is misplaced and give the regulatory agency a less significant role. The onus of safety rests with the electrical utility which must set its own standards (with guidance from the regulatory body) and supervise its own safety procedures. This should produce a safer plant, though there is some confusion as to how safety standards are actually set and who decides that certain minimum levels of safety are acceptable. So while the locus of responsibility in the US is public, in the UK it is private. In Sweden it is a mixture of the two with considerable emphasis being placed on special commissions of inquiry and the de facto veto power of the local government in the locality in which a nuclear plant may be sited. Here the locus of responsibility rests on a formal combination and political judgment (where it must lie whatever country is involved) coupled with independent studies by specially commissioned groups of experts.

The West Germans, like the British, take the view that the combination of independence and experise is unlikely to be found in any one individual, and in West Germany the final judgment is left to the courts utilizing the trained legal mind. In Britain the public inquiry is often a de factor court, for evidence is assembled and cross-examined by senior lawyers. However, the public inquiry is not always held (though each new reactor type requires a special ministerial consent which usually follows a public inquiry) and in any case it is only an administrative procedure. The final decision is left with the Energy Secretary who may take into account not only the findings of the inquiry but relevant government policy as well. The policy dimension is not examinable at the inquiry. In Canada the locus of responsibility is still governmental though the commission of inquiry, if held, can be remarkably democratic and, in political terms, influential.

In general one can observe that the interface between science and politics which is an essential aspect of risk analysis is recognized in all countries and that more often than not the intervention of the legal mind is called upon to clarify choices for the politicians. In some respects, this is not particularly remarkable given that this is part of the lawyer's brief; and all western democracies recognize the rigors of legal analysis.

3. Risk studies abound in the context of energy policy generally. What has emerged from these papers is that one cannot separate the role of risk/safety from the wider political context in which nuclear energy is promoted. Hence all the authors look at the relative position of the nuclear industry in the structure of the national energy supply picture and some try to analyze how national governments respond to the nuclear industry in the light of commitments to other energy supply industries, notably coal, oil, and gas. There is no doubt that where indigenous supplies of hydrocarbon fuels are poor, the nuclear option, risks and all, is viewed more favorably than in cases where other supply options exist. Thus Ontario is the only Canadian province that has gone nuclear because it had no reliable alternative source. Similarly France was far more pro-nuclear than the UK with its vast reserves of coal or Sweden with its scope for hydropower and solar based supplies. Generally speaking, the nuclear industry world wide is in the doldrums largely because new plants are not required in great numbers and safety regulations are constantly being reviewed. So the uncertainty is great.

What does begin to emerge from these studies is a feeling that where there are real political commitments to nuclear power, the problem of incorporating risk into policy and practice gives a government many headaches. Risk studies, therefore, have a much greater potilical dimension, in the sense that powerful interests want to know how nuclear plants can be made publicly acceptable, than a mere scientific one.

Another factor to emerge is the increasingly responsible role taken by the nuclear utilities. They know they cannot afford another major accident, and they also believe they must be seen to be responsive and receptive to public opinion. They must, therefore, set standards of safety which go far beyond the standards set for any other industry even when the costs are astronomic (especially if retrofitting is required). They also must be seen to be listening to public opinion and to respond to safety demands that may go well beyond what is 'necessary' in engineering terms. So there is a public dimension to risk standards which seem to surpass both the economic and scientific elements of rationality. Yet the industry must accede to these tough requirements (which seem to be getting tougher all the time) in order to court public and political popularity, even when there is no scientific basis for the standard set. This may prove to be its undoing, but it has little choice. Thus the industry is trapped in a difficult dilemma. It must bear the burden of proof of safety, yet when doing so it is vulnerable to all manner of accusations. The rapidity with which the industry world wide responded to the TMI event is a good demonstration of this.

The next phase of this study is to discover what are the connections between government economic policy generally, its energy strategy and finally its coal-nuclear investment plans in the light of these other two considerations. Only then can a proper assessment be made of the manner in which risk analyses are thought through and finally executed. This phase will also look at national responses to the risks associated with radioactive waste disposal. This may well prove to be the most intractable issue of all the fuel cycle risk studies, unless dumping sites can be found where there are no local inhabitants (e.g., deep sea disposal). Needless to say this is an area which anti-nuclear activists can exploit mercilessly, since here is the high point of the assymetry between expert judgment and lay judgment of risk. No country has yet solved this dilemma, which could, conceivably, stop the nuclear industry in its tracks.

DR. BAILOR

In sequence - sometimes in very rapid sequence - this entire conference has been stimulating, confusing, exhilarating, frustrating, and a good many other things; and I really see no good way to summarize. What I will do instead is to read to you some of my haphazard notes that I originally began jotting down, as I often do to relieve my frustrations in a socially acceptable way. I make no claim that these very sketchy notes are comprehensive or perceptive or even consistent. They just give my reactions to various points during these precedings:

1. In assessing and managing risk, bias is likely to be far more common than deliberate deception and far more pernicious. Is anyone here unbiased? Only one, each of us might answer.

2. The discussions have been punctuated by comments that fundamental objections to nuclear power are almost unrelated to risk assessments, yet most risk assessments do not go beyond picking at accidents and natural disasters. The public is worried, and I'm worried, about nonaccidental matters such as theft, sabotage, catastrophic problems in reprocessing cycles, and proliferation of technology, as well as major problems on the other side, too - like war over limited fossil fuel supplies. The public is concerned, and I am concerned, that we have no way to know whether we have been fully, objectively, and honestly informed about these matters from the right or from the left.

3. The title of our conference suggests that we should understand the difference between actual risk and perceived risk; but I am also interested in the difference between the perception of risk and acceptability of risk. Perhaps we can come back to this in some future year.

4. There seems to be an underlying assumption that the public is not rational. If the word rational applies to correct reasoning from specific starting points, I suspect rather that the public may on the whole be entirely rational; but they start from different premises and rather naturally come to different conclusions.

5. One speaker said, in what I assume was a slip of the tongue, that we must "take a closer look at the risk we represent." There may be some wisdom hidden in that.

6. There seems to be no reason to think that anyone has suffered physical harm from the TMI incident, but the lack of direct damage may be irrelevant to the main concerns. How much did the utility and the NRC contribute to the public perception of the problem? How sure can we now be that there really was no likelihood of serious consequences, or is there any residual possibility that we have been systematically misinformed about a very narrow escape? What does the bungling of various involved groups tell us about possible bungling in future accidents? Aren't the two questions of loss of trust and bungling much more important objectively, as well as in the public view, than the second or third significant digit in biased or incomplete risk assessments?

7. With respect to the differences between perceptions of the public and perceptions of the experts, what groups of persons are most expert about such matters such as our military strength, or the doctrines promoted by various religions, or the goals and capabilities of one or another political party, or whatever? Do we automatically and instinctively trust the "experts" in these fields? Clearly not, so why should we trust experts in the nuclear industry, or the tobacco industry, or the medical industry, or even the risk assessment industry? It even seems possible that the widespread

trust of scientists in past decades was the anomaly, not the present distrust.

8. One of our labor leaders once said "Never overestimate the knowledge of a worker - or underestimate his intelligence." What does that say to us as risk assessors when we communicate with the public?

9. What is an acceptable probability of death? It depends on whether it's my death or yours.

10. In a democratic society, if I don't want something in my backyard, or my community, or my country, do I need to express a reason? Do I even need to have a reason, or is my preference enough to guide my actions?

11. I am uneasy about recurring references to a distinction between rational decisions and political decisions. Politics is one of the most fiercely rational of all human endeavors. The scientist who forgets this will soon be in trouble - the politician who forgets it will be in bigger trouble, faster.

12. Have we raised any issues here that won't be raised again at the second meeting of this society, or the third, or the fourth, or the nth? Have we settled any? If not, is it perhaps because we have not yet put them in a form that can be settled, and could we do any better in this respect?

13. The word "overreact" has been used several times to describe the public response to nuclear power plants. There seems to be little understanding that at other times and places the dominant theme and concern would be "underreact."

14. Recurring references to Love Canal have skipped over the question of evidence of direct physical harm. Is there such evidence? And if so, how good is it? More generally, is the evidence of human health effects of toxic waste dumps any better or any worse than the evidence of such effects from nuclear power plants, or agricultural pesticides, or other things? I'm worried about hazardous waste dumps, but we would do nobody any favors by pretending that the evidence of risk is stronger than it really is. Precisely the same skepticism seems appropriate for problems with contaminated ground water. If we can't find the sick or the dead, then the indirect data had better be stronger than what I have yet seen.

15. It somtimes seems that the assessment and management of risk is almost synonymous with the assessment and management of uncertainty. Experts are generally self-identified by their interest groups if not by themselves. Each time I have heard here about loss of trust in experts, I think "That's me! I've lost my trust, too!"

Why should I believe the self-proclaimed experts on politics, or what we are pleased to call "national defense," or science, or even the experts on risk analysis?

16. And my last note: With privilege goes responsibility; with knowledge goes bias.

Thank you.

DR. SCHNEIDERMAN

I find myself in a peculiar position. I'm here at the far right, a position to which I am unaccustomed. On the other hand, if the look at me from out there, I'm on the far left. Which is actual and which is perceived?

I want to thank the organizers of this meeting, for inviting me and permitting me to participate and for inviting so many bright and competent people that I might learn a great many things. I hope that these meetings will continue, and I hope that I will be able to attend them and continue to learn. Much of what I wanted to talk about has already been said so much better, that I need make only some brief remarks.

I want to start with a British author in honor of my colleague sitting at my left. I remind you of Tom Stoppard's Rosencranz and Gildenstern. Remember, they were characters out of Hamlet, but they saw the world entirely differently than did Hamlet and the rest of the Danish court. Seeing the world differently epitomizes what we have been dealing with here in the problem of actual vs. perceived risk. I think perceived risks are real. Perceived risks are the ones on which people take action - and in that sense perceived risks are not only actual, but also activating. They are more actual than the so-called observed risks on which people don't do anything.

I look at Dr. O'Riordan's schema here, and I am intrigued. I was asking myself where do I fit in his particular groupings, but I don't know the answer yet. In looking at that schema I am pushed to ask, "where do the experts come from?" It seems to me that most of the time - not all the time - but very often, the experts come from the technocentric side. Now if the experts are largely drawn from the technocentric side because there they are the persons who have concerned themselves with the technologic issues, then we are faced with the problems that Dr. Bailar and Dr. Upton raised. Is there a conflict of interest, or perhaps unconscious, bias that leads the technical experts to support where they are, or what they are doing or what they have done or what they might want to continue to do? If there is this tendency, is a generally nonexpert group, which will not be so technocentrically oriented, grossly in error in having some distrust of the experts or not believing the experts or being concerned about what the experts say? I really don't know how to deal with this possible bias - possible distrust complex. I hope continuing discussions will lead us to how we can use expert opinion better.

I was concerned in the course of this meeting with the frequent reference to the loss of trust, loss of credibility, loss of confidence in the experts. I am not surprised, but I am concerned. I am not surprised because in this country we have had four presidents in a row who have campaigned for President by running against the federal government and, of course, by implication, the experts in the federal government. Much of our expertise in many of these areas - and certainly in the regulatory area - resides in the federal government. So if the President of the United States thinks that his staff is not very good why should the rest of the people in the United States think his staff is any good? I don't know when this fashion began of running against the "bureaucracy," but running against the bureaucracy means running against your own experts. I don't know when this is going to stop, but if it doesn't stop, I see a continuing erosion of the belief in experts. This is a self-perpetuating process. If we have a disbelief in the experts from any other social organization including universities and private operations, it will lead, and it has led, to the cynical approach - which scares the dickens out of me - of describing people as "Hertz-rent-a-scientist." The argument goes: If you want a point of view, you'll find someome who will give you a point of view, and you'll give them enough of a consultant's fee and they'll tell your story. This says to me that there is an actual (as well as a perceived) basis for the public perception that mtakes for a lack of belief in the experts. This perception has been strongly supported and fortified and reinforced by behavior that has gone on in this country for at least the last 20 years.

How do we reinstitute, reconstitute, bring back belief in experts and belief in institutions, because erosion of belief in experts is part of erosion in belief in institutions? On a first level, I think we have to go to the kind of process that Dr. Upton indicated. We have to have much more public participation in activities in which experts are involved. I believe that it is possible for lay persons who are interested to learn enough about what I am doing to make some judgments about the consequences of what I am doing. I don't expect them to know in detail the kind of technical things that I known, but I do expect persons who are interested to learn enough to examine what I am doing and to decide whether I am doing it appropriately or inappropriately. I think we have to have more public members on presidential commissions and things of that sort. I believe it was Clemenceau who said that war was too complex to be left to the generals. Perhaps health and safety, and nuclear power are also too complex to be left only to the experts. I think we also have to recognize that there is a distinction between private and public expertise, and I think that got blurred here many times during the course of this meeting. I think we have to have open and rapid and honest speaking out by authorities, by government, and the press. Yesterday, Mrs. Trunk was very critical of the press. I raised Mrs. Trunks charges with Ms. Omang, who is from the press. I reported (I hope accurately) to Ms. Omang that the lady who spoke here yesterday was very critical about what the press did with respect to Three Mile Island. She said the press misreported; the press faked news; the press put out scare statements. You will recall that Ms. Omang replied, "I don't think we misreported; the statements we put out were statements made to us by either the industry people or the federal government's Nuclear Regulatory-Commission people. And we reported what people said to us, about putting fake 'for sale' signs." She continued, "I've heard people charge we put out those fake 'for sale' signs. Yet I've not yet found any one person who could tell me about the fake 'for sale' signs and who knew this information first hand. People only knew it second hand. They were told somebody put out fake 'for sale' signs and took pictures of them." I don't know who is faking what in this particular argument. (Maybe Mrs. Trunk has inadvertently put out fake reports of fake press reports.) But I think we have to have much more public participation in the kinds of activities we are involved in and then we honestly and openly and completely have to listen to each other. I don't like the we/them kind of approach. You known, that goes this way: "We, the experts, <u>know</u>. And they are out there and maybe they'll get to know; but they really aren't terribly interested or concerned." They are often sufficiently interested and concerned to disagree with us strenuously.

This brings me then to two other authors. I have here a modified quote from T. S. Elliott. (I have mangled the quote, of course. All famous quotations are mangled. That's the definition of a famous quotation; it gets mangled when it's cited.) Elliott asked the questions: "Where is the wisdom we've lost in knowledge? And where is the knowledge we have lost in information?" (I put those lines up on my blackboard and one of my colleagues adeed, "Where is the information we've lost in data?" And then my computer man added "where is the data we've lost?"). I think we have to go back to Elliott's first line, "Where is the wisdom we have lost in knowledge?" Do we sometimes lose wisdom in an over-elaboration of the knowledge? I have been concerned at times, for example, with controlled clinical trials, just as John Bailar has. And we are now concerned with the problem of the fully informed consent of the patient. Do we really want to teach the patient all there is in the textbook of medicine and the recent papers that appeared in the Journal of the National Cancer Institute before we ask that the patient consent to a particular activity? No we don't. What we want is for the patient to have a decent understanding of what is involved and what will go on and the mind set that will allow him (her) to ask questions, easily, and so that he believes he has enough basis on which to make an adult decision. I see public participation in the kinds of issues we are talking about here to be exactly on the same level. I don't want to make a mathematical statistician out of someone in order that this person might understand in exquisite detail what I've done in making my risk assessment, but I do want this person to know what my assumptions are, what I have plugged in, and what I have thought about. Affected people have a perfect right to know this.

I hope that we can restore confidence through extensive public participation, through openness, through invitation of responsible community leaders to participate with us and help us. I mean the local minister who is well thought of, the school teacher who has been teaching in this community for 30 years and who has a lot of students out there who are still fond of her. I think that we have to approach these issues with less arrogance than, "I am the expert and I really know what's right; and when I compare what I have done and what you have done, I am the standard and your perception is wrong." I am not sure the expert's perception always corresonds to actuality.

Cliff Grobstein said at a Brooking Institution meeting not long ago that there are three kinds of science: basic science, applied science, and the science for regulatory purposes. I think we have to recognize that the science for regulatory purposes has to be, in certain circumstances, far better, far more open, far more challengeable, far more understandable than any of the others. As experts, we may have the feeling that our not being believed very much any longer may have brought us to "Paradise Lost." Remember Milton also wrote another book "Paradise Regained." Maybe we can achieve that with a little more message and a lot less mystique. And the next President I vote for is going to be a man (or woman) who has faith and confidence in his (her) experts - and says so.

DR. UPTON

No one wants a reactor next door, and no one wants a waste-disposal dump next door! Yet an equitable distribution of risks and benefits is impossible. With nuclear power, of course, we are not dealing merely with an energy system. There are those who see in nuclear energy the threat of diverting weapons materials, through sabotage or other means, and hence an increase in the risk of a nuclear holocaust.

There is also at issue the centralization of economic and political power - the notion that small is beautiful - that the interdependence among individuals is becoming too great, and that the nuclear approach simply accentuates an undesirable socio-political trend. Philosophical, ethical, political, and religious values are all inextricably interwosen into these perceptions.

Another issue that seriously complicates the assessment of risk comes in connection with the scientific uncertainty that is attached to any extrapolation we must make. Particularly when risks are not readily verified, there will be scientific controversy, which confuses the public and leads to a loss of credibility. We have in the nuclear situation, scientists like Dr. Sternglass, who repeatedly make assertions, which are repudiated by the scientific community (NAS, 1972, 1980)*, but who nevertheless capture public attention and contribute to apprehension.

In connection with the toxic substances problem, such as in relation to chemical wastes, much of our information today is derived from experiments with animals, or even unicellular model systems. There again, there are extrapolation problems and inevitable scientific uncertainties, which are grounds for controversy, public confusion, and distrust. It was the same originally with tobacco: the first suggestions that cigarette smoke was a carcinogen aroused the same denials and controversies.

Obviously, for the answer to the question posed by the regulators, who asked several times yesterday afternoon how to cope with the situation, we must deal with the perceived risks as well as the actual risks, and there is a necessity to arrive at some concensus about them. This will require agreements among technologists - convergence if you will into some recognizable confidence intervals. One must define the boundaries of uncertainty as Dr. O'Riordon pointed out in assessing the risk of low-level radiation. There does appear to be a remarkable convergence, although not yet without some contention.

Here there needs to be a new science - a new process developed - and I am immensely encouraged from the discussion to see that Dr. Slovic, Dr. Fischhoff, and their colleagues at Decision Research; Dr. Hohenemeser and his colleagues at Clark University; and Dr. Rasmussen and his associates are all developing a similar approach: the framework in which to try to arrive at quantitative, or reasonably quantifiable, estimates of perceived risk and to develop a scientific rationale for arriving at estimates of perceived risk. I agree that this will inevitably require much better communication among scientists - "hard" scientists and "soft" scientists, physical

---

*The Effects on Populations of Exposure to Low Levels of Ionizing Radiation. Committee on the Biological Effects of Ionizing Radiation, National Academy of Sciences-National Research Council, Washington, D.C. 1972, 1980.

scientists and social scientists - and between the scientific community and the public at large. I agree with Dr. O'Riordan that we are dealing with a highly heterogeneous population when it comes to perceptions of risk.

I think that to have the credibility and trust that this approach to risk assessment deserves will require the constant participation of the public itself. This will involve new kinds of relationships with the public and new frameworks of communication. Responsibility on the part of the experts is also important here. I referred a moment ago to Dr. Sternglass and his distortions. We heard yesterday from Mrs. Trunk that the media too presented a distorted picture of the reactions of the population around Three Mile Island. So responsibility on the part of all communicators needs to be stressed and needs to be developed further, as a means of influencing behavior in a more responsible way.

I certainly don't profess to see how to develop the process that is needed. Suggestions come to mind, such as the Science Court or the Natonal Academy of Sciences itself; the Kemeney Commission, or other independent commissions of scientists, technicians, and nontechnical people; and obviously the federal-rule making process, which is a multi-step procedure involving public input. One of the difficulties, of course, is that the perceived risk of an option is often, as a matter of fact, much larger than the actual risk. If for example, the ratio of perceived risks of nuclear power vs. coal power, is 9000 to 1, as Dr. Rasmussen emphasized, the problem for the policy-maker is much more difficult, because the ratio may lead to the election of options or alternative strategies involving greater actual risks than those one seeks to avoid. Hence there may be a large foregone benefit.

Again, as Joanne Omang said this morning, if one had been able to predict that 50,000 people a year would die on the highway as a result of traffic accidents, one would never have allowed automobiles to have been built in the first place. That's an actual risk and not a perceived risk; but arriving at rational, reasonable, and adequate comparisons among alternative strategies is much more difficult in light of differences in perceived risks. Therefore, I think that a meeting such as this is highly encouraging in its promise for the development of needed strategies for assessment of perceived, as well as actual, risks.

DR. RALL

We have heard some elegant discussions of the general principles of actual and perceived risks. We heard something about elegantly derived mortality numbers. Those numbers may have been elegantly derived, but I think if one had a confidence interval around them, we would find that the number itself may have been fine, but it had a very large error band. I haven't heard much discussion about estimated vs. actual risks. I've got a sneaking suspicion that sometimes the perceived risks might be as close to the actual risks as the estimated risks and that bothers me a little bit. I think it's certainly true in toxic chemical waste dumps that this is a very newly perceived problem.

I don't think anybody will know how serious a problem the Love Canal was or will be. There are two bits of evidence that something happened; first, I think it's agreed that there was a higher incidence of miscarriages and abortions in rings one and two, that is, those families closest to the old canal and in the so-called swale areas. Now a swale is simply a dry stream which fills up when it rains. There was some thought that the swales might allow for the movement of chemicals out of the dump along the swales into the basements of the houses that sat on the swales - that turns out very likely not to be true. What happened is very much simpler and I think very typical of the sort of problem one faces. Swales are depressions, and one wants to fill in depressions. There is a whole bunch of dump stuff over there; so to fill in the depressions one takes a truck, digs into the canal where the wastes are located, dumps the material into the depression and levels it off - it's very nice. So houses around the canal appeared to contain a higher incidence of miscarriages and did have a higher incidence of low birth weight babies. Those are the only two hard facts, as far as I know, of health effects. On the other hand, we are dealing with a population of something like three to four hundred people. It's very hard, unless it is a very unusual occurrence, to get any sort of good statistics on that, particularly since there was never a good epidemological study. Fortunately, I think Love Canal will not be typical. It was enormously emotionalized because these were the houses which contained the financial equity of a number of lower-middle-class people. This was their only sort of chance to achieve a little financial independence, and they found that that was being threatened by a whole variety of very strange things that happened decades before, such as a company's waste disposal practices and the Board of Education buying for $1 a track of land. I think we must be alert the next time we see something like a Love Canal to try to get very much better information. I don't really know about the seriousness of the contaminated water supplies, but I know one thing: If we don't look, we're not going to find out. If we don't analyze some of those water supplies, we won't know; if we don't try to set up studies of the effects of those chemicals in animals and some epidemological studies of the populations exposed, clearly 20 years from now we won't know. So I think there are some real problems along these lines.

I think chemical waste dumps in retrospect were perhaps not as usual for this conference as it could have been. I think it's a fascinating problem and one that we will spend much time on in the future.

I think there is perhaps only one other comment I would like to make. We do tend to focus, as so often happens at these meetings, on cancer mortality; and we really do have to realize, although relatively little attention is paid to it, that chemicals do many other things - for example, they damage the kidneys; and the direct cost of the United States in terms of medicare for instage renal disease dialysis is a billion dollars a year and a large fraction of that probably is associated with chemical exposures. So I think this has been a fascinating conference. Maybe in 20 years we will know enough about Love Canal and the like for toxic chemical waste dumps to be a useful adjunct to the conference.

DR. RUPERT

We are, of course, somewhat the victims of our own knowledge. If we did not know some of the things that we know today, no one would worry about them, and there would be no need for conferences or confrontations. While it's good to know, to have control over matters, to try to foresee problems before they exist and do something about them before they become serious, this creates its own burden of responsibility and anxiety. Sometimes the effort to foresee makes trouble.

You might remember that a few years ago, the people involved in recombinant DNA research rather innocently set out to do what scientists always tend to do: get together and talk about perplexing possibilities in their work. The moment they started, they brought down a storm of public wrath on their heads, much to their amazement. They had created a problem, simply by creating perception of a problem where none had existed before. One can't say they shouldn't have done this, but it's in the nature of the world that a certain amount of chaos and tension accompany any such awakening.

In its own realm, Science provides effective methods for dealing with the unknown and the uncertain, for pooling informations and insights, and for moving toward resolution of disagreements. While scientists as individuals are just as ornery as anybody else, they do manage, within the framework of their profession, to add to each others' insight and information more often than they cancel out. Usually, however, this process requires that all participants keep track of a lot of small details painted in various shades of gray. As soon as people who cannot follow all those details become involved, the entire process changes. Matters then fall into the simpler black-and-white, true-or-false, good-or-bad, guilty-or-not-guilty categories characteristic of adversary proceedings. Shades of gray are no longer detected. This tends to be the case with citizens'

action groups which are often ill-equipped to deal with complicated
options and simply have to be for something or against it. The scientists' taste for weighing fine shadings and factual complexities,
such as have been discussed in this meeting, has some difficulty
adapting to that situation. Yet we will have to deal with it, if we
are going to interact with the public, rather then merely talk to
outselves and each other.

Hopefully, one can try to frame the discussions in ways that
lead to effective examination of the alternative choices. The public
naturally wishes for safety rather than nonsafety, and - when it gets
concerned enough over an issue to become active about it - calls for
a zero risk, or zero pollution. But we seldom have the option of
simply avoiding all actions that might be objectionable. What we
have is a chance to choose between actions, each carrying different
risks, along with different costs and benefits, and one has to try
to put the matter into such a context.

Take, for instance, the chlorofluoromethanes held up here as a
likely cause of stratospheric ozone depletion. These substances
represent, by the way, something of a Greek tragedy in the chemical
world. The properties making them desirable for industry are in-
separable from the reason one worries about them. They have a high
chemical inertness, closely associated with the fact they are color-
less, odorless, tasteless, noncorrosive, nontoxic, and nonflammable -
in other words, ideal for use around human beings. They are, consequently, used for degreasing electronic parts, as blowing agents in
manufacture of plastic insulating foams, for fast freezing of foods,
and in refrigeration and air conditioning machinery, with little
worry about people's exposure. Their one flaw is that, because of
their inertness, nothing much can happen to them and, released into
the atmosphere, they remain. The principle, perhaps the only, way
out is a diffusion into the stratosphere, where they are photochem-
ically dissociated, and enter into the complex cycle of ozone-de-
stroying reactions - a worrisome business.

But the uses to which these substances are put will not just
disappear; at the very least some different materials will replace
them. And, unless we are terribly fortunate, replacement is likely
to be by substances less noncorrosive, less nontoxic, less nonflammable, etc., leading to different occupational and environmental
troubles we know not of - a new set of risks created in eliminating
the old. One has to always think in getting rid of a thing, what
are you going to use in its place? We are always involved in some
kind of trade-off.

It may be simpler to deal with the public if one can get this aspect of the matter into the front of the discussion. Identification
of the alternatives, and examination of what each entails is more to
a scientist's operating mode than adversary proceedings. But, then,

let us not be too romantic either. In the long run, we will be faced with the necessity of dealing very much with the public on the public's own, not-so-scientific terms, hopefully with adequate patience and with full integrity in the bargain.

DR. NELSON

Because we are moving very rapidly into the time of anticipated plane departures and evening traffic, we can entertain a very small number of urgent questions and comments from the floor. Yes, please identify yourself.

SALLY WORTS, PSYCHOLOGIST

Dr. Nelson, I'm Sally Worts, a psychologist, who lives in the society at large. I'm a member of the lay public, not a participant at this meeting, and I would like to ask your permission to speak.

DR. NORTON NELSON

I am afraid that we do not have time for a statement. If you have comments that you can make let's say in two or three or four minutes, that's fine.

SALLY WORTS

I can.

DR. NORTON NELSON

Good.

SALLY WORTS

I wish to make my bias clear from the beginning: I am a feminist, I am a peace worker, and I am an activist. I'd like to offer my perception of this meeting.

I see you experts who have very much to say in this country as having a very great affect on public opinion. I see you as largely in power beholding to those in power, receiving your salaries from those in power. I perceive the National Academy of Sciences and people who are admitted to these halls to be sometimes on the high technology bandwagon that this country is devoted to and committed to at the present time, with a bias to continue in that direction. We of the public listen to what you say. The format of this meeting does not allow people from the public to say what they think; it allows you to say what you think, and that's the way most public hearings are. I feel that in your presentations you have overlooked the fact that human suffering cannot be quantified. I don't see that you have made an effort to quantify human suffering. I would like to add to what's been said about nuclear power. I wish each of you who has spoken of nuclear power would make it a point within the next month to meet three victims of nuclear power or nuclear weapons production. There is a woman here in Washington who is suffering from cancer of the face through excess exposure to medical radiation. Her father was a Hanford Waste Site Inspector, he contracted cancer and committed suicide. She visited the Hanford Waste Site when she was two months pregnant, when the lungs and the genitals of her child were forming, and her 20 year old son now has a deformed penis and lungs that do not function correctly. One individual became sterile after watching the nuclear blast in the Southwest. Another watched 17 of blasts in the Bikini Islands, and he was sworn to secrecy by the military, and he has only in the last three years allowed, what has basically driven him crazy to be spoken through his lips.

It's hard for me to sit in this room and watch your slides of figures, because I've talked with the nuclear victims. And there is suffering and there is tragedy and I want - I wish somehow I could appeal to your hearts to be active in your research, as well as your minds. I feel that you, as risk analysts, are probably particularly cautious people, and that in order to get to the point you have reached in your careers you have had to be very careful in what you say, and I imagine that most of the men in this room haven't had a lot of practice in going out on a limb and taking a chance on being wrong, or taking a chance on taking a stand against what is the accepted view. I am sorry if I am wrong about that. But I feel the absence of women in this room (there are some women present), the absence of blacks, the absence of Indians - the uranium coming out of the ground in our Southwest is giving the Navahoes lung cancer.

The last point I would like to make is that the greatest risk of all, which I perceive is nuclear war, has been left off your agenda. I am aware that the thought of it leads us all to despair, and our first thought would be: let us not think of it. But if you are our risk analysts there are despair workshops that you can go to and face up to the fear and the anger that you probably have as well as I that that is the end that is foreseen for us and perhaps offer something to the public in that regard.

DR. NELSON

Thank you. I take this as a statement rather than a question, and I am sure there are comments and appreciative comments that could be made from everyone on this panel. I would simply personally like to say that you have hit on an issue that has been repeatedly mentioned throughout this meeting, namely, the distrust of all experts regardless of what their known or unknown biases may be. This is a paralyzing issue that confronts us, and I think in my own mind, that this may be the most serious problem we confront. It's not so much perhaps the inadequacy of the science - although, God help us, science in this is terribly inadequate - it's making use of what science we do have in a way that is effective. That is probably our biggest challenge here in this particular area. I appreciate your comments and I think you have touched on an issue that has been, as a thread, running through almost every presentation here. Did you just come or were you here throughout all the meeting?

SALLY WORTS

I was here this afternoon and for part of yesterday's afternoon presentation.

DR. NELSON

Had you been here throughout, I think you would have recognized that this was a continuing and reoccurring concern. Are there other questions or comments.

RUDOLPH YAKSICK

National Academy of Sciences
Board on Mineral and Energy Resources

I guess I would like to follow up on the previous comment by saying that I think the problem is not whether or not you reduce uncertainty. Dr. Schneiderman raised several possible sources of gap between actual and perceived risks because of lack of information about a process which is meaningful and open to both experts and nonexperts. I wonder if the sources are more fundamental, and that the scientists have "a rational way of looking at the world." That there are certain assumptions about the nature of rational choices is from classical economic theory, and I wonder if the problem is that certain people (particularly on the spectrum that Dr. O'Riordan showed) may have evidence of an intuitive and an effective way of looking at the world as opposed to a cognitive way. The debate is over whether or not the risk is 10^{-8} with a particular confidence interval. It's a proxy for a more fundamental debate over whether the world is viewed through cognitive or intuitive modes. I guess it raises a question for risk analysts as to whether or not one's efforts to try to get a better handle on 10^{-7} or 10^{-8} risk of a catastrophic event is futile if the real issue is the way people look at events in the world.

DR. UPTON

Mr. Chairman, I tried to make this same point, namely that the risk in question is the perceived risk, and that this is a complicated entity. One can't divorce the issues that you speak about from the quantification of an extrapolated risk of 10^{-7} or 10^{-8}. One really has to get all the issues on the agenda; and for that reason, the dialogue must be broadened very much more than it has in the past. So I would agree with you.

DR. TARDIFF
National Academy of Sciences

I hope my comments are not premature, and I don't want to foreclose any additional question; but before adjournment I did want to express on behalf of the Organizing Committee and the Society for Risk Analysis our great indebtedness to Professor Nelson and to the members of the panel for what, I believe, has been a very enlightening and highly beneficial three days. I think there is really no other way to describe the program except as a "new frontier," one that clearly has a significant amount of relevance to societal decision making, one that is going to become ever more prominent within the next few years. Very clearly, you gentlemen are the pioneers in this area; and I hope that everyone will agree that as we look back upon this conference, we will realize that we placed a well founded trust in individuals who could address this in what has been a most objective perspective.

DR. NORTON NELSON

Are there other comments or questions. If not, I'm not going to try to summarize or analyze the proceedings. However, I would like to just quickly comment on what I see as a series of issues which have been referred to repeatedly. Number one, in my opening remarks, I expressed some concern about difficulties of communication among those diverse backgrounds - I haven't seen that at all. Certainly there is a need possibly to tidy up some terminological differences, but communication, I think is a difficulty that I thought might be present was not present. I think that there may have been much disagreement, and indeed there was; but I really don't think there have been any serious misunderstandings between us because of our failure to understand the others' points of view. There are some points that have emerged again and again. One of these has to do with the issue of looking at benefits at the same time that one looks at risks and also the intruding issue of the acceptability of whatever equation is directed in this risky business of risk analyses. I think what has emerged is the possibility of treating the risk issue in a temporary isolation from the benefits. I think we all recognize, however, that the risk analysis is useless without going on to that stage - it could be a parallel stage and often is - that is a benefit analysis and finally the third stage of acceptability of risks. I am still a simple minded boy scout at heart in the sense that I think that it is important to separate issues in order to analyze them clearly and to keep our motivations clear. And I do think that, whereas the analysis of risk can be regarded as a largely technical issue, the analysis of benefits can also be regarded as largely as a technical issue. However, the acceptability of whatever decision comes out of this analysis is by no means a technical issue. The determination of acceptability is wholly nontechnical, in which human values and human concerns have to be the dominating issues. My point is merely that even though they become entangled in the analysis, I think it's urgent to attempt conceptually to keep these issues separate, in somewhat separate compartments, and not confuse them. As a scientists, I make only the contribution of another layman if you will, to the determination of acceptability. But I hope that I can add some extra expertise to very narrow issues of risk analysis. We've had much discussion, much concern, and much difficulty with the issue of the perceptions of risks and it's clear that perception almost invariably drives the final societal responses.

It's remarkable that there has been very little effort to attempt to measure perceptions; and perhaps such measurement is not possible. Realistically, the perception of risk may in fact be of most consequence in many instances. We have some very serious problems with this; for example, the mere announcement of concern is certainly going to produce apprehension and result in a perception of risk, the degree of which we cannot determine. We heard a paper yesterday in which it was said that the media, perhaps deliberately or perhaps through ignorance, in some instances produced exaggerated concerns. Whether that's true or not, those concerns were real and lead to anguish and concern and require consideration of a different sort, perhaps equal to the finding of physical injury. How are we going to balance of people's right to know with the transmission of uncertain - sometimes speculative data? I suppose we must accept some degree of apprehension and perhaps some misplaced perception of concern, and at the same time try to minimize arousing needless concerns in producing "riskogenic" disease - disease which is real, in the sense of apprehension, but which grows out of announcements or warnings based on ignorance. I see this as a very difficult problem. But I think that we need to regard this as an inevitable consequence of working in this particular area. In the same way the experts are going to have to recognize that they have to be living more and more in a very open arena - side by side with many more confronting situations than they have been accustomed to in their quiet academic or scientific circles or laboratories. We must learn a kind of communication in which perhaps we've been largely unsuccessful. It is necessary to keep in perspective the wide range of concerns that are a legitimate part of the decisions that grow out of these issues and in which we're only limited players.

I think this has been excellent meeting, a very informative one; I've learned a great deal and I am sure others have as well. I think the Society for Risk Analysis has made a fine step forward in calling this meeting, and I'd like to thank the organizers for it and all of the participants in this affair - thank you very much.

ACTUAL VS. PERCEIVED RISK: A POLICY-RELATED BIBLIOGRAPHY*

Vincent T. Covello and Mark Abernathy

INTRODUCTION

During the past decade, the literature on actual and perceived risk has grown from a handful of articles and books to a formidable collection of material. The objective of this bibliography is to provide a listing of this material, with emphasis on works of generic interest to researchers and policymakers. The bibliography is divided into two sections: 1) Actual Risk: A Theoretical and Methodological Overview; 2) Perceived Risk: A Theoretical and Methodological Overview. Based on judgments about the subject matter of the work, several references appear in both sections.

Aside from inadvertent omissions and some minor exceptions, the following types of material were intentionally excluded from the bibliography: works dealing narrowly with a specific technology or hazard; technical or methodological works dealing exclusively with risk estimation or the measurement of probabilities and uncertainties; articles from mass circulation magazines; book reviews and foreign language works; unpublished dissertations and theses; and unpublished symposium, conference, or workshop papers. Because numerous sources were consulted in compiling the bibliography, it was not possible to provide complete reference information (such as page numbers) in all cases.

---

*The views expressed in this paper are those of the authors and do not necessarily reflect the views of the National Science Foundation.

ACTUAL RISK: A THEORETICAL AND METHODOLOGICAL OVERVIEW

Ackerman, B., Rose-Ackerman, S., Sawyer, J. W., and Henderson, D. W., The Uncertain Search for Environmental Quality, New York, Free Press (1974).
Aharoni, A., The No Risk Society, Old Greenwich, Connecticut, Chatham Press (1981).
Altzinger, E., Brook, M., Wilbert, J., Chernick, M. R., Elsner, B., and Foster, W. V., Special Publication No. 4, Compendium of Risk Analysis Techniques, Aberdeen, Maryland, U.S. Army Material Systems Agency (1972).
Apostolakis, G. E., Mathematical Methods of Probabilistic Safety Analysis (UCLA-ENG-7464), Los Angles, University of California (September 1974).
Apostalakis, G. E., "Probability and Risk Assessment, The Subjectivistic Viewpoint and Some Suggestions," Nuclear Safety 22 (January-February, 1981):1.
Apostalakis, G. E., Garribba, S., and Volta, G., eds., Synthesis and Analysis Methods for Safety and Reliability Studies, NATO Advanced Study Institute Series, New York, Plenum Press (1980).
Apostalakis, G. E., and Kaplan, S., "Pitfalls in Risk Calculations," Reliability Engineering, 2:135-145 (1981).
Ashby, E., "The Risk Equations: The Subjective Side of Assessing Risks," New Scientist, 74:398-400 (May 19, 1977).
Ashford, N. A., Crisis in the Workplace: Occupational Injury and Disease, Cambridge, Massachusetts, MIT Press (1976).
Ayers, F. T., "The Management of Technological Risk," Research Management, 24-28 (November 1977).
Bazelon, D., "Risk and Responsibility," Science, 205:277-280 (July 20, 1979).
Beauchamp, D. E., "Public Health and Individual Liberty," Annual Review of Public Health, 1:121-136 (1980).
Berg, G. G., and Maillie, H. D., eds., Measurement of Risks, New York and London, Plenum Press (1981).
Besuner, P. M., Tetelman, A. S., Eagen, G. R., and Rau, C. A., "The Combined Use of Engineering and Reliability Analysis in Risk Assessment of Mechanical and Structural Systems," Risk-Benefit Methodology and Application (UCLA-ENG-7598), D. Okrent, ed., Los Angeles, University of California (1975).
Bogen, K. T., "Public Policy and Technological Risk," IDEA: The Journal of Law and Technology, 21:37-74 (1980).
Bowen, J., "The Choice of Criteria for Individual Risk," Risk-Benefit Methodology and Application (UCLA-ENG-7589), D. Okrent, ed., Los Angeles, University of California (December 1975).
Bunker, J. P., Barnes, B. A., and Mosteller, F., eds., Costs, Risks, and Benefits of Surgery, New York, Oxford University Press (1977).
Burton, I., Kates, R., and White, G., The Environment as Hazard, London, Allen and Unwin (1978).

Cairns, J., Cancer: Science and Society, San Francisco, W. H. Freeman (1978).
Cairns, J., "Estimating Hazard," BioScience, 30:101-107 (February 1980).
Carter, L., "Dispute Over Cancer Risk Quantification," Science, 203: 1324-1325 (March 30, 1979).
Carter, L., "How to Assess Cancer Risks," Science, 204:811 (May 25, 1979).
Cederlof, R., et al., "Air Pollution and Cancer: Risk Assessment Methodology and Epidemiological Evidence," Environmental Health Perspectives, 22:1-13 (1978).
Chiang, C. L., "A Stochastic Model of Competing Risks of Illness and Competing Risks of Death," Biology, J. Gurland, ed., Madison, Wisconsin, University of Wisconsin Press (1964).
Clark, E. M., and Van Horn, A. J., "Risk-Benefit Analysis and Public Policy, A Bibliography," Updated and extended by L. Hedal and E. A. C. Crouch, Cambridge, Massachusetts, Energy and Environmental Policy Center, Harvard University (1978).
Clark, W., "Managing the Unknown," Managing Technological Hazard: Research Needs and Opportunities, R. Kates, ed., Boulder, Institute of Behavioral Science, University of Colorado (1977), pp. 109-142.
Clark, W., "Witches, Floods, and Wonder Drugs: Historical Perspectives on Risk Management," Societal Risk Assessment: How Safe is Safe Enough? R. Schwing and W. Albers, eds., New York, Plenum Press (1980), pp. 287-312.
Cohen, B., and Sing Lee, I., "A Catalog of Risks," Health Physics, 36:707-722 (June 1979).
Commission of the European Communities, Nuclear and Non-Nuclear Risk - An Exercise in Comparability (EUR 6417EN), Brussels, Belgium, Commission of the European Communites (1980).
Conrad, J., "Society and Risk Assessment: An Attempt at Interpretation," Society, Technology and Risk Assessment, J. Conrad, ed., London, Academic Press (1980), pp. 241-276.
Conrad, J., ed., Society, Technology, and Risk Assessment, London, Academic Press (1980).
Coppola, A., and Hall, R. E., A Risk Comparison (NUREG/CR-1916), Upton, New York, Brookhaven National Laboratory, prepared for the Nuclear Regulatory Commission (February 1981).
Council for Science and Society, The Acceptability of Risk, London, Barry Rose, Ltd. (1977).
Covello, V., "Technological Hazards, Risk, and Society: A Perspective on Risk Analysis Research," Risk Benefit Analysis in Water Resources Planning and Management, Y. Haimes, ed., New York, Plenum Press (1981).
Covello, V., and Menkes, J., "Issues in Risk Analysis," Risk in the Technological Society, C. Hohenemser and J. Kasperson, eds., Boulder, Colorado, Westview Press (1981).
Crandall, R., and Lave, L., eds., The Scienfific Basis of Health and Safety Regulations, Washington, D.C., Brookings Institution (1981).

Cumming, R. B., "Is Risk Assessment a Science?", Risk Analysis: An International Journal, 1:1-4 (March 1981).

de Neufville, R., and Pate, M. E., "A Conceptual Risk Assessment Procedure," Two Conceptual Approaches to Health Risk Assessment for Alternative National Ambient Air Quality Standards, Washington, D.C., U.S. Environmental Protection Agency (September 1980).

DeSchmukh, S. S., "Risk Analysis," Chemical Engineering, 81:141-144 (June 24, 1974).

Dierkes, M., "Assessing Technological Risks and Benefits," Technological Risk: Its Perception and Handling in the European Community, M. Dierkes, S. Edwards, and R. Coppock, eds., Cambridge, Massachusetts, Oelgeschlager, Gunn, and Hain, Publishers, Inc., (1980), pp. 21-30.

Dierkes, M., Edwards, S., and Coppock, R., eds., Technological Risk: Its Perception and Handling in the European Community, Cambridge, Massachusetts, Oelgeschlager, Gunn and Hain, Publishers, Inc. (1980).

Dunster, H. J., "The Assessment of Risk - Its Value and Limitations," Nuclear Engineering International, 24:23-25 (August 1979).

Dunster, H. J., "The Risk Equations: Virtue in Compromise," New Scientist, 74:454-456 (May 26, 1977).

Dunster, H. J., and McLean, A. S., "The Use of Risk Estimates in Setting and Using Basic Radiation Protection Standards," Health Physics (July 1970).

Elster, J., "Risk, Uncertainty, and Nuclear Power," Social Science Information, London and Beverly Hills, Sage Publications (1979), pp. 371-400.

Epstein, S. S., "Information Requirements for Determining the Benefit-Risk Spectrum," Perspective on Benefit-Risk Decision Making. Edited by the Committee on Public Engineering Policy, Washington, D.C., National Academy of Engineering (1972).

Erickson, L. E., Issues and Experiences in Applying Benefit Cost Analysis to Health and Safety Standards, Report to the U.S. Nuclear Regulatory Commission, Richland, Washington, Battelle Pacific Northwest Laboratories (September 1977).

Fagnani, F., "Role and Function of Risk Assessment," Society, Technology, and Risk Assessment, J. Conrad, ed., London, Academic Press (1980), pp. 165-172.

Fairley, W., "Assessment of Catastrophic Risks," Risk Analysis: An International Journal, 1:3 (1982).

Farmer, F. R., "Experience in the Reduction of Risk," Proceedings of the Symposium on Major Loss Prevention in the Process Industries, London, The Institution of Chemical Engineers (1971).

Farmer, F. R., "Some Considerations of Major Non-Nuclear Hazards," International Atomic Energy Agency Bulletin 20:13-20 (December 1978).

Ferreira, J., and Slesin, L., Observations on the Social Impact of Large Accidents (Technical Report No. 122), Cambridge, Massachusetts, Operations Research Center, Massachusetts Institute of Technology (October 1976).

Fischhoff, B., Hohenemser, C., Kasperson, R., and Kates, "Handling Hazards," Environment, 20:16-37 (September 1978).
Fischhoff, B., Hohenemser, C., Kasperson, R., and Kates, R., "Handling Hazards," Environment, 20:16-37 (September 1978).
Fischhoff, B., Slovic, P., and Lichtenstein, S., "Weighing the Risks," Environment, 21:17-38 (May 1979).
Fischhoff, B., and Whipple, C., "Assessing Health Risks Associated with Ambient Air Quality Standards," Four Conceptual Approaches to Health Risk Assessments, Washington, D.C., U.S. Environmental Protection Agency (August 1980).
ENG-7598), D. Okrent, ed., Los Angeles, University of California (December 1975).
Goetz, A. A., "Health Risk Appraisal: The Estimation of Risk," Public Health Reports, 95:119-126 (1980).
Goodman, G. T., and Rowe, W. D., eds., Energy Risk Management, New York, Academic Press (1979).
Green, H. P., "The Risk-Benefit Calculus in Safety Determinations," George Washington Law Review, 43:791-808 (March 1975).
Greer-Motten, B., "Context, Concept, and Consequence in Risk Assessment Research: A Comparative Overview of North American and Europen Approaches in the Social Sciences," Society, Technology and Risk Assessment, J. Conrad, ed., London, Academic Press (1980), pp. 67-104.
Griffiths, R., ed., Dealing with Risk: The Planning, Management, and Acceptability of Technological Risk, New York, Halsted Press (1981).
Haddon, W. H., Jr., Suchman, E. A., and Klein, D., Accident Research: Methods and Approaches, New York, Harper and Row (1964).
Haimes, Y., ed., Risk Benefit Analysis in Water Resources Planning and Management, New York, Plenum Press (1981).
Hall, W. K., "Why Risk Analysis Isn't Working," Long Range Planning 25-29 (December 1975).
Hammer, W., Product Safety Management and Engineering, Englewood Cliffs, New Jersey, Prentice-Hall (1980).
Handler, P., "A Rebuttal: The Need for a Sufficient Scientific Base for Government Regulation," George Washington Law Review, 43:808-813 (March 1975).
Handler, P., "Some Comments on Risk Assessment," 1979 - Current Issues and Studies, Washington, D.C., National Research Council, National Academy of Sciences (1979).
Harriss, R., Hohenemser, C., and Kates, R., "Our Hazardous Environment," Environment, 20 (April 1978).
Hartley, H. O., Manton, K. G., and Woodbury, M. A., "Estimation of Risk of Adverse Health Effects Associated with Air Quality Standards for Pollutants," Four Conceptual Approaches to Health Risk Assessment, Prepared for U.S. Environmental Protection Agency (August 1980).
Health and Safety Executive, Canvey: An Investigation of Potential Hazards from Operations in the Canvey Island/Thurrock Area, London, H.M.S.O. (1978).

Herbert, J. H., Swanson, C., and Reddy, P., "A Risky Business: Energy Production and the Inhaber Report," Environment, 20: 28-33 (1979).

Heuser, F. W., and Homke, P., "Reliability Analysis and Its Application for Safety Assessment of Nuclear Plants," Risk Benefit Methodology and Application (UCLA-ENG-7598), D. Okrent, ed., Los Angeles, University of California (December 1975).

Hoel, D., A Representation of Mortality Data by Competing Risks," Biometrics, 28:475-488 (1972).

Hoffman, S. D., Unreasonable Risk of Injury Revisited, Chicago, Underwriters Laboratories (1976).

Hohenemser, C., and Kasperson, J., eds., Risk in the Technological Society, Boulder, Colorado, Westview Press (1981).

Holdren, J., Smith, K., and Morris, G., "Energy: Calculating the Risks," Science, 204:564-568 (May 11, 1979).

Holdren, J., et al., Risk of Renewable Energy Sources: A Critique of the Inhaber Report (ENG 79-3), Berkeley, Energy and Resources Group, University of California, and Honolulu, Resource Systems Institute, East-West Center (June 1979).

Howard, N., "What Price Safety? The 'Zero-Risk' Debate," Dun's Review, 114:49-51, 53, 57 (September 1979).

Inhaber, H., "Risk from Conventional and Nonconventional Energy Sources," Science, 203:718-723 (February 23, 1979).

Inhaber, H., Risk of Energy Production (Report AECB-rev. 3), Ottawa, Atomic Energy Control Board (1979).

Interagency Regulatory Liaison Group, Scientific Basis for Identification of Potential Carcinogens and Estimation of Risk, Washington, D.C., Interagency Regulatory Liaison Group (February 1979).

Irwin, J., and Stoner, G. D., Facets of Biohazard Control Program - Agent Registration, Risk Assessment and Computerization of Data," American Journal of Public Health, 66:372-374 (1976).

Johnson, R., "The Characteristics of Risk Assessment Research," Society, Technology and Risk Assessment, J. Conrad, ed., London, Academic Press (1980), pp. 105-122.

Kaplan, S., and Garrick, B. J., "On the Quantitative Definition of Risk," Risk Analysis: An International Journal, 1:11-28 (March 1981).

Kasperson, R. E., "Societal Management of Technological Hazards," Managing Technological Hazard: Research Needs and Opportunities, R. Kates, ed., Boulder, Institute of Behavioral Science, University of Colorado (1977), pp. 49-80.

Kates, R., "Assessing the Assessor: The Art and the Ideology of Risk Assessment," AMBIO, 6:247-252 (1977).

Kates, R., Risk Assessment of Environmental Hazard, New York, John Wiley and Sons (1978).

Kates, R., ed., Managing Technological Hazard: Research Needs and Opportunities, Boulder, Institute of Behavioral Science, University of Colorado (1977).

Kates, R., and Hohenemser, C., eds., Technological Hazard Management, Cambridge, Massachusetts, Oelgeschlager, Gunn, and Hain Publishers, Inc. (1981).

Keeney, R. L., Siting Energy Facilities, New York, Academic Press (1980).

Keeney, R. L., and Raiffa, H., "A Critique of Formal Analysis in Public Decision Making," Analysis of Public Systems, A. W. Drake, R. L. Keeney, and P. M. Morse, eds., Cambridge, Massachusetts, MIT Press (1972).

Keeney, R. L., and Robilliard, G. A., Assessing and Evaluating Environmental Impacts at Proposed Nuclear Power Plant Sites (IIASA PP-76-3), Laxenburg, Austria, International Institute for Applied Systems Analysis (February 1976).

Kemeny, J. G., "Saving American Democracy: The Lessons of Three Mile Island," Technology Review, 65-75 (June/July(1980).

Kirschman, J. C., "Toxicology - The Exact Use of an Inexact Science," Food, Drug, and Cosmetic Law Journal, 31:455-461 (1976).

Kranzberg, M., "Prospects for Change," Societal Risk Assessment: How Safe is Safe Enough?, R. Schwing and W. Albers, ed., New York, Plenum Press (1980), pp. 319-332.

Lathrop, J., "Measuring Social Risk and Determining Its Acceptability," Two Blowouts in the North Sea: Managing Technological Disaster, D. W. Fischer, ed., Oxford, Pergamon Press (1980).

Lathrop, J., Measuring Societal Risk and Determining Its Acceptability (UCRL-81060), Berkeley, California, Lawrence Livermore Laboratory (1978).

Lave, L., and Seskin, E. P., "Epidemiology, Causality, and Public Policy," American Scientist, 67 (March-April 1979).

Lawless, E., Technology and Social Shock, New Brunswick, New Jersey, Rutgers University Press (1977).

Lewis, H. W., et al., Risk Assessment Review Group Report to the U.S. Regulatory Commission (NUREG/CR-0400), Washington, D.C., Nuclear Regulatory Commission (1978).

Libby, L. M., Technological Risk Versus Natural Catastrophes (P-4602), Santa Monica, California, Rand Corporation (March 1971).

Linnerooth, J., "Methods for Evaluating Mortality Risk," Futures, 8:293-304 (August 1976).

Lovins, A. B., "Cost-Risk-Benefit Assessments in Energy Policy," George Washington Law Review, 45:911-943 (August 1977).

Lowrance, W., "The Nature of Risk," Societal Risk Assessment: How Safe is Safe Enough?, R. Schwing and W. Albers, eds., New York, Plenum Press (1980), pp. 5-14.

Lowrance, W., Of Acceptable Risk: Science and the Determination of Safety, Los Altos, California, W. Kaufman, Inc. (1976).

Lowrance, W., "Probing Societal Risks," Chemical and Engineering News, 59:13-20 (July 6, 1981).

Mann, N. R., Shafer, R. E., and Singpurwalla, N. O., Methods for Statistical Analysis of Reliability and Life Data, New York, John Wiley and Sons (1974).

Mattson, R., Ernst, M., Minners, W., and Spangler, M., "Concepts, Problems, and Issues in Developing Safety Goals and Objectives for Commercial Nuclear Power," Nuclear Safety, 21:6 (November-December 1980):703-716.

Mazur, A., "Disputes Between Experts," Minerva, 2:243-262 (1973).

Mazur, A., "Societal and Scientific Causes of the Historical Development of Risk Assessment," Society, Technology and Risk Assessment, J. Conrad, ed., London, Academic Press (1980), pp. 151-158.

Meadows, D. H., Meadows, D. L., Randers, J., and Behrens, W. W., The Limits to Growth, New York, Signet (1972).

Mehlman, M. A., Shapiro, R. E., and Blumenthal, H., eds., Advances in Modern Toxicology, Vol. 1, New Concepts in Safety Evaluation, New York, John Wiley and Sons (1979).

Mendeloff, J. Regulating Safety: An Economic and Political Analysis of Occupational Safety and Health Policy, Cambridge, Massachusetts, MIT Press (1979).

Merkhofer, M. W., "A Procedure Based on Decision Analyses for Assessing the Health Risks Associated with Alternative Ambient Air Quality Standards," Four Conceptual Approaches to Health Risk Assessment, Washington, D.C., U.S. Environmental Protection Agency (August 1980).

Miller, B., and Hall, R. E., Index of Risk Exposure and Risk Acceptance Criteria (NUREG/CR-1930), Prepared for the U.S. Nuclear Regulatory Commission, Upton, New York, Brookhaven National Laboratory (February 1981).

MITRE Corporation, Risk Assessment and Governmental Decisionmaking: Symposium/Workshop on Nuclear and Nonnuclear Energy Systems (MTR-79W-00335), McLean, Virginia, MITRE Corporation (1979).

Moll, K. D., and Tihansky, D. P., "Risk Benefit Analysis for Industrial and Social Needs," American Industrial Hygiene Association Journal, 38:153-161 (1977).

Moreau, D., Quantitative Assessment of Health Risks by Selected Federal Regulatory Agencies, Washington, D.C., U.S. Environmental Protection Agency, Office of Air Quality Planning and Standards (October 1980).

Morgan, G., "Choosing and Managing Technology-Induced Risk," IEEE Spectrum, 53-60 (December 1981).

Muelhause, C. O., "Risk-Benefit Analysis," ASTM Standardization News, 8-27 (February 1973).

National Academy of Engineering, Perspectives on Benefit Risk Decision Making, Washington, D.C., National Academy of Engineering, Committee, on Public Engineering Policy (1972).

National Academy of Sciences, Energy in Transition 1985-2010: Final Report of the Committee on Nuclear and Alternative Energy Systems, San Franciso, W. H. Freeman (1979).

National Academy of Sciences, Committee for a Study on Saccharin and Food Safety Policy, Food Safety Policy: Scientific and Societal Considerations, Washington, D.C., National Academy of Sciences (1979).
National Academy of Sciences, Food Safety Policy: Scientific and Social Considerations, Part 2, Washington, D.C., National Academy of Sciences (March 1979).
National Academy of Sciences, Committee on Environmental Decision Making, Decision Making in the Environmental Protection Agency, Washington, D.C., National Academy of Sciences (1977).
National Academy of Sciences, Committee on Principles of Decision Making for Chemicals in the Environment, Decision Making for Regulating Chemicals in the Environment, Washington, D.C., National Academy of Sciences (1975).
National Academy of Sciences, Committee on Prototype Explicit Analyses for Pesticides, "Risk Assessment and Benefit Assessment," Regulatory Pesticides, Washington, D.C., National Academy of Sciences (1980), pp. 65-130.
National Academy of Sciences, Committee on the Biological Effects of Ionizing Radiation, The Effects on Populations of Exposure to Low Levels of Ionizing Radiation, Washington, D.C., National Academy of Sciences (1980).
National Academy of Sciences, Safe Water Drinking Committee, "Problems of Risk Estimation," Drinking Water and Health, 3:25-65 (1980).
National Academy of Sciences/National Research Council, Advisory Committee on Biological Effects of Ionizing Radiation, Considerations of Health Benefit Cost Analysis for Activities Involving Ionizing Radiation and Alternatives, Washington, D.C., National Academy of Sciences (1977).
National Safety Council, Accident Facts, 1980 Edition, Chicago, National Safety Council (1980).
National Science Foundation, Risk Assessment and Catastrophe Management Policies: The Role of Science and Technology, Papers Commissioned as Inputs to Annual Science and Technology Report (ASTR), 2nd ed., Washington, D.C., National Science Foundation (1980).
Nehnevajsa, J., and Menkes, J., "Technology Assessment and Risk Analysis," Technological Forecasting and Social Change, 19: 245-255 (1981).
Office of Technology Assessment, "Methods for Assessing Health Risks," Environmental Contaminants in Food, Washington, D.C., U.S. Congressional Office of Technology Assessment, 59-70 (December 1979).
Okrent, D., "Comment on Societal Risk," Science, 208:372-375 (April 25, 1980).
Okrent, D., A General Evaluation Approach to Risk-Benefit for Large Technological Systems and Its Application to Nuclear Power (UCLA-ENG-7777), Los Angeles, University of California (December 1977).

Okrent, D., "Risk-Benefit Evaluation for Large Technological Systems," Nuclear Safety, 20:148-164 (March-April 1979).

Okrent, D., ed., Risk-Benefit Methodology and Application: Some Papers Presented at the Engineering Foundation Workshop, September 22-26, 1975, Asilomar, California (UCLA-ENG-7598), Los Angeles, University of California (1975).

Okrent, D., and Whipple, C., An Approach to Societal Risk Acceptance Criteria and Risk Management (UCLA-ENG-7746), Los Angeles, University of California, School of Engineering and Applied Sciences (1977).

Otway, H. J., and Edwards, W., Application of a Simple Multi-Atribute Rating Technique to Evaluation of Nuclear Waste Disposal Sites: A Demonstration (IIASA-RM-77-31), Laxenburg, Austria, International Institute for Applied Systems Analysis (1977).

Otway, H.J., and Pahner, P. D., "Risk Assessment," Futures, 8:122-134 (April 1976).

Page, T., "A Generic View of Toxic Chemicals and Similar Risks," Ecology Law Quaterly, 7:207-243 (1978).

Parry, G. W., "Characterization and Evaluation of Uncertainty in Probabilistic Risk Analysis," Nuclear Safety, 22:28-42 (1981).

Philipson, L. L., Investigation of the Feasibility of the Delphi Technique for Estimating Risk Analysis Parameters (TES-20-74-5), Prepared for the Office of Hazardous Materials, Department of Transportation, Los Angeles, University of Southern California (1974).

Poulinquen, L. Y., Risk Analysis in Project Appraisal, World Bank Staff Occasional Papers, No. 11, International Bank for Reconstruction and Development, Baltimore, Maryland, Johns Hopkins University Press (1970).

Raiffa, H., Schwartz, W. B., and Weinstein, M. C., "Evaluating Health Effects of Societal Decisions and Programs," Decision Making in the Environmental Protection Agency: Selected Working Papers, Washington, D.C., National Academy of Sciences (1977).

Raiffa, H., Schwartz, W., and Weinstein, M., On Evaluating Health Effects of Societal Programs, Cambridge, Massachusetts, Harvard University (October 1976).

Raiffa, H., and Zeckhauser, R., "Reporting of Uncertainties in Risk Analysis," Two Conceptual Approaches to Health Risk Assessment for Alternative National Ambient Air Quality Standards, Washington, D.C., U.S. Environmental Protection Agency (September 1980).

Rasmussen, N., Kleitman, D. J., Stewart, R. B., and Yellin, J., "Nuclear Power: Can we Live With It?", Technology Review, 81:32-47 (1979).

Richmond, C. R., The Science of Risk Assessment (ORNL/PPA-80/2), Oak Ridge, Tennessee, Oak Ridge National Laboratory (1980).

Rivard, J. B., "Risk Minimizing by Optimum Allocation of Resources Available for Risk Reduction," Nuclear Safety, 12:305-309 (July/August 1971).

Roschin, A. V., and Timofeevsaya, L. A., "Chemical Substance in the Work Environment: Some Comparative Aspects of U.S.S.R. and U.S. Hygienic Standards," AMBIO, 4:30-33 (1975).

Ross, M., "Quantitative Decision Making," Risk vs. Benefit: Solution or Dream?", H. J. Otway, ed., Los Alamos, New Mexico, Los Alamos National Laboratory (1971), pp. 33-36.

Rothschild, N. M. V., "Coming to Grips with Risk," The Wall Street Journal (March 13, 1979).

Rowe, W. D., An Anatomy of Risk, New York, John Wiley and Sons (1977).

Rowe, W. D., "Development of Approaches for Acceptable Levels of Risk," Risk-Benefit Methodology and Application (UCLA-ENG-7598), D. Okrent, ed., Los Angeles, University of California (December 1975).

Rowe, W. D., "Risk Assessment Approaches and Methods," Society, Technology and Risk Assessment, J. Conrad, ed., London, Academic Press (1980), pp. 3-29.

Sage, A., and White, E., "Methodologies for Risk and Hazard Assessment: A Survey and Status Report," IEEE Transactions on Systems, Man and Cybernetics, 6:10 (August 1980):425-445.

Samuels, S., "Role of Scientific Data in Health Decisions," Environmental Health Perspectives, 32:301-307 (October 1979).

Sather, H. N., Biostatistical Aspects of Risk Benefit: The Use of Competing Risk Analysis. Prepared for the National Science Foundation, Los Angeles, University of California, School of Engineering and Applied Science (September 1974).

Schaeffer, R., What Are We Talking About When We Talk About Risk? A Critical Survey of Risk and Risk Preference Theories (RM-78-69), Laxenburg, Austria, International Institute for Applied Systems Analysis (December 1978).

Schneider, S. H., "Comparative Risk Assessment of Energy Systems," Energy 4:919-932 (1979).

Schneiderman, M., "The Uncertain Risk We Run: Hazardous Materials," Societal Risk Assessment: How Safe is Safe Enough? R. Schwing and W. Albers, eds., New York, Plenum Press (1980), pp. 19-38.

Schwing, R., and Albers, W. A., Jr., eds., Societal Risk Assessment: How Safe is Safe Enough? New York, Plenum Press (1980).

Singer, M., "How to Reduce Risks Rationally," Public Interest, 51: 93-112 (1978).

Slovic, P., et al., "Rating the Risks," Environment, 21:14-39 (1979).

Slovic, P., Fischhoff, B., and Lichtenstein, S., "Risk Assessment: Basic Issues," Managing Technological Hazard: Research Needs and Opportunities, R. Kates, ed., Boulder, Colorado, Institute of Behavioral Science, University of Colorado (1977), pp. 81-108.

Smalley, R. D., "Risk Assessment: An Introduction and Critique," Coastal Zone Management Journal, 7:133-162 (1980).

Solomon, K. A., and Abraham, S. C., "The Index of Harm: A Useful Measure for Comparing Occupational Risk Across Industries," Health Physics, 38:375-391 (March 1980).

Starr, C., et al., Energy and the Environment: A Risk-Benefit Analysis, New York, Pergamon Press (1977).

Starr, C., and Whipple, C., "Risks of Risk Decisions," Science, 208: 1114-1119 (June 6, 1980).

Tannenbaum, S. R., "Relative Risk Assessment," Advances in Modern Toxicology, Vol. 1: New Concepts in Safety Evaluation, M. A. Mehlman, R. F. Shapiro, and H. Blumenthal, eds., New York, John Wiley and Sons (1979).

Union of Concerned Scientists, The Risks of Nuclear Power Reactors, A Review of the NRC Reactor Safety Study (WASH-1400, NUREG-75/014), Cambridge, Massachusetts, Union of Concerned Scientists (1977).

U.S. Congress, House Committee on Science and Technology, Subcommittee on Science, Research, and Technology, Comparative Risk Assessment, 96th Congress, Washington, D.C., U.S. Government Printing Office (1980).

U.S. Congress, House Committee on Sciences and Technology, Subcommittee on Science, Research and Technology, Risk/Benefit Analysis in the Legislative Process: Joint Hearings, July 24, 25, 1979, 96th Congress, 1st Session (No. 71), Washington, U.S. Government Printing Office (1980).

U.S. Congress, Office of Technology Assessment, Assessment of Technologies for Determining Cancer Risks from the Environment, Washington, D.C., U.S. Government Printing Office (1981).

U.S. Council on Environmental Quality, and the U.S. Department of State, The Global 2000 Report to the President: Entering the Twenty-First Century, Volume I, The Summary Report, Washington, D.C., U.S. Government Printing Office (1980).

U.S. Council on Environmental Quality, and the U.S. Department of State, The Global 2000 Report to the President: Entering the Twenty-First Century, Volume II, The Technical Report, Washington, D.C., U.S. Government Printing Office (1980).

U.S. Council on Environmental Quality, and the U.S. Department of State, The Global 2000 Report to the President: Entering the Twenty-First Century, Volume III, The Government's Global Model, Washington, D.C., U.S. Government Printing Office (1980).

U.S. Department of Health, Education, and Welfare, National Institute of Environmental Health Sciences, Human Health and the Environment: Some Research Needs, Washington, D.C., U.S. Government Printing Office (1977).

U.S. Environmental Protection Agency, Considerations of Health Benefit-Cost Analysis for Activities Involving Ionizing Radiation Exposure and Alternatives, Report of the Advisory Committee on the Biological Effects of Ionizing Radiation (EPA 520/4-77-003), Washington, D.C., U.S. Environmental Protection Agency (1977).

U.S. Interagency Regulatory Liaison Group, Work Group on Risk Assessment, Scientific Bases for Identifying Potential Carcinogens and Estimating Their Risks, Washington, D.C. (February 1979).

U.S. Nuclear Regulatory Commission, Comparative Risk-Cost-Benefit Study of Alternative Sources of Electrical Energy (WASH-1224), Washington, D.C., U.S. Atomic Energy Commission (December 1974).

U.S. Nuclear Regulatory Commission, Executive Seminar on the Future Role of Risk Assessment and Reliability Engineering in Nuclear Regulation, Washington, D.C., National Technical Information Service (1981).

U.S. Nuclear Regulatory Commission, Fault Tree Handbook (NUREG-0492), Washington, D.C., U.S. Nuclear Regulatory Commission (January 1981).

U.S. Nuclear Regulatory Commission, Plan for Developing a Safety Goal (NUREG-9735), Washington, D.C., U.S. Nuclear Regulatory Commission (August 1980).

U.S. Nuclear Regulatory Commission, Reactor Safety Study: An Assessment of Accidental Risks in U.S. Commercial Nuclear Power Plants (WASH-1400, NUREG 75/014), Washington, D.C., U.S. Atomic Energy Commission, Nuclear Regulatory Commission (October 1975).

U.S. President's Commission for a National Agenda for the Eighties, Panel on Science and Technology: Promises and Dangers, Washington, D.C., U.S. Government Printing Office (1980).

Van Horn, A., and Wilson, R., The Status of Risk-Benefit Analysis, Cambridge, Massachusetts, Harvard University Energy and Environmental Policy Center (November 1976).

Vaupel, J. W., and Graham, John D., "Egg in Your Bier?", The Public Interest, 58:3-17 (Winter 1980).

Vesely, W. E., "Estimating Common Cause Failure Probabilities in Reliability and Risk Analyses: Marshall-Olkin Specializations," Journal of the Society for Industrial and Applied Mathematics, 314-341 (1977).

Weatherwax, R. K., "Virtues and Limitations of Risk Analysis," Bulletin of the Atomic Scientists, 31:29-32 (1975).

Webre, A. L., and Liss, P. H., The Age of Cataclysm, New York, Putnam (1974).

Weinberg, A. M., "Reflections on Risk Assessment," Risk Analysis: An International Journal, 1:5-8 (March 1981).

Wilson, R., "Analyzing the Daily Risks of Life," Technology Review, 81:40-46 (February 1979).

Wilson, R., "Examples in Risk-Benefit Analysis," Chemical Technology, 6:604-607 (October 1975).

Winkler, R. L., and Sarin, R. K., "A Risk Assessment Methodology for Environmental Pollutants," Four Conceptual Approaches to Health Risk Assessment, Washington, D.C., U.S. Environmental Protection Agency (August 1980).

PERCEIVED RISK: A THEORETICAL AND METHODOLOGICAL OVERVIEW

Atkinson, J. W., "Motivational Determinants of Risk-Taking Behavior," Psychology Review, 64:359-372 (1957).
Baldweicz, W., Haddock, G., Lee, Y., Prajoto, O., Whitley, R., and Denny, V., Historical Perspective on Risk for Large Scale Technological Systems (UCLA-ENG-7485), Los Angeles, University of California (December 1974).
Becker, G. M., and McClintock, C. G., "Value: Behavioral Decision Theory," Annual Review of Psychology, 18:239-286 (1967).
Bem, D. J., Wallach, M., and Kogan, N., "Group Decision Making Under Risk of Aversive Consequences," Journal of Personality and Social Psychology, 1:453-460 (1965).
Bettman, O. L., The Good Old Days - They Were Terrible! New York, Random House (1974).
Bowen, J., "The Choice of Criteria for Individual Risk, for Statistical Risks, and for Public Risk," Risk-Benefit Methodology and Application (UCLA-ENG-7598), D. Okrent, ed., Los Angeles, University of California (December 1975).
Bowman, C. H., et al., The Prediction of Voting Behavior on a Nuclear Energy Referendum (IIASA RM-78-8), Laxenburg, Austria, International Institute for Applied Systems Analysis Research (February 1978).
Buttel, F., and Flinn, W., "The Politics of Environmental Concern: The Impacts of Party Identification and Political Ideology on Environmental Attitudes," Environment and Behavior, 10:17-35 (March 1978).
Cole, G., and Withey, S., "Perspectives on Risk Perceptions," Risk Analysis: An International Journal, 1:2 (1982).
Combs, B., and Slovic, P., "Causes of Death: Biased Newspaper Coverage and Biased Judgments," Journalism Quaterly, 56:837-843 (1979).
Craik, K. H., "Environmental Psychology," New Directions in Psychology, T. M., Newcomb, ed., New York, Holt, Rinehart, and Winston (1970).
Crowe, M. J., "Toward a 'Definitional Model' of Public Perceptions of Air Pollution," Journal of the Air Pollution Control Association, 18:154-157 (March 1968).
de Boer, C., "The Polls: Nuclear Energy," Public Opinion Quaterly, 402-411 (Fall 1977).
Delcoigne, G., "Education and Public Acceptance of Nuclear Power Plants," Nuclear Safety, 20:655-664 (November-December 1979).
Downs, A., "Up and Down with Ecology - The Issue Attention Cycle," The Public Interest, 28:38-50 (1972).
Edwards, W., "Behavioral Decision Theory," Annual Review of Psychology, 12:473-498 (1961).
Edwards, W., and Tversky, A., Decision Making, Selected Readings, Middlesex, Penguin Books (1967).

Englemann, P. A., and Renn, O., "On the Methodology of Cost-Benefit Analysis and Risk Perception," Directions in Energy Policy, B. Kursunoglu and A. Perlmutter, eds., Cambridge, Massachusetts, Ballinger Publishing Company (1979), pp. 357-364.

Falk, H., "The Effect of Personal Characteristics on Attitudes Toward Risk," Journal of Risk and Insurance, 43:215-241 (June 1976).

Ferkiss, V. C., Technological Man: The Myth and the Reality, New York, New American Library (1969).

Fischhoff, B., "Behavioral Aspects of Cost Benefit Analysis," Impacts and Risks of Energy Strategies: Their Analysis and Role in Management, G. Goodman, ed., London, Academic Press (1979).

Fischhoff, B., "Hindsight/Foresight: The Effect of Outcome Knowledge on Judgment Under Uncertainty," Journal of Experimental Psychology: Human Perception and Performance, 1:288-299 (1975).

Fischhoff, B., "Informed Consent in Societal Risk - Benefit Decisions," Technological Forecasting and Social Change, 13:347-357 (May 1979).

Fischhoff, B., Hohenemser, C., Kasperson, R., and Kates, R., "Handling Hazards," Environment, 20:16-37 (September 1978).

Fischhoff, B., Slovic, P., and Lichtenstein, S., "Labile Values: A Challenge for Risk Assessment," Society, Technology, and Risk Assessment, J. Conrad, ed., London, Academic Press (1980), pp. 57-66.

Fischhoff, B., Slovic, P., Lichtenstein, S., Read, S., and Combs, B., "How Safe is Safe Enough? A Psychometric Study of Attitudes Toward Technological Risks and Benefits," Policy Sciences, 9: 127-152 (1978).

Fishbein, M., Readings in Attitude Theory and Measurement, New York, John Wiley and Sons (1967).

Fishbein, M., and Ajezen, I., Belief, Attitude, Intention and Behavior: An Introduction to Theory and Research, Reading, Massachusetts, Addison-Wesley (1975).

Fishburn, P. C., Utility Theory for Decision Making, New York, John Wiley and Sons (1970).

Flanders, J. P., and Thistlewaite, D. L., "Effects of Familiarization and Group Discussion Upon Risk Taking," Journal of Personality and Social Psychology, 5:91-97 (1967).

Foreman, H., ed., Nuclear Power and the Public, Minneapolis, University of Minnesota Press (1970).

Friedman, M., and Savage, L. J., "The Utility Analysis of Choices Involving Risks," Journal of Political Economy, 56:279-304 (1948).

Gould, L., and Walker, C. A., eds., Too Hot to Handle: Public Policy Issues in Nuclear Waste Management, New Haven, Connecticut, Yale University Press (1981).

Green, C. H., "Revealed Preference Theory: Assumptions and Presumptions," Society, Technology and Risk Assessment, J. Conrad, ed., London, Academic Press (1980), pp. 49-56.

Green, C. H., "Risk: Attitudes and Beliefs," Behaviour in Fires, D. V. Canter, ed., Chichester, Milay (1980).

Green, C. H., and Brown, R. A., "Counting Lives," Journal of Occupational Accidents (1978).
Green, C. H., and Brown, R. A., Life Safety: What Is It and How Much Is It Worth" (CP52/78), Borehamwood, Hertfordshire, U.K., Department of the Environment, Building Research Establishment (1978).
Green, C. H., and Brown, R. A., Metrics for Societal Safety (Note N 144/78), Borehamwood, Hertfordshire, U.K., Department of the Environment, Building Research Establishment (1978).
Green, C. H., and Brown, R. A., Perceived Safety as an Indifference Function (Note N 156/78), Borehamwood, Hertfordshire, U.K., Department of the Environment, Building Research Establishment (1978).
Green, C. H., and Brown, R. A., The Perception of, and Attitudes Towards, Risk: Final Report, Vol. 2, Measure of Safety (FRO/028/68), Dundee, Scotland, School of Architecture, Duncan of Jordanstone College of Art, University of Dundee (April 1977).
Green, C. H., and Brown, R. A., The Perception of, and Attitudes Towards, Risk, Final Report, Vol. 3, Stability of Perception under Time and Data (FRO/028/68), Dundee, Scotland, School of Architecture, Duncan of Jordanstone College of Art, University of Dundee (April 1977).
Green, C. H., and Brown, R. A., The Perception of, and Attitudes Towards, Risk, Final Report, Vol. 4, Initial Experiments on Determining Satisfaction With Safety Levels (FRO/028/68), Dundee, Scotland, School of Architecture, Duncan of Jordanstone College of Art, University of Dundee (April 1977).
Green, C. H., and Brown, R. A., Problems of Valuing Safety (Note N 70/78), Borehamwood, Hertfordshire, U.K., Department of the Environment, Building Research Establishment (1978).
Greenberg, P. F., "The Thrill Seekers," Human Behavior, 6:17-21 (April 1977).
Hammond, K. R., and Adleman, L., "Science, Values and Human Judgment," Science, 194:389-396 (October 22, 1976).
Harris, Louis and Associates, Inc., Harris Perspective 1979: A Survey of the Public and Environmental Activists on the Environment (59), New York, Louis Harris and Associates (1979).
Harris, Louis and Associates, Inc., Risk in a Complex Society, Chicago, Marsh and McClennan Public Opinion Survey (1980).
Harris, Louis and Associates, Inc., A Second Survey of Public and Leadership Attitudes Toward Nuclear Power Development in the United States, New York EBASCO (1976).
Kahan, J. P., How Psychologists Talk About Risk (P-6403), Santa Monica, California, Rand Corporation (October 1979).
Kahneman, D., and Tversky, A., "Availability: A Heuristic View for Judging Frequency and Probability," Cognitive Psychology, 5: 207-232 (1973).
Kahneman, D., and Tversky, A., "On the Psychology of Prediction," Psychological Review, 80:237-251 (July 1973).

Kahneman, D., and Tversky, A., "Prospect Theory: An Analysis of Decision Under Risk," Econometrica, 47:263-291 (March 1979).
Kasper, R. G., "Perceived Risk: Implications for Policy," Impacts and Risks of Energy Strategies: Their Analysis and Role in Management, London, Academic Press (1979).
Kasper, R., "Perceptions of Risk and Their Effects on Decision-Making," Societal Risk Assessment: How Safe is Safe Enough? R. Schwing and W. Albers, eds., New York, Plenum Press (1980), pp. 71-80.
Kasperson, R. E., Berk, G., Pijaka, D., Sharaf, A., and Wood, J., "Public Opposition to Nuclear Energy: Retrospect and Prospect," Science, Technology and Human Values, 5:11-23 (Spring 1980).
Keeney, R. L., and Kirkwood, C. W., "Group Decision Making Using Cardinal Social Welfare Functions," Management Science, 22: 430-437 (1975).
Keeney, R. L., and Raiffa, H., Decisions with Multiple Objectives: Preferences and Value Tradeoffs, New York, John Wiley and Sons (1976).
Klausner, S., ed., Why Man Takes Chances: Studies in Stress-Seeking, Garden City, New York, Doubleday (1968).
Kogan, N., and Wallach, M. A., "Risk Taking as a Function of the Situation, the Person, and the Group," New Directions in Psychology III, New York, Holt, Rinehart, and Winston (1967).
La Porte, T., "Public Attitudes Toward Present and Future Technology," Social Studies of Science, 5:373-391 (1975).
LaPorte, T. R., and Metlay, D., "Technology Observed: Attitudes of a Wary Public," Science, 188:121-127 (April 11, 1975).
Laporte, T. R., and Metlay, D., They Watch and Wonder: Public Attitudes Toward Advanced Technology, University of California Institute of Governmental Studies, Berkeley (1975).
Lerch, I., "Risk and Fear," New Scientist, 185:8-11 (January 3, 1980).
Lichtenstein, S., Fischhoff, B., and Phillips, L. D., "Calibration of Probabilities: The State of the Art," Decision Making and Change in Human Affairs, H. Jungermann and G. de Zeeuw, eds., Dordrecht, Reidel (1977).
Lichtenstein, S., Slovic, P., Fischhoff, B., Layman, M., and Combs, B., "Judged Frequency of Lethal Events," Journal of Experimental Psychology: Human Learning and Memory, 4:551-578 (1978).
Lowrance, W., Of Acceptable Risk: Science and the Determination of Safety, Los Altos, California, Kaufman (1976).
Maderthaner, R., Guttman, G., Swaton, E., and Otway, H. J., "Effect of Distance on Risk Perception," Journal of Applied Psychology 63:3 (1978):380-382.
Maderthaner, R., Pahner, P., Guttman, G., and Otway, H. J., "Perceptions of Technological Risks: The Effect of Confrontation: (RM-76-53), Laxenburg, Austria, International Institute for Applied Systems Analysis (1976).
Marquis, D. G., and Reitz, H. J., "Effects of Uncertainty on Risk Taking in Individual and Group Decisions," Behavioral Sciences, 14:281-288 (July 1969).

Maynard, W. S., Nealey, S. M., Hebart, J. A., and Lindell, M. K., Public Values Associated with Nuclear Waste Disposal (BNWL-1997), Seattle, Battelle Human Affairs Research Centers (1976).

Mazur, A., "Disputes Between Experts," Minerva, 11:55-81 (1973).

Mazur, A., "Opposition to Technological Innovation," Minerva, 13: 58-81 (1975).

McEnvoy, J., "The American Concern with the Environment," Natural Resources and the Environment, W. R. Burch, Jr., et al., eds., New York, Harper and Row (1972).

Melber, B. D., Nealey, S. M., Hammersla, J., and Rankin, W. L., Nuclear Power and the Public: Analysis of Collected Survey Research (PNL-2430), Seattle, Battelle Human Affairs Resarch Centers (1977).

Mitchell, R. C., Public Opinion on Environmental Issues: Results of a National Public Opinion Survey, Washington, D.C., Council on Environmental Quality, Department of Agriculture, Department of Energy, and Environmental Protection Agency (1980).

Mitchell, R. C., "Silent Spring/Solid Majorities," Public Opinion, 2 (August-September 1979).

National Council on Radiation Protection and Measurements, Perceptions of Risk: Proceedings of the Fifteenth Annual Meeting, March 14-15, 1979, Washington, D.C., National Council on Radiation Protection and Measurements (March 1980).

Nelkin, D., Jetport, New Brunswick, New Jersey, Transactional Books (1975).

Nelkin, D., Nuclear Power and Its Critics, Ithaca, New York, Cornell University Press (1971).

Nelkin, D., "The Political Impact of Technical Expertise," Social Studies of Science, 5:35-54 (1975).

Nelkin, D., "Some Social and Political Dimensions of Nuclear Power: Examples from Three Mile Island," American Political Science Review, 75:132-145 (March 1981).

Nelkin, D., Technological Decisions and Democracy, Beverly Hills, California, Sage Publications (1977).

Nelkin, D., and Pollack, M., "Political Parties and the Nuclear Energy Debate in France and Germany," Comparative Politics (January 1980).

O'Hare, M., "Not On My Block You Don't: Facility Siting and the Strategic Importance of Compensation," Public Policy, 25 (Fall 1977).

Okrent, D., and Whipple, C., An Approach to Societal Risk Acceptance Criteria and Risk Management (UCLA-ENG-7746), Los Angeles, University of California, School of Engineering and Applied Science (June 1977).

Opinion Research Corporation, "Public Attitudes Toward Environmental Trade-Offs," ORC Public Opinion Index, 33:1-8 (August 1975).

Otway, H. J., "The Perception of Technological Risks: A Psychological Perspective," Technological Risk: Its Perception and Handling in the European Community, M. Dierkes, S. Edwards, and R. Coppock, eds., Cambridge, Massachusetts, Oelgeschlager, Gunn and Hain, Publishers, Inc. (1980), pp. 35-45.

Otway, H. J., A Review of Research on the Identification of Factors in Influencing Social Response to Technological Risks (IAEA-CN-36/4), Proceedings of the IAEA Conference on Nuclear Power and Its Fuel Cycle, Salzburg, Austria (1977).

Otway, H. J., Risk Assessment and Societal Choices (RM-75-2), Laxenburg, Austria, International Institute for Applied Systems Analysis (February 1975).

Otway, H. J., et al., "On the Social Aspects of Risk Assessment," Journal of the Society for Industrial and Applied Mathematics (1977).

Otway, H. J., and Cohen, J. J., Revealed Preferences: Comments on the Starr Benefit-Risk Relationships (RM-75-5), Laxenburg, Austria, International Instituted for Applied Systems Analysis (1975).

Otway, H. J., and Fishbein, M., Public Attitudes and Decision Making (RM-77-54), Laxenburg, Austria, International Institute for Applied Systems Analysis (1977).

Otway, H. J., and Fishbein, M., The Determinants of Attitude Formation: An Application to Nuclear Power (RM-76-80), Laxenburg, Austria, International Institute for Applied Systems Analysis (1976).

Otway, H. J., Maderthaner, R., and Guttman, G., Avoidance Response to the Risk Environment: A Cross-Cultural Comparison (RR-75-14), Laxenburg, Austria, International Institute for Applied Systems Analysis (1975).

Otway, H. J., Maurer, D., and Thomas, K., "Nuclear Power, The Question of Public Acceptance," Futures, 10:109-118 (April 1978).

Otway, H. J., Pahner, P. D., and Linnerooth, J., Social Values in Risk Acceptance (RM-75-54), Laxenburg, Austria, International Institute for Applied Systems Analysis (November 1975).

Pahner, P. D., "The Psychological Displacement of Anxiety: An Application to Nuclear Energy," Risk-Benefit Methodology and Application (UCLA-ENG-7598), D. Okrent, ed., Los Angeles, University of California (December 1975).

Pahner, P. D., A Psychological Perspective of the Nuclear Energy Controvery (RM-76-67), Laxenburg, Austria, International Institute for Applied Systems Analysis (1976).

Payne, J. W., "Relation of Perceived Risk to Preferences Among Gamblers," Journal of Experimental Psychology: Human Perception and Performance, 104:86-94 (1975).

Pearce, D. W., "The Nuclear Debate is About Values," Nature, 274:200 (1978).

Pearce, D. W., "The Preconditions for Achieving Consensus in the Context of Technological Risk," Technological Risk: Its Perception and Handling in the European Community, M. Dierkes, S. Edwards, and R. Coppack, eds., Cambridge, Massachusetts, Oelgeschlager, Gunn, and Hain, Publishers, Inc. (1980), p. 58.

Powers, W. T., Behavior: The Control of Perception, Chicago, Aldine (1973).

Pratt, J. W., Raiffa, H., and Schlaifer, R., "The Foundations of Decision Under Uncertainty," The American Statistical Association Journal, 59:353-376 (1964).

Raiffa, H., Decision Analysis: Introductory Lectures on Choices Under Uncertainty, Reading, Massachusetts, Addison-Wesley (1968).

Rapoport, A., and Wallsten, T. S., "Individual Decision Behavior," Annual Review of Psychology, 23:131-175 (1972).

Ravetz, J. R., "Public Perceptions of Acceptable Risks as Evidence for Their Cognitive, Technical, and Social Structure," Technological Risk: Its Perception and Handling in the European Community, M. Dierkes, S. Edwards, and R. Coppack, eds., Cambridge, Massachusetts, Oelgeschlager, Gunn, and Hain, Publishers, Inc. (1980), pp. 46-57.

Rowe, W. D., An Anatomy of Risk, New York, John Wiley and Sons (1977).

Sapolsky, H. M., "Science, Voters, and the Fluoridation Controversy," Science, 162:427-433 (1968).

Sjoberg, L., "Risk Generation and Risk Assessment in a Social Perspective," Foresight, the Journal of Risk Management, 3:4-12 (1978).

Sjoberg, L., "Strength of Belief and Risk," Policy Sciences, 2:39-52 (August 1979).

Slovic, P., "Assessment of Risk-Taking Behavior," Psychological Bulletin, 61:220-233 (1964).

Slovic, P., "Choice Between Equally Valued Alternatives," Journal of Experimental Psychology: Human Perception and Performance, 1:280-287 (1975).

Slovic, P., and Fischhoff, B., "Cognitive Process and Societal Risk Taking," Cognition and Societal Behavior, J. S. Carroll and J. W. Payne, eds., Potomac, Maryland, Lawrence Erlbaum Associates (1976).

Slovic, P., Fischhoff, B., and Lichtenstein, S., "Behavioral Decision Theory," Annual Review of Psychology, 28:1-39 (1977).

Slovic, P., Fischhoff, B., and Lichtenstein, S., "Characterizing Perceived Risk," Technological Hazard Management, R. W. Kates and C. Hohenemser, eds., Cambridge, Massachusetts, Oelgeschlager, Gunn, and Hain (1981).

Slovic, P., Fischhoff, B., and Lichtenstein, S., Expressed Preferences, Eugene, Oregon, Decision Research Report, 80-1 (1980).

Slovic, P., Fischhoff, B., and Lichtenstein, S., "Facts and Fears: Understanding Perceived Risk," Societal Risk Assessment: How Safe is Safe Enough? R. Schwing and W. Albers, Jr., eds., New York, Plenum (1980), pp. 181-216.

Slovic, P., Fischhoff, B., and Lichtenstein, S., "Informing People about Risk," Product Labeling and Health Risks (Banbury Report 6), L. Morris, M. Marsis, and I. Barofksy, eds., Cold Spring Harbor, New York, Cold Spring Harbor Laboratory (1980).

Slovic, P., Fischhoff, B., and Lichtenstein, S., "Perceived Risk," Societal Risk Assessment: How Safe is Safe Enough? R. Schwing and W. Albers, eds., New York, Plenum Press (1980).

Slovic, P., Fischhoff, B., and Lichtenstein, S., "Rating the Risks," Environment, 21:3 (April 1979):14-39.
Slovic, P., Fischhoff, B., and Lichtenstein, S., "Risky Assumptions," Psychology Today, 14:44-45, 47-48 (June 1980).
Slovic, P., Fischhoff, B., Lichtenstein, S., Corrigan, B., and Combs, B., Preference for Insuring Against Probable Small Losses: Insurance Implications," Journal of Risk and Insurance, 45:237-258 (June 1977).
Slovic, P., Kunreuther, H., and White, G., "Decision Processes, Rationality and Adjustments to Natural Hazards," Natural Hazards: Local, National, and Global, G. F. White, ed., New York, Oxford Press (1974).
Slovic, P., Lichtenstein, S., and Fischhoff, B., "Images of Disaster: Perception and Acceptance of Risks from Nuclear Power," Energy Risk Management, G. Goodman and W. Rowe, eds., London, Academic Press (1979).
Slovic, P., and Tversky, A., "Who Accepts Savage's Axiom?" Behavioral Science, 19:368-372 (1974).
Spangler, M. B., "Risks and Psychic Costs of Alternative Energy Sources for Generating Electricity," The Energy Journal (January 1981).
Starr, C., "Social Benefit versus Technological Risk," Science, 165:1232-1238 (September 19, 1969).
Starr, C., "Some Comments on the Public Perception of Personal Risk and Benefit," Risk vs. Benefit: Solution or Dream? H. J. Otway, ed., Los Alamos, New Mexico, Los Alamos National Laboratory (1971).
Starr, C., and Whipple, C., "Risk of Risk Decisions," Science, 208:1114-1119 (June 1980).
Stumpf, S. E., "Culture, Values, and Food Safety," BioScience, 28:186-190 (March 1978).
Swaton, E., Maderthaner, R., Pahner, P. D., Guttman, G., and Otway, H. J., The Determinants of Risk Perception: A Survey (RM-76-XX), Laxenburg, Austria, International Institute for Applied Systems Analysis (1976).
Swelm, R. O., "Utility Theory Insights into Risk Taking," Harvard Business Review, 123-136 (November-December 1966).
Tamerin, T., and Resnick, L. P., "Risk Taking by Individual Option - Case Study: Cigarette Smoking," Perspectives on Benefit-Risk Decision Making, Washington, D.C., National Academy of Engineering (1972), pp. 73-84.
Thomas, K., Maurer, D., Fishbein, M., Otway, H., Hinkle, R., and Simpson, D., A Comparative Study of Public Beliefs About Five Energy Systems (RR 80-1), Laxenburg, Austria, International Institute for Applied Systems Analysis (1979).
Thomas, K., Swaton, E., Fishbein, M., and Otway, H., Nuclear Energy: The Accuracy of Policy Maker's Perceptions of Public Beliefs (RR 80-2), Laxenburg, Austria, International Institute for Applied Systems Analysis (1979).

Thompson, M., "Aesthetics of Risk: Context or Culture?" Societal Risk Assessment: How Safe is Safe Enough? R. Schwing and W. Albers, eds., New York, Plenum Press (1980), pp. 273-286.

Thorngate, W., "Efficient Decision Heuristics," Behavioral Science, 25:219-225 (1980).

Tubiana, M., "One Approach to the Study of Public Acceptance," Directions in Energy Policy, B. Kursunoglu and A. Perlmutter, eds., Cambridge, Massachusetts, Ballinger Publishing Company (1979).

Tversky, A., "Elimination by Aspects: A Theory of Choice," Psychological Review, 79:281-299 (1972).

Tversky, A., and Kahneman, D., "The Framing of Decisions and the Psychology of Choice," Science, 211:1453-1458 (1981).

Tversky, A., and Kahneman, D., "Judgment Under Uncertainty: Heuristics and Biases," Benefit-Cost and Policy Analysis, R. Zeckhauser, et al., eds., Chicago, Aldine Publishing Company (1975).

Tversky, A., and Kahneman, D., "Judgment Under Uncertainty: Heuristics and Biases," Science, 185:1124-1131 (September 27, 1974).

Tversky, A., and Sattath, S., "Preferences Tress," Psychological Review, 86:542-573 (1979).

Velimirovic, H., An Anthropological View of Risk Phenomena (RM-75-XX), Laxenburg, Austria, International Institute for Applied Systems Analysis (1975).

Vlek, C., and Stallen, P. J., "Judging Risks and Benefits in the Small and in the Large," Organizational Behavior and Human Performance, 28 (October 1981).

Vlek, C., and Stallen, P. J., "Rational and Personal Aspects of Risk," Acta Psychologica, 45 (1980).

Van Neuman, J., and Morgenstern, O., Theory of Games and Economic Behavior, Princeton, Princeton University Press (1944).

Von Winterfeldt, D., Edwards, W., Anson, J., Stillwell, W., and Slovic, P., Development of a Methodology to Evaluate Risks From Nuclear Electric Power Plants: Phase I - Identifying Social Groups and Structuring Their Values and Concerns, Final Report to Sandia National Laboratories, Albuquerque, New Mexico (May 1980).

Von Winterfeldt, D., and Rios, M., "Conflicts about Nuclear Power Safety: A Decision Theoretic Approach," Proceedings of the ANS/ENS Topical Meeting on Thermal Reactor Safety, M. H. Fontana and D. R. Patterson, eds., Springfield, Virginia, National Technical Information Service (1980), pp. 696-709.

Wendt, D., and Vlek, C. A. J., eds., Subjective Probability, Utility and Human Decision Making, Dordrecht, Reidel (1974).

White, G. F., "Formation and Role of Public Attitudes," Environmental Quality in a Growing Economy, H. Jarret, ed., Baltimore, Johns Hopkins University Press (1966).

INDEX

Abbreviated Injury Scale (AIS), 2
Adriamycin, 164
Air bags (see Passenger restraints)
Air pollution, 252
AIS (see Abbreviated Injury Scale)
Albinism, 282
Animal health, 232
Antinuclear sentiment, 229
Anti-smokers, 190
   antismoking campaign, 184, 185, 186, 190, 191, 194, 200, 210
     messages, 188
     publicity, 188
Atomic bomb, 228
Automobile, 1

Behavioral effects, 246
Benefits, 223, 240, 251-257, 269, 270
Bias, 236, 245
Bioaccumulation of chemicals, 260, 263
Blue-collar workers, 204, 208, 210, 211
   occupations, 206

Cancer
   breast, 166
   in humans, 281
   latency period, 262
   lung, 157, 158, 162, 183, 208
   non-melanoma, 282

Cancer (continued)
   rates, 223, 230, 243, 262, 270, 271
Carbon monoxide, 173, 175, 178, 279
Carcinoma
   bronchogenic, 157
Centers for Disease Control, 260
Chemical
   carcinogens, 242, 261-263, 267-271
     environmental, 169
     low dose, 169
   dumps, 263
   risks, 259-263, 267-271
Chemicals
   migration, 261
   persistent, 260, 262
Chemotherapy, 158, 159, 162, 164
   and cancer, 159
   chemotherapeutic program, 163
   treatment, 158
Chest acceleration, 6
Chlordane, 271
Chlorofluorocarbons, 281
Chlorofluoromethanes, 279
Chronic toxicity, 260-263
Cigarettes (in U.S. and Canada), 175
   consumption of, 184, 188, 189
     actual, 186
     aggregate, 188, 196
     daily, 195
     per capita, 189, 191, 199
     per-smoker, 191
   filtered, 210
   filter-tipped, 186, 191

Cigarettes (continued)
  hole-blocking, 176, 177, 179, 180, 181
    of ventilated filters, 175
  less hazardous, 183, 199
  low-nicotine, 195
  low-yield, 175, 180, 181
  ultra-low tar, 179
  unfiltered, 191, 195, 210
  ventilated filter, 175, 177, 179, 180, 181
Coal, 253
Collision Deformation Classification, (CDC), 3
Compensation for added risk, 254
Conflicting values, 240
Crashworthiness standards, 15

DDT, 269, 270
Death
  from disease, 162
  from treatment, 162
Dose-response relationships, 287
Driving behavior, 15

Ecocentricism, 327
Energy supply, 251
EPA, 226, 271
Expert opinion, 235-247, 270
Extrapolation methods
  linear, 170
  non-linear, 170

Fairness Doctrine, 183, 188, 198
Fetotoxicity, 261
Food contamination, 260, 263
Fossil fuels, 252
Friends and Family of TMI, 230

Governmental protection, 236

Hazards
  description, 58, 324
Head Injury Criteria (HIC), 6
Health
  and hazards, 193
    work related, 211
  threat to, 194
Hematologic problems, 159

Hexamethylmelamine, 164
Hydro-electric power
  in Sweden, 69-71
Hydrogen bubble, 228

Ignition-interlock, 6, 13
Infant mortality, 232
Inhaling, 194, 196
Injury
  severity of, 1
  severity score (ISS), 3
Insurance premiums, 254
ISS (*see* Injury, severity score)

Journalism, 267-271

Karnofsky Performance Status Criteris, 161

Landfills, 259
Lentigo Maligna Melanoma, 284
Log normal distribution, 215
Love Canal, 260, 342

Malignant melanoma, 282
Mantel-Bryan probit method, 170
Maternal nutritional factors, 208
Media Institute, 230
Meltdown ("China Syndrome"), 225, 228, 255
Methotrexate, 164
Metropolitan Edison Co., 226, 231
Milk contamination, 229, 230
Moral convictions, 203
Mortality
  among radiologists, 209
  ratio, 186
  risks, 208
  standard tables, 209
  statistics on smoking, 208
  surveys, 208
Multistage methods, 170
Mutagenesis, 261, 262

National risk strategy, 251-257, 263
Nausea, 159
News media, 227, 228, 243, 245, 267-271

INDEX

Nicotine, 173, 175, 177, 178, 179, 193
  compensation, 195, 196, 199
  compensation hypothesis, 190
  content of cigarettes, 199
  regulation, 200
  regulation hypothesis, 199
Non--smokers
  wives, 204
  women, 206
NRC, 229, 230, 231, 256
Nuclear power, 226, 230, 239, 241, 244, 251-257, 270, 271
  in Sweden
    licensing procedures, 73-74
    risk assessment studies, 74-75, 76-78
  in the United Kingdom
    cross-national comparisons, 115
    management, 109-115
    political and economic considerations, 101-105
    regulation, 113-114
    risk assessment studies, 117-123
  in the United States
    Kemeny commission, 134-143
    management, 130-131, 141-143
    risk assessment studies, 129
    siting and energency preparedness, 149-153
  in West Germany
    judicial review, 88-89
    licensing procedure, 84-88
    public acceptance, 82, 94-98
    regulation, 82-84
    risk assessment studies, 91-94
Nuclear war, 244
Nuclear weapons, 244
Nutritional deficiencies, 208

Objective tumor response, 164
Occupational
  distribution, 210
  dusts, 207
  groups, 204
  hazards, 204

Oil, 253
Organic halogens, 274
Ozone, 281
  stratospheric, 273, 281, 303

Passenger restraints, 1, 11
  air bags
    cost, 32
    effectiveness, 22, 33, 45-47
    history, 21, 24-28
    performance, 30-31
    public acceptance, 29-30
  air cushions, 16
  child, 11
  lap and shoulder belt, 7
  seat belts, 11
    cost, 32
    effectiveness, 22, 33, 45-47
    history, 21, 24-28
    performance, 30-31
    public acceptance, 29-30
Penicillin, 270
Pennsylvania Power and Light, 226
Phosphates, 269
"Police-scale," 2
Political process, 236, 242, 244, 256
Political stability, 252
Post-accident analyses, 246
Post-war baby boom, 199
Psychological stress, 229, 246, 247
Public fears, 228, 229, 230, 240-243, 256, 268
Public Health Service surveys, 195
Public opinion, 235-247

Radiation
  ultraviolet, 276, 281
  UV-B, 303
Radioactive gas release, 227
Reactor safety, 251-257
Reactor Safety Study (WASH-1400), 223
Risk
  acceptance, 220, 221, 230, 240, 253, 254, 269
  actual
    correlation with perceived risk, 64, 66

Risk, actual (continued)
  definition, 49
 assessment, 158, 210, 237–247, 262, 263
 aversion, 215
 awareness, 235, 267–271
 of bronchogenic carcinoma, 166
  smoking, 185
  toxic wastes, 259–263
 comparisons, 164, 218, 219, 254, 255
 Conversion Factors (RFCs), 213–223
 definition, 51
 delayed, 215, 223, 260–263
 determinants of, 159
 estimation, 351
 factor, 183
 hypothetical, 253
 immediate, 213, 215, 223
 information, 242, 246
 involuntary, 239
 local, 254
 management, 235, 238, 241, 251–257, 263
 man-caused, 222
 modification of, 166
 negligible, 254, 255, 256
 nuclear, 222, 223, 225–232 251–257
 occupational, 217, 242, 259, 263, 268
 perceived
  correlation with actual risk, 64, 66
  definition, 49
 perceptions of, 158, 198, 213, 229, 230, 232, 235–247, 257, 267–271
 physical, 158
 primary, 162
 psychological, 158
 radiation, 228, 242
 terminology, 238
 voluntary, 163

Saccharin, 240, 245
Salt mines, 257

Seat belts, (see Passenger restraints)
Self-regulation, 256
Shale oil, 252
Smokers
 behavior of, 179
 current, 184, 206
 heavy, 206
 light, 196
 modern, 199
Smoking
 and health, 169
  Surgeon General's report on, 183, 185, 188, 191, 193, 196, 199
 and health messages, 184
 behaviors, 183, 186, 195, 196, 199, 200, 206
 compensatory, 181
 cigarette (see Cigarettes)
  rate of, 190
 deterrents to, 198
 habits, 198, 206
 machine, 177, 179
 parental, 205
 rates, 190
 related illnesses, 200
 sociology of, 210

Tar yields, 175, 177, 178, 179, 181, 193
Technocentricism, 327
Teratogenesis, 261
Terrorism, 253
Therapy
 aggressiveness of, 163
 intensive, 164
Three-Mile Island, 225–232, 253, 255, 256, 338
Threshold level values, 261
Tobacco consumers, 181
Toxicity data, 261–263
 wildlife toxicity, 260
Toxic wastes, 259–263, 269, 271
 explosions, 260
 exposure to, 260–263
 flammability, 260

Underreporting, 198

Veterans data (U.S.), 210
"Virtually safe" dose, 169

Waste
    incineration, 259
    storage, 253, 260-263
    treatment, 259